Source and Components Analysis of Aerosols in Air Pollution

Source and Components Analysis of Aerosols in Air Pollution

Guest Editors

Jian Gao

Jiaqi Wang

MDPI

Basel • Beijing • Wuhan • Barcelona • Belgrade • Novi Sad • Cluj • Manchester

Guest Editors
Jian Gao
State Key Laboratory of Environmental Criteria and Risk Assessment
Chinese Research Academy of Environmental Sciences
Beijing
China

Jiaqi Wang
State Key Laboratory of Environmental Criteria and Risk Assessment
Chinese Research Academy of Environmental Sciences
Beijing
China

Editorial Office
MDPI AG
Grosspeteranlage 5
4052 Basel, Switzerland

This is a reprint of the Special Issue, published open access by the journal *Toxics* (ISSN 2305-6304), freely accessible at: https://www.mdpi.com/journal/toxics/special_issues/M5DIKT02AY.

For citation purposes, cite each article independently as indicated on the article page online and as indicated below:

Lastname, A.A.; Lastname, B.B. Article Title. *Journal Name* **Year**, *Volume Number*, Page Range.

ISBN 978-3-7258-7713-3 (Hbk)
ISBN 978-3-7258-7714-0 (PDF)
https://doi.org/10.3390/books978-3-7258-7714-0

Contents

About the Editors

Jian Gao

Jian Gao, Research Fellow/PhD Supervisor, Deputy Director of the Institute of Atmospheric Environment, Chinese Research Academy of Environmental Sciences (CRAES). His main research interests include comprehensive urban air pollution control, air pollution source apportionment, and R&D and application of air environmental monitoring and information management technologies. He has published more than 230 research papers, including over 150 SCI-indexed papers, with an h-index of 53 and more than 13,000 citations. He has had 16 national invention patents authorized and won 1 National Science and Technology Progress Award, 4 Science and Technology Awards of the Ministry of Ecology and Environment, 1 Science and Technology Award of the Ministry of Education, and 1 National Patent Award.

Jiaqi Wang

Jiaqi Wang, Associate Research Fellow, Chinese Research Academy of Environmental Sciences (CRAES). Her main research interests include atmospheric environment observation, formation mechanisms of complex air pollution, and sources and sinks of atmospheric reactive nitrogen compounds. She has published more than 10 papers in journals such as *Environmental Science & Technology*, *Atmospheric Chemistry & Physics*, and *Science of the Total Environment*. She has participated in several national and local government cooperative projects, including the National Natural Science Foundation of China and the National Key Research and Development Program.

toxics

MDPI

Article

Characteristic Analysis and Health Risk Assessment of $PM_{2.5}$ and VOCs in Tianjin Based on High-Resolution Online Data

Yanqi Huangfu [1,2], Feng Wang [1,2,3], Qili Dai [1,2], Danni Liang [1,2], Guoliang Shi [1,2] and Yinchang Feng [1,2,*]

1 State Environmental Protection Key Laboratory of Urban Ambient Air Particulate Matter Pollution Prevention and Control, College of Environmental Science and Engineering, Nankai University, Tianjin 300350, China; hfyq1993@hotmail.com (Y.H.); wfeng@email.tjut.edu.cn (F.W.); daiql@nankai.edu.cn (Q.D.); nksgl@nankai.edu.cn (G.S.)

2 CMA-NKU Cooperative Laboratory for Atmospheric Environment-Health Research (CLAER/CMA-NKU), Tianjin 300350, China

3 School of Environmental Science and Safety Engineering, Tianjin University of Technology, Tianjin 300384, China

* Correspondence: fengyc@nankai.edu.cn

Abstract: This study leveraged 2019 online data of particulate matter ($PM_{2.5}$) and volatile organic compounds (VOCs) in Tianjin to analyze atmospheric pollution characteristics. $PM_{2.5}$ was found to be primarily composed of water-soluble ions, with nitrates as the dominant component, while VOCs were predominantly alkanes, followed by alkenes and aromatic hydrocarbons, with notable concentrations of propane, ethane, ethylene, toluene, and benzene. The receptor model identified six major sources of $PM_{2.5}$ and seven major sources of VOCs. The secondary source is the main contribution source, while motor vehicles and coal burning are important primary contribution sources in $PM_{2.5}$. And, industrial processes and natural gas volatilization were considered major contributors for VOCs. A health risk assessment indicated negligible non-carcinogenic risks but potential carcinogenic risks from trace metals As and Cr, and benzene within VOCs, underscoring the necessity for focused public health measures. A risk attribution analysis attributed As and Cr in PM to coal combustion and vehicular emissions. Benzene in VOCs primarily originates from fuel evaporation, and industrial and vehicular emissions. These findings underscore the potential for reducing health risks from PM and VOCs through enhanced regulation of emissions in coal, industry, and transportation. Such strategies are vital for advancing air quality management and safeguarding public health.

Keywords: $PM_{2.5}$; VOCs; online data; source apportionment; health risk assessment

1. Introduction

With the swift progression of urbanization, an array of pollutants, including particulate matter and VOCs, have been introduced into the atmosphere, exacerbating air pollution issues in numerous urban centers across China [1]. In response, the initiation of the Air Pollution Prevention and Control Action Plan (APPCAP) in 2013 marked a significant regulatory effort to mitigate these emissions [2,3]. Consequently, there has been a notable reduction in anthropogenic atmospheric particulates and inhalable particulate matter, as indicated by substantial decreases in particulate matter concentrations [2,4,5]. Despite the overall particulate pollution having been significantly improved, the problem of secondary pollution has gradually become prominent. Additionally, the emergence of ozone pollution has introduced a new environmental challenge. VOCs have been identified as key precursors in the formation of O_3 and secondary organic aerosols (SOAs), which are integral to fine particulate matter and significant contributors to haze events [6,7]. The public health implications of particulates and VOCs have risen to prominence as a pivotal area of research [8–10].

VOCs originate from diverse sources, spanning both anthropogenic activities and natural processes [11,12]. The intensification of human activities has led to a continuous increase in the anthropogenic emissions of VOCs. Research on the impact and health risks of VOCs emitted from specific industries is a burgeoning field, with the latest references indicating a growing focus on this area [13–15]. In recent years, there has been a surge in studies aimed at tracing the sources of VOCs in the atmospheric environment. The Positive Matrix Factorization (PMF) model has been extensively applied in studies aimed at identifying sources of VOCs [16]. Chronic exposure to certain toxic VOCs at high concentrations poses significant health risks, including sensory irritation, fatigue, neurological damage, and carcinogenesis [17–21]. Researchers have utilized methodologies aligned with USEPA standards to assess VOC-related health risks in certain regions of India [22,23], Canada [24], and China [25,26].

The concentration of atmospheric particulate matter has seen a marked decline, with current pollution primarily attributed to secondary pollutants. During periods of severe pollution, secondary pollutants can account for approximately half of the total $PM_{2.5}$ load. The health risks associated with particulate matter are largely due to trace metal elements such as Pb, Mn, Ni, Cr, Cd, and Hg. These elements possess pathophysiological toxicity and are linked to a range of severe health hazards, including respiratory inflammation, cardiovascular and pulmonary diseases, and DNA damage [27]. They have also been shown to induce severe cardiovascular and central nervous system disorders [28,29]. Cd toxicity is associated with lung cancer, hepatic and renal pathologies, osteocalcin, and dermatological conditions [30]. Ni and V, emitted through fuel combustion, have a positive correlation with human cardiovascular and respiratory diseases [31–33]. Excessive Ni intake can lead to respiratory ailments [34], while Pb overexposure can impair the nervous system, endocrine and immune functions, skeletal development, and enzymatic cycles [35]. Mn has been identified as a potential neurotoxin and nephrotoxin.

This research delves into an integrated analysis of PM and VOCs in Tianjin for the year 2019, leveraging the online monitoring data from the advanced Super Observation Station located in Jinnan District. It presents a detailed characterization of the concentration and compositional profiles of PM and VOCs, offering insights into the chemical signatures of air pollutants within the region. The principal contributors of PM and VOC pollution have been identified in this area based on the widely used PMF model. Furthermore, this study conducts a thorough assessment of the health impacts associated with the identified PM and VOC components. By quantifying the exposure levels and evaluating the associated health risks, it pinpoints the predominant sources contributing to health risks, providing critical insights into the environmental and public health challenges posed by these pollutants.

2. Materials and Methods

2.1. Sampling and Chemical Analysis

Data collection was conducted at the Nankai University Air Quality Research Supersite (NKAQRS), situated at 38.99° N latitude and 117.33° E longitude, within the Jinan District of the Tianjin campus. The monitoring period spanned from 1 January to 31 December 2019. The NKAQRS is adjacent to a student dormitory area, with a low-traffic road located 20 m to the north. The winds predominantly blew from the east and southeast throughout this study. This study recorded a mean wind speed (WS) of 1.5 m/s; an average temperature of 14.5 °C; and a mean relative humidity (RH) that fluctuated between 14.1% and 98.7%, averaging 58.4% for the year 2019.

Despite the synchronization of the majority of data across the years, occasional gaps were observed due to instrument malfunctions or unforeseen events. The $PM_{2.5}$ mass concentration was continuously tracked using a BPM-200 Beta Particulate Monitor (Focused Photonics Inc., Hangzhou, China). Meteorological parameters, such as Temp, RH, and solar radiation, were captured by a compact weather station (WS600-UMB, LUFFT Inc., Fellbach, Germany).

All particulate matter instruments were integrated with a $PM_{2.5}$ inlet to ensure accurate and targeted sampling of fine particulate matter. Concentrations of elements, including K, Ca, Ti, Cr, Mn, Fe, Ni, Cu, Zn, As, Ba, and Pb were meticulously monitored on an hourly basis utilizing an in situ X-ray fluorescence analyzer (AMMS-100, Focused Photonics Inc., Hangzhou, China). Concurrently, water-soluble ions such as $SO_{4}{}^{2-}$, $NO_{3}{}^{-}$, $NH_{4}{}^{+}$, Cl^{-}, Na^{+}, K^{+}, Mg^{2+}, and Ca^{2+} were precisely quantified through an in situ ion chromatography system (URG 9000D, Thermo Fisher Scientific Inc., Waltham, MA, USA). Additionally, Organic Carbon (OC) and Elemental Carbon (EC) were continuously assessed hourly with a sophisticated thermal-optical carbon analyzer (OCEC-100, Focused Photonics Inc., Hangzhou, China). Elements K and Ca were selected to be included when constructing the $PM_{2.5}$ dataset.

In total, 54 VOC species across four sub-categories—27 alkanes, 10 alkenes, 16 aromatic hydrocarbons, and 1 alkyne (acetylene)—were recognized as photochemical precursors by the United States Environmental Protection Agency (US EPA). The analyses were executed using a photoionization detector (PID) and a flame ionization detector (FID), ensuring high sensitivity and effective identification (GC955-611/811, Synspec B.V., Groningen, Netherlands). It should be noted that due to the large range of missing values of styrene, although it has a more important indicative role, it was still removed from the overall system.

Quality assurance and quality control (QA/QC) measures were implemented based on previous studies [36–38] to ensure the reliability of the collected data. For $PM_{2.5}$ monitoring, it was meticulously maintained through monthly cleaning and calibration of the cutting head, as well as calibration of the sampling flow rate. The allowable deviation between the instrument's indicated flow rate and that of the flowmeter was within ±2%. Single-point calibration was performed monthly, with multi-point calibration conducted semi-annually for the VOC analysis. The calibration curve had to demonstrate a correlation coefficient greater than 0.95 for at least 95% of the species. Ion calibration was adjusted every 1~2 months based on changes in the peak retention time, employing a multi-point calibration across six concentration gradients. The resulting standard curves for anions exhibited a correlation coefficient above 0.99, while for cations, all except sodium ions demonstrated a correlation coefficient above 0.95. The elemental analysis was subjected to annual verification against standard membrane filters, with the verification outcomes ensuring that the concentration discrepancies for all elements were within ±2% of the standard membrane filter concentrations. OC/EC calibration was performed annually or following the replacement of the helium–methane calibration gas. A multi-point calibration was executed with five points, where the correlation coefficient exceeded 0.95.

2.2. PMF Model

The EPA PMF 5.0 model served as the analytical framework for the source apportionment of PM2.5 and VOCs. Mathematically, the PMF model is represented as follows [39,40]:

$$X_{ij} = \sum_{k=1}^{p} g_{ik} \cdot f_{kj} + e_{ij} \tag{1}$$

In this equation, X_{ij} denotes the concentration of the jth species in the ith sample, g_{ik} signifies the contribution of the kth source to the ith sample, f_{kj} is the profile of the jth species in the kth source, e_{ij} is the residual error for the jth species in the ith sample, and p refers to the total number of sources identified. The source contribution matrix and source profile matrix are the set of all g_{ik} and f_{kj} together.

The objective of employing the PMF model is to minimize the objective function $Q(E)$, defined as follows:

$$Q(E) = \sum_{i=1}^{n} \sum_{j=1}^{m} \left(\frac{e_{ij}}{\sigma_{ij}} \right)^2 \tag{2}$$

σ_{ij} means the uncertainty for the jth species in the ith sample. Here, uncertainty is an important part of the input of the PMF model. Zeroes and negatives are not permitted for either the detection limit or the percent uncertainty. If the concentration is less than or equal to the Method Detection Limit (MDL) provided, the uncertainty is calculated using a fixed fraction of the MDL (Equation (3)). If the concentration is greater than the MDL provided, the calculation is based on a user-provided fraction of the concentration and MDL (Equation (4)).

$$Unc = \frac{5}{6} \times MDL \tag{3}$$

$$Unc = \sqrt{(Error\ fraction \times Conc.)^2 + (0.5 \times MDL)^2} \tag{4}$$

where the $Error\ fraction$ (EF) encapsulates the aggregate deviations that occur throughout the entire experimental procedure, commencing with instrumental sampling and culminating with completion of the sample analysis. This encompasses potential sources of error such as instrumental instability, operational variability, and other elements that could compromise data precision. While specific EF values are often not explicitly stated in the majority of studies, a general range of 0.1 to 0.5 is commonly adopted as a reference [41].

Factor identification stands as a critical phase in the PMF analysis process, entailing the correlation of model-resolved factors with tangible sources of pollution. Employing marker species for this purpose is a method broadly recognized and utilized across the field. Our research approaches factor identification by drawing upon and consolidating insights gleaned from existing scholarly works.

2.3. Health Risk Evaluation of VOCs

The potential inhalation risks were quantified by integrating toxicity values with exposure estimates. The US EPA Inhalation Unit Risk (IUR) and Reference Concentrations (RfCs) were utilized to assess the carcinogenic and non-carcinogenic risks associated with each VOC species, providing insights into the actual exposure levels for workers in operational environments [42].

$$EC = \frac{CA \times ET \times EF \times ED}{AT} \tag{5}$$

where EC represents the exposure concentration, CA is the ambient VOC concentration, ET means the daily exposure duration, EF refers to the exposure frequency per year, ED denotes the exposure duration in years, and AT is the average time for exposure calculations. The values of ET, EF, ED, and AT are 8 h/day, 365 day/year, 70 yr, and (70 × 365 × 24) h, respectively.

The non-carcinogenic risk was evaluated using the hazard quotient (HQ):

$$HQ = \frac{EC}{RfC \times 1000} \tag{6}$$

The RfC values for specific VOCs were sourced from databases such as the Integrated Risk Information System (IRIS), the United States Agency for Toxic Substances and Disease Registry, or the California Office of Environmental Health Hazard Assessment (OEHHA) (Table S1).

The Hazard Index (HI), which is the sum of HQs for multiple pollutants, was used to assess the overall non-carcinogenic risk:

$$\mathrm{HI} = \sum HQ_i \tag{7}$$

For carcinogenic risks, the risk was calculated as follows:

$$Risk = EC \times IUR \tag{8}$$

Here, IUR represents the unit risk for a specific VOC species, which was also sourced from IRIS and OEHHA.

2.4. Health Risk Evaluation of Trace Metals

Drawing upon the research findings from the USEPA IRIS and the International Agency for Research on Cancer (IARC), health risks were categorized into carcinogenic and non-carcinogenic effects [43]. The health risk assessment model recommended by USEPA was employed, with the following computational approach:

$$ADD(LADD) = \frac{C_i \times IR \times EF \times ED}{BW \times AT} \tag{9}$$

where ADD is the average daily dose for non-carcinogenic metals; $LADD$ is the lifetime average daily dose for carcinogenic metals; C_i is the concentration of trace elements in PM fractions; IR refers to the ingestion rate; EF represents exposure frequency; ED means the exposure duration; BW denotes the average body weight; and AT is the average time for exposure. The values of IR, EF, ED, BW, and AT are 20 m^3/day, 350 day/year, 24 yr, 70 kg, and 24 $\times$ 365 days for non-carcinogens and 70 $\times$ 365 days for carcinogens.

Health risks were quantified using the hazard quotient (HQ) for non-carcinogenic effects and the incremental lifetime cancer risk ($ILCR$) for carcinogenic effects:

$$HQ = \frac{ADD}{RfD} \tag{10}$$

$$ILCR = LADD \times CSF \tag{11}$$

Here, RfD denotes the reference dose, and CSF means the cancer slope factor. The parameters for these equations were obtained from the USEPA (Table S2). An HQ value greater than one for individual trace elements or the cumulative HQ indicates a potential for non-carcinogenic effects due to exposure. The acceptable risk level for carcinogenic effects was set at 10^{-6}, representing a one in one million lifetime cancer risk.

2.5. Health Risk of Sources

Firstly, we needed to obtain the concentration contributions of every factor in the PMF. Subsequently, these contributions were multiplied by every species with a health risk present within the profiles of each factor. These results elucidated the specific contributions of various factors to the concentration of health risk species.

Second, with the obtained factor-specific contribution concentrations of health risk species, the health risk assessment formulas delineated in Sections 2.3 and 2.4 were integrated. This integration facilitated the quantification of health risks attributable to the presence of these species within each factor. The aggregate of these health risks across all factors for a particular species was the total health risk of the species. Furthermore, the ratio of the health risk attributable to each factor relative to the total health risk needed to be calculated.

Finally, we obtained the source apportionment of health risk for each health risk species building upon the outcomes of the preceding step. These results could be used to evaluate the health risk of the sources.

3. Results and Discussion

3.1. Characteristics of $PM_{2.5}$ and VOCs in Tianjin

The year 2019 witnessed the delineation of total volatile organic compound (TVOC) concentrations and their respective proportions, as meticulously detailed in Table S3 and illustrated in Figure 1a,b. The average TVOC concentration peaked at 24.2 parts per billion by volume (ppbv), accompanied by a standard deviation of 22.5 ppbv. The mean concentrations for alkanes, alkenes, aromatics, and alkynes were recorded as 15.0, 4.8, 2.3, and 2.2 ppbv, respectively. These figures correspond to proportions of 61.9%, 19.8%, 9.4%, and 8.9%. of the total TVOC, respectively. Alkanes notably dominated with a significant edge in both concentration and proportion over other VOC subclasses. Propane and ethane emerged as the predominant alkane species, with concentrations of 4.6 ppbv

(30.4%) and 2.9 ppbv (19.1%), respectively. Their origins can be primarily traced back to the evaporation and incomplete combustion of liquefied petroleum gas (LPG) and natural gas (NG), underscoring the substantial influence of NG or LPG in the area. The diagnostic ratio of isobutane to n-butane, standing at 0.7, highlights the notable impact of natural gas emissions, while the iso-pentane to n-pentane ratio of approximately 1.7 signaling the significant influence of gasoline evaporation. Among the alkenes, ethylene was found to have the highest proportion, accounting for 54.8%. In the aromatic category, toluene and the combined proportion of m-/p-xylene were the most prevalent, representing 28.9%, 29.0%, and 20.0%, respectively. The collective concentration of benzene, toluene, ethylbenzene, and xylene (BTEX) reached 2.1 ppbv, which encompasses 95.3% of the total aromatic content. This prevalence likely mirrors the substantial influence of solvent usage and vehicular emissions within the region. The toluene-to-benzene (T/B) ratio, with a value of 1.0, indicates that emissions from combustion processes are a source that merits considerable attention. In this study, the alkanes claimed the largest share of the VOC composition, succeeded by alkenes and aromatics. These findings are congruent with prior research conducted by Liu et al. [44]. and Li et al. [45] in the Tianjin and Beijing–Tianjin–Hebei (BTH) regions. However, it is important to acknowledge the marked variability in the primary VOC species across different regions within China. In some locales, aromatics hold a significant, if not dominant, position, as reported by Zhang et al. [46]. This underscores the geographically dependent nature of air pollutant composition and the necessity for localized air quality management strategies.

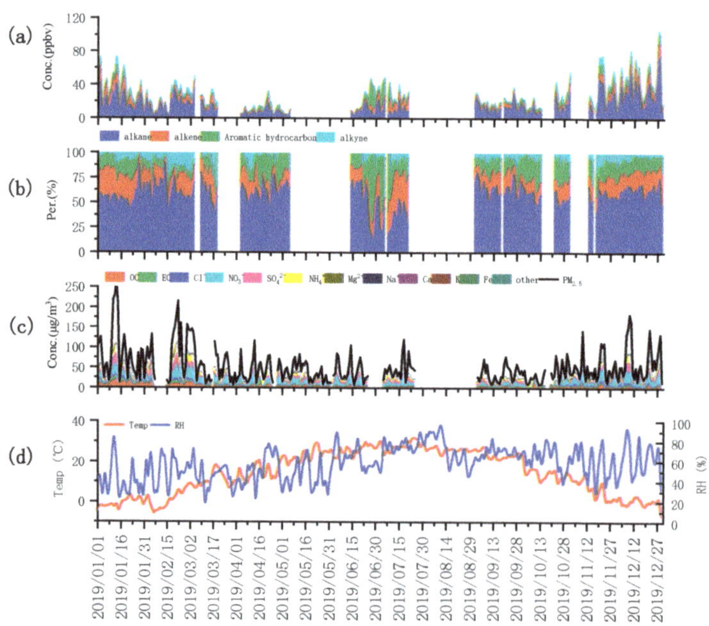

Figure 1. Time series of concentrations (**a**), percentages (**b**) of VOCs, aerosol concentrations and compounds (**c**) and meteorological parameters (**d**).

This 2019 study presents a detailed analysis of particulate matter concentrations and their chemical composition within the research area, as elegantly illustrated in Figure 1c. The effective online monitoring revealed a particulate matter concentration of 54.7 ± 50.7 μg/m^3. Within this matrix, water-soluble ions predominated, constituting 39% of the particulate load, trailed by carbon components at 12%, elemental constituents at 4%, and a diverse array of other components that account for 45%. NO_{3^-} was identified as the principal component of the particulate matter, with a concentration of 8.7 μg/m^3. The remaining constituents are sequentially ordered by their abundance, including NH_{4^+}, $SO_{4^{2-}}$, OC, EC, Cl^-, Na^+, K, Fe, Ca, Zn, Mn, and others. The compositional analysis underscores the predominance of secondary particulate matter-related components, highlighting the imperative to address secondary pollution, particularly the formation of secondary nitrates. OC and EC are correlated with combustion sources, indicating that emissions from motor vehicles and coal combustion continue to be pivotal contributors to primary emissions. Tianjin, situated as a coastal city, exhibits elevated levels of Cl^-, likely attributed to the influence of marine aerosols. Moreover, chloride emissions have also been associated with combustion processes. The presence of elements such as Fe, Ca, and Mn may be indicative of crustal dust contributions, while these elements, along with others, also suggest a linkage to industrial emissions.

3.2. Source Apportionment of $PM_{2.5}$ and VOCs

The PMF model was used to analyze the sources of VOCs and particulate matter. The species included in the model needed to be screened, guided by stringent criteria: (1) the presence of species unique to each source profile, (2) their prevalence at elevated concentrations, (3) the capacity to compensate for potential sample volatilization, (4) the exclusion of highly reactive species that could skew results, and (5) an adequate representation of species to authentically reflect the actual source apportionment. The results are shown in Figure 2.

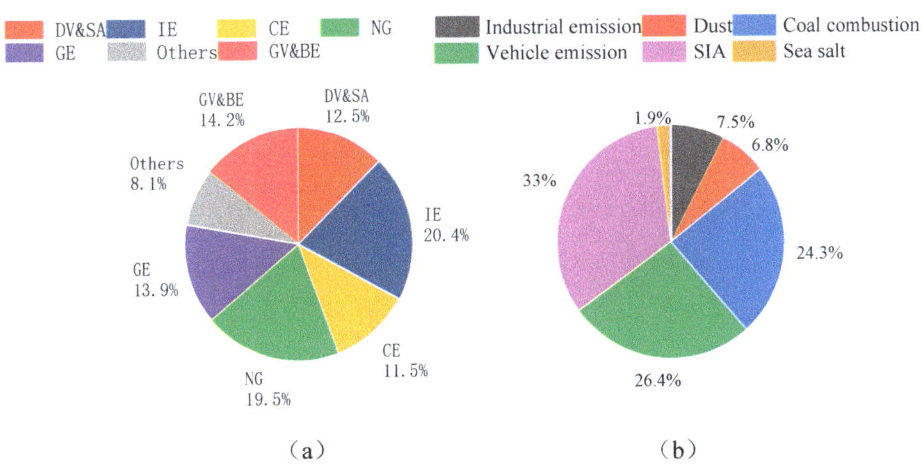

Figure 2. Source contribution of VOCs (**a**) and $PM_{2.5}$ (**b**) in Tianjin.

For VOCs, we meticulously selected 31 VOCs for the source contribution analysis. Following a meticulous evaluation process, we identified seven distinct factors for the PMF calculations, with their profiles for the year 2019, elegantly depicted in Figures 2a and S1. Factor 1 is distinguished by its significant loading of aromatic hydrocarbons, including o-xylene, m-/p-xylene, toluene, ethylbenzene, and 1,2,4-trimethylbenzene. Notably, substantial concentrations of n-nonane, n-heptane, and n-hexane were also detected. C_6 aromatics such as 1,2,4-trimethylbenzene and 1,3,5-trimethylbenzene are predominantly traced back to organic solvents utilized in the coating industry [47]. Trimethylbenzenes are established as principal constituents in a vast array of solvents employed across manufacturing sectors [48]. Toluene and ethylbenzene are byproducts of organic solvent applications [49]. The presence of long-chain alkanes, namely n-nonane, n-heptane, and n-hexane, in diesel exhaust fumes positions them as quintessential markers for diesel vehicle emissions [50]. Consequently, Factor 1 is delineated as an amalgam of diesel vehicle and solvent application (DV&SA). In Factor 2, we observed pronounced levels of ethylene and propylene, alongside n-pentene. Ethylene and propylene are emblematic of the petrochemical industry [51]. n-pentane as a key ingredient in the production of polystyrene foam, further substantiates the industrial linkage [49]. Thus, Factor 2 is demarcated as an industrial emission (IE). Factor 3 is emblematic of acetylene, recognized for its signature association with combustion sources [52]. Acetylene's presence is intrinsically linked to the phenomenon of incomplete combustion predominantly emanating from the exhaust of internal combustion engines [53]. Accordingly, Factor 3 is categorized under combustion emission (CE). Factor 4 is characterized by ethane, a species quintessential to natural gas, thereby being classified as an NG volatilization (NG). Factor 5 exhibits a high concentration ratio of isobutane, 2-methylpentane, and 3-methylpentane, along with 1-butene. 2-methylpentane as a species is predominantly associated with gasoline-fueled vehicles, 1-butene primarily to gasoline volatilization [54], and C_4-C_7 alkanes as the principal volatiles in the gasoline headspace [50], thereby cementing the identification of Factor 5 as gasoline evaporation (GE). Factor 6 is marked by elevated concentrations of methylcyclohexane, accompanied by a notable presence of alkanes, alkenes, and aromatic hydrocarbons. While methylcyclohexane is recognized as a significant species for diesel volatilization, the industrial fingerprint of the accompanying species remains ambiguous. Thus, Factor 6 is posited as an 'others' category that includes diesel evaporation. Factor 7 is predominantly associated with isoprene, which constitutes 60% of its profile and stands as a significant bio-emission indicator [55]. Consequently, it is classified as a biogenic emission source. The high percentage of C_4-C_6 low-carbon alkane species, including pentane, butane, 3-methylhexane, 1,3-butadiene, 2,3-dimethylbutane, and cyclohexane, further corroborates the vehicular exhaust signature [56]. Therefore, Factor 7 is identified as a composite source of gasoline vehicle emissions and biogenic emission (GV&BE). In general, IE was responsible for the largest fraction of TVOC mass on average (20%), followed by NG (19%), GV&BE (14%), GE (14%), DV&SA (13%), CE (12%), and others (8%).

Twenty strictly screened components were incorporated into the PMF model for particle source analysis, and six factors were analyzed (Figure S2). The source contributions are elegantly portrayed in Figure 2b. The first factor, composed of Mn, Fe, Zn, Pb, OC, and Cl^-, is attributed to industrial sources. Fe and Mn originate from emissions during the steel smelting process [57], while Zn and Pb are associated with non-ferrous metal processing, such as emissions from galvanization [2]. Cl^- is likely emitted from the production and use of chlorine-containing products or combustion emission. After all, Tianjin has long been renowned for its developed industry. The second factor consists of Ca, Ti, Fe, Mn, and Ba. These components are widely present in crustal dust; thus, this factor is attributed to urban dust resuspension sources. The third factor includes OC, EC, NH_4^+, $SO_{4^{2-}}$, As, Fe, Cu, and Zn and is assigned to coal combustion. Coal is the primary energy source for space heating in northern China during the cold season, and its combustion typically results in emissions of OC and EC. Studies have found a close relationship between As and coal combustion [58]. Previous research by Dai et al. [59] reported that coal burned in domestic stoves during winter is a major source of environmental sulfates, with sulfate reacting with ambient ammonia to form ammonium salts. The fourth factor, traffic emissions, is primarily composed of OC and EC, along with minor amounts of NH_4^+, Cl^-, Mg^{2+}, K, Ca, and Pb. Cl and K have been reported to come from vehicular emissions [60]. The online data lack components of Al and Si, and Ca can be considered to come from road dust [61]. Studies have indicated that motor vehicles may be a significant source of NH_4^+ in urban areas [62]. The fifth factor, characterized by high concentrations of NO_{3^-}, $SO_{4^{2-}}$, and NH_4^+, is identified as secondary inorganic aerosol (SIA), consistent with the typical mixed SIA factor reported in other studies [63,64]. The sixth factor comprises components such as Na^+, Mg^{2+}, Ti, Ni, EC, OC, Cl^-, and $SO_{4^{2-}}$. Na^+, Mg^{2+}, Cl^-, and $SO_{4^{2-}}$ are typical components of sea salt [65], and the Bohai Gulf is located approximately 30 km southeast of the site. Sea salt has previously been considered a source of particulate matter for Tianjin and the surrounding coastal and inland areas. Ti, Ni, EC, and OC are likely derived from the mixing and aging with other particles along the transport pathways. In general, SIA was responsible for the largest fraction of $PM_{2.5}$ mass on average (32.7%), followed by traffic emission (26.2%), coal combustion (24.1%), industry emissions (7.5%), dust (6.7%), and sea salt (1.9%). On average, traffic emission appears to be the predominant primary source of $PM_{2.5}$ for the measurement campaign.

3.3. Health Risk of $PM_{2.5}$ and VOCs

Figure 3 presents an assessment of the health risks posed by particulate matter and VOCs in Tianjin City for the year 2019. The non-carcinogenic risk, quantified by the hazard quotient (HQ), associated with trace metals in particulate matter—namely As, Cr, Ni, Cu, Pb, Zn, and Mn—is detailed in Figure 3a. The HQ values for these metals are ordered as follows: Cr > Mn > As > Cu > Pb > Zn > Ni, ranging from 1.7×10^{-4} to 9.4×10^{-2}. With all individual HQ values significantly below unity and the cumulative HQ also below the threshold of 1, it is evident that the non-carcinogenic risks are well within safe limits, indicating no cumulative non-carcinogenic effects from exposure to these trace metals. The carcinogenic risk, expressed as the incremental lifetime cancer risk (ILCR), for the selected trace metals (As, Cr, and Ni) in particulate matter is depicted in Figure 3b. The ILCR values for Cr, As, and Ni are 3.8×10^{-5}, 3.8×10^{-6}, and 9.9×10^{-7}, respectively, with chromium and arsenic presenting a higher risk in the order of Cr > As > Ni. The ILCR benchmarks indicate that a value below 1×10^{-6} is considered to carry no carcinogenic risk, a value between 1×10^{-4} and 1×10^{-6} suggests a potential carcinogenic risk, and a value exceeding 1×10^{-4} indicates a definite carcinogenic risk. The ILCR for chromium and arsenic falls within the range of potential risk, while nickel's ILCR is below the threshold for negligible risk. This signifies a latent carcinogenic risk to the residents of Tianjin from exposure to chromium and arsenic. The aggregate ILCR, positioned between 1×10^{-4} and 1×10^{-6}, underscores a certain degree of health hazard and a potential carcinogenic threat due to long-term exposure via the respiratory pathway.

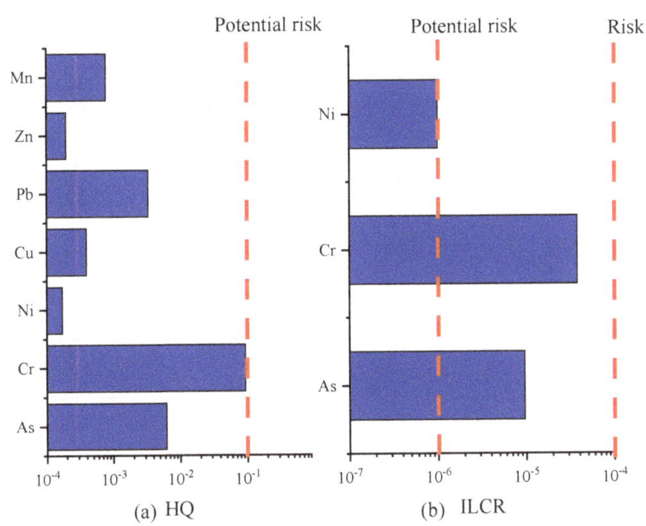

Figure 3. Health risks of trace metals in Tianjin. The result of hazard quotient (**a**) and incremental lifetime cancer risk (**b**).

In the realm of VOCs, a health risk assessment was conducted on harmful species such as n-hexane, cyclohexane, and aromatic hydrocarbon compounds, with the non-carcinogenic risk (HQ) portrayed in Figure 4. The HQ values for these species fluctuate between 1.5×10^{-5} and 3.6×10^{-2}, culminating in a total Hazard Index (HI) of 0.08. An

HI value below the safety threshold of 1 demarcates the absence of a cumulative non-carcinogenic impact. Furthermore, the carcinogenic risk of benzene, a prominent VOC, was calculated. The carcinogenic threshold of VOCs is the same as that of heavy metals. Benzene's carcinogenic risk, at 6×10^{-6}, lies within the bracket of potential risk, signifying a latent carcinogenic threat to human health due to long-term environmental exposure.

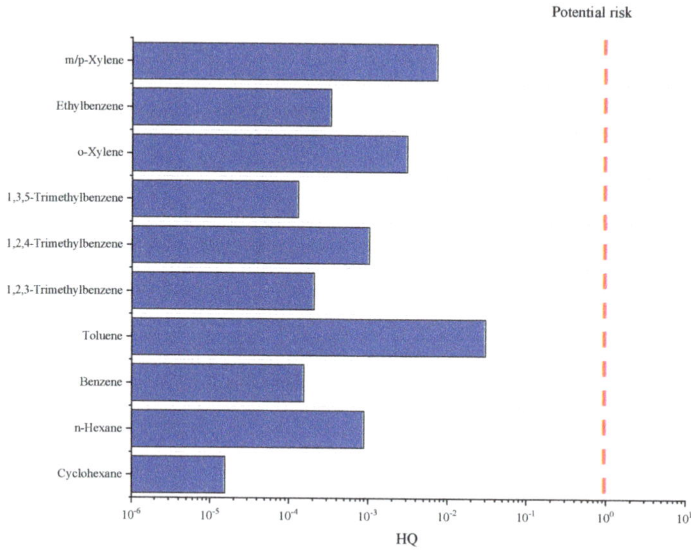

Figure 4. Hazard quotient of VOCS in Tianjin.

3.4. Health Risk of Source

Drawing upon the outcomes of source apportionment analyses for $PM_{2.5}$ and VOCs, this study has conducted a comprehensive assessment of the non-carcinogenic and carcinogenic health risk potentials associated with As and Cr, as well as benzene. The delineation of these risk sources is depicted in the accompanying Figures 5 and 6.

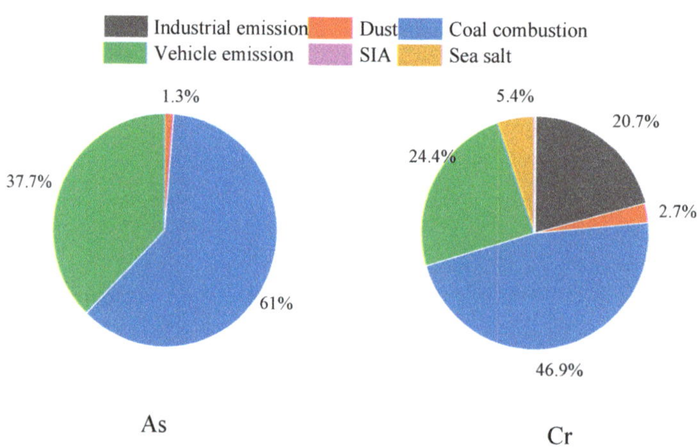

Figure 5. Sources of trace metals with health risk (As and Cr).

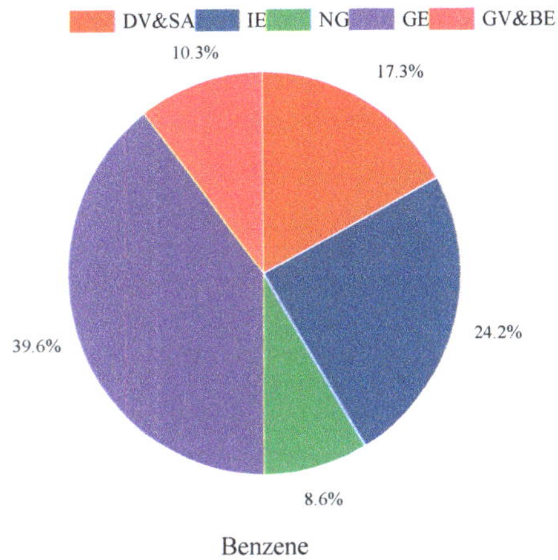

Figure 6. Sources of VOCs with health risk (benzene).

The non-carcinogenic health risk potential for arsenic is predominantly linked to emissions from coal combustion, which accounts for a substantial 61% of the total. Vehicle emissions also contribute significantly, with a 37.7% share. Notably, contributions from dust sources are minimal, at approximately 1.3%, while the impacts of industrial emissions, secondary formation, and marine aerosols are found to be negligible. In the case of chromium, the non-carcinogenic health risk potential is primarily driven by coal combustion, which leads with a 46.9% contribution. Vehicle emissions follow closely, constituting 24.4% of the total risk. Industrial emissions are a significant factor, contributing 20.7%. The influence of marine aerosols and dust sources is comparatively minor, at 5.4% and 2.7%, respectively. It is evident that the control of emissions from coal combustion and vehicular sources is pivotal in mitigating the non-carcinogenic health risk potentials of heavy metals such as As and Cr in particulate matter.

Turning our attention to benzene, a known carcinogen, its health risk potential is attributed to several key sources (Figure 6). The most significant contributor is the volatilization of fuels, which encompasses 48.2% of the total risk, with GE accounting for 39.6% and natural gas (NG) for 8.6%. Emissions from vehicles and the use of solvents are also substantial, contributing 27%, with DV&SA collectively accounting for 17.3%, and GV&BE contributing 10%. Additionally, IE still contributes a notable 24.2%. The findings underscore the critical need for targeted strategies to control the volatilization of fuels, curtail industrial emissions, and regulate vehicular and solvent emissions. Such measures are essential in reducing the carcinogenic health risk potential of benzene in VOCs, thereby safeguarding public health and contributing to the broader goals of air quality management.

4. Conclusions

We conducted an analysis of the atmospheric pollution characteristics in Tianjin City using a 2019 online dataset of $PM_{2.5}$ and VOCs. The atmospheric particulate matter in Tianjin is predominantly composed of water-soluble ions, complemented by carbonaceous components, with nitrates emerging as the most significant constituent. Within the VOC profile, alkanes are the most abundant, succeeded by alkenes and aromatic hydrocarbons, with compounds such as propane, ethane, ethylene, toluene, and benzene exhibiting notably high concentrations.

Utilizing receptor modeling techniques, we resolved the pollution sources of particulate matter and VOCs into distinct categories. For particulate matter, six major source types were identified—industrial emission, vehicle emissions, coal combustion, dust, secondary sources, and sea salt—with secondary sources, vehicular emissions, and coal combustion identified as the principal contributors. For VOCs, seven major source types were delineated—industrial emission, combustion emission, NG volatilization, gasoline evaporation, gasoline vehicles and biogenic emission, diesel vehicle and solvent application, and others—with industrial processes, NG volatilization, gasoline evaporation, and gasoline vehicles and biogenic emission recognized as the primary source categories.

In conclusion, we assessed the health impacts of the particulate matter and VOC components within the region. The non-carcinogenic risks associated with particulate matter and VOCs were determined to be negligible. Nevertheless, certain trace metals, namely As and Cr, along with benzene among the VOCs, were found to pose a potential carcinogenic risk, indicating the need for targeted monitoring and mitigation strategies to safeguard public health. This study conducted a risk attribution analysis for components in $PM_{2.5}$ and VOCs that pose health risks. The findings reveal that the primary sources of the hazardous constituents As and Cr in $PM_{2.5}$ are coal combustion and vehicular emissions, which account for a striking 99% and 70% of the total contributions to As and Cr, respectively. For the VOCs, the source of the high-risk component benzene is predominantly attributed to fuel evaporation, industrial emissions, and vehicular exhaust. These sources are identified as the principal contributors to the health risks associated with benzene exposure. The implications of these findings are clear: by intensifying the regulation and control of toxic and harmful components in coal combustion, industrial processes, and vehicular emissions, there is a significant potential to reduce the health risks posed by $PM_{2.5}$ and VOCs. Such measures are essential for safeguarding public health and achieving broader air quality management objectives.

Supplementary Materials: The following supporting information can be downloaded at https://www.mdpi.com/article/10.3390/toxics12090622/s1, Figure S1: Factor profile of VOCs; Figure S2: Factor profile of $PM_{2.5}$; Table S1: The values of RfC and IUR of the VOCs; Table S2: The values of RfD and CSF of the trace metals; Table S3. The concentrations and percentages of the VOC species.

Author Contributions: Conceptualization, Y.H., Q.D. and Y.F.; methodology, Y.H., F.W. and G.S.; software, Y.H., Q.D. and G.S.; formal analysis, Y.H. and Y.F.; investigation, D.L.; data curation, D.L.; writing—original draft preparation, Y.H.; writing—review and editing, F.W., Q.D. and Y.F.; supervision, Y.F.; project administration, Y.F. and G.S.; funding acquisition, Y.F. All authors have read and agreed to the published version of the manuscript.

Funding: This research was funded by the National Natural Science Foundation of China, grant number 42177085.

Institutional Review Board Statement: Not applicable.

Informed Consent Statement: Not applicable.

Data Availability Statement: The data are provided in tables and figures directly within the manuscript, and raw data are available via e-mail upon request to the corresponding author.

Conflicts of Interest: The authors declare no conflicts of interest.

References

1. Zheng, S.; Xu, X.; Zhang, Y.; Wang, L.; Yang, Y.; Jin, S.; Yang, X. Characteristics and sources of VOCs in urban and suburban environments in Shanghai, China, during the 2016 G20 summit. *Atmos. Pollut. Res.* **2019**, *10*, 1766–1779. [CrossRef]
2. Dai, Q.; Liu, B.; Bi, X.; Wu, J.; Liang, D.; Zhang, Y.; Feng, Y.; Hopke, P.K. Dispersion Normalized PMF Provides Insights into the Significant Changes in Source Contributions to $PM_{2.5}$ after the COVID-19 Outbreak. *Environ. Sci. Technol.* **2020**, *54*, 9917–9927. [CrossRef]
3. Wang, Q.; Fang, J.; Shi, W.; Dong, X. Distribution characteristics and policy-related improvements of PM2.5 and its components in six Chinese cities. *Environ. Pollut.* **2020**, *266*, 115299. [CrossRef] [PubMed]
4. Gao, M.; Liu, Z.; Zheng, B.; Ji, D.; Sherman, P.; Song, S.; Xin, J.; Liu, C.; Wang, Y.; Zhang, Q.; et al. China's emission control strategies have suppressed unfavorable influences of climate on wintertime PM 2.5 concentrations in Beijing since 2002. *Atmos. Chem. Phys.* **2020**, *20*, 1497–1505. [CrossRef]
5. Huang, X.-F.; Zou, B.-B.; He, L.-Y.; Hu, M.; Prévôt, A.S.H.; Zhang, Y.-H. Exploration of PM 2.5 sources on the regional scale in the Pearl River Delta based on ME-2 modeling. *Atmos. Chem. Phys.* **2018**, *18*, 11563–11580. [CrossRef]
6. Liu, Y.; Li, X.; Tang, G.; Wang, L.; Lv, B.; Guo, X.; Wang, Y. Secondary organic aerosols in Jinan, an urban site in North China: Significant anthropogenic contributions to heavy pollution. *J. Environ. Sci.* **2019**, *80*, 107–115. [CrossRef]

7. Wu, W.; Zhao, B.; Wang, S.; Hao, J. Ozone and secondary organic aerosol formation potential from anthropogenic volatile organic compounds emissions in China. *J. Environ. Sci.* **2017**, *53*, 224–237. [CrossRef]
8. Bari, M.A.; Kindzierski, W.B. Ambient volatile organic compounds (VOCs) in Calgary, Alberta: Sources and screening health risk assessment. *Sci. Total Environ.* **2018**, *631-632*, 627–640. [CrossRef]
9. Gao, J.; Zhang, J.; Li, H.; Li, L.; Xu, L.; Zhang, Y.; Wang, Z.; Wang, X.; Zhang, W.; Chen, Y.; et al. Comparative study of volatile organic compounds in ambient air using observed mixing ratios and initial mixing ratios taking chemical loss into account – A case study in a typical urban area in Beijing. *Sci. Total Environ.* **2018**, *628-629*, 791–804. [CrossRef]
10. Wu, S.; Tang, G.; Wang, Y.; Yang, Y.; Yao, D.; Zhao, W.; Gao, W.; Sun, J.; Wang, Y. Vertically decreased VOC concentration and reactivity in the planetary boundary layer in winter over the North China Plain. *Atmos. Res.* **2020**, *240*, 104930. [CrossRef]
11. Hui, L.; Liu, X.; Tan, Q.; Feng, M.; An, J.; Qu, Y.; Zhang, Y.; Deng, Y.; Zhai, R.; Wang, Z. VOC characteristics, chemical reactivity and sources in urban Wuhan, central China. *Atmos. Environ.* **2020**, *224*, 117340. [CrossRef]
12. Mozaffar, A.; Zhang, Y.-L.; Fan, M.; Cao, F.; Lin, Y.-C. Characteristics of summertime ambient VOCs and their contributions to O_3 and SOA formation in a suburban area of Nanjing, China. *Atmos. Res.* **2020**, *240*, 104923. [CrossRef]
13. Li, J.; Qin, Y.; Zhang, X.; Shan, B.; Liu, C. Emission Characteristics, Environmental Impacts, and Health Risks of Volatile Organic Compounds from Asphalt Materials: A State-of-the-Art Review. *Energy Fuels* **2024**, *38*, 4787–4802. [CrossRef]
14. Tran, H.-T.; Binh, Q.A.; Van Tung, T.; Pham, D.T.; Hoang, H.-G.; Hai Nguyen, N.S.; Xie, S.; Zhang, T.; Mukherjee, S.; Bolan, N.S. A critical review on characterization, human health risk assessment and mitigation of malodorous gaseous emission during the composting process. *Environ. Pollut.* **2024**, *351*, 124115. [CrossRef] [PubMed]
15. Wang, Y.; Ding, L.; Shi, Q.; Liu, S.; Qian, L.; Yu, Z.; Wang, H.; Lei, J.; Gao, Z.; Long, H.; et al. Volatile organic compounds (VOC) emissions control in iron ore sintering process: Recent progress and future development. *Chem. Eng. J.* **2022**, *448*, 137601. [CrossRef]
16. Mo, Z.; Shao, M.; Lu, S.; Niu, H.; Zhou, M.; Sun, J. Characterization of non-methane hydrocarbons and their sources in an industrialized coastal city, Yangtze River Delta, China. *Sci. Total Environ.* **2017**, *593-594*, 641–653. [CrossRef]
17. Maung, T.Z.; Bishop, J.E.; Holt, E.; Turner, A.M.; Pfrang, C. Indoor Air Pollution and the Health of Vulnerable Groups: A Systematic Review Focused on Particulate Matter (PM), Volatile Organic Compounds (VOCs) and Their Effects on Children and People with Pre-Existing Lung Disease. *Int. J. Environ. Res. Public Health* **2022**, *19*, 8752. [CrossRef]
18. Pál, L.; Lovas, S.; McKee, M.; Diószegi, J.; Kovács, N.; Szűcs, S. Exposure to volatile organic compounds in offices and in residential and educational buildings in the European Union between 2010 and 2023: A systematic review and health risk assessment. *Sci. Total Environ.* **2024**, *945*, 173965. [CrossRef]
19. Paterson, C.A.; Sharpe, R.A.; Taylor, T.; Morrissey, K. Indoor PM2.5, VOCs and asthma outcomes: A systematic review in adults and their home environments. *Environ. Res.* **2021**, *202*, 111631. [CrossRef]
20. Yang, D.-L.; Zhang, Z.-N.; Liu, H.; Yang, Z.-Y.; Liu, M.-M.; Zheng, Q.-X.; Chen, W.; Xiang, P. Indoor air pollution and human ocular diseases: Associated contaminants and underlying pathological mechanisms. *Chemosphere* **2023**, *311*, 137037. [CrossRef]
21. Zheng, H.; Kong, S.; Yan, Y.; Chen, N.; Yao, L.; Liu, X.; Wu, F.; Cheng, Y.; Niu, Z.; Zheng, S.; et al. Compositions, sources and health risks of ambient volatile organic compounds (VOCs) at a petrochemical industrial park along the Yangtze River. *Sci. Total Environ.* **2020**, *703*, 135505. [CrossRef] [PubMed]
22. Masih, A.; Lall, A.S.; Taneja, A.; Singhvi, R. Exposure levels and health risk assessment of ambient BTX at urban and rural environments of a terai region of northern India. *Env. Pollut.* **2018**, *242*, 1678–1683. [CrossRef] [PubMed]
23. Tamrakar, A.; Pervez, S.; Verma, M.; Majumdar, D.; Pervez, Y.F.; Candeias, C.; Dugga, P.; Mishra, A.; Verma, S.R.; Deb, M.K.; et al. BTEX in Ambient Air of India: A Scoping Review of their Concentrations, Sources, and impact. *Water Air Soil Pollut.* **2022**, *233*, 411. [CrossRef]
24. Bari, M.A.; Kindzierski, W.B. Concentrations, sources and human health risk of inhalation exposure to air toxics in Edmonton, Canada. *Chemosphere* **2017**, *173*, 160–171. [CrossRef]
25. Li, L.; Li, H.; Zhang, X.; Wang, L.; Xu, L.; Wang, X.; Yu, Y.; Zhang, Y.; Cao, G. Pollution characteristics and health risk assessment of benzene homologues in ambient air in the northeastern urban area of Beijing, China. *J. Environ. Sci.* **2014**, *26*, 214–223. [CrossRef]
26. Zhou, J.; You, Y.; Bai, Z.; Hu, Y.; Zhang, J.; Zhang, N. Health risk assessment of personal inhalation exposure to volatile organic compounds in Tianjin, China. *Sci. Total Environ.* **2011**, *409*, 452–459. [CrossRef]
27. Gavett, S.H.; Haykal-Coates, N.; Copeland, L.B.; Gilmour, J.H.I. Metal composition of ambient PM2.5 influences severity of allergic airways disease in mice. *Environ. Health Perspect* **2003**, *111*, 1471–1477. [CrossRef]
28. Xu, H.; Han, S.; Bi, X.; Zhao, Z.; Zhang, L.; Yang, W.; Zhang, M.; Chen, J.; Wu, J.; Zhang, Y.; et al. Atmospheric metallic and arsenic pollution at an offshore drilling platform in the Bo Sea: A health risk assessment for the workers. *J. Hazard. Mater.* **2016**, *304*, 93–102. [CrossRef]
29. Joseph, T.; Dubey, B.; McBean, E.A. Human health risk assessment from arsenic exposures in Bangladesh. *Sci. Total Environ.* **2015**, *527-528*, 552–560. [CrossRef]
30. Żukowska, J.; Biziuk, M. Methodological Evaluation of Method for Dietary Heavy Metal Intake. *J. Food Sci.* **2008**, *73*, R21–R29. [CrossRef]
31. Lippmann, M.; Ito, K.; Hwang, J.-S.; Maciejczyk, P.; Chen, L.-C. Cardiovascular effects of nickel in ambient air. *Environ. Health Perspect* **2006**, *114*, 1662–1669. [CrossRef] [PubMed]

32. Zhao, M.; Zhang, Y.; Ma, W.; Fu, Q.; Yang, X.; Li, C.; Zhou, B.; Yu, Q.; Chen, L. Characteristics and ship traffic source identification of air pollutants in China's largest port. *Atmos. Environ.* **2013**, *64*, 277–286. [CrossRef]
33. Zwolak, I. Vanadium carcinogenic, immunotoxic and neurotoxic effects: A review of in vitro studies. *Toxicol. Mech. Methods* **2013**, *24*, 1–12. [CrossRef]
34. Palmer, K.T. Inflammatory responses to the occupational inhalation of metal fume. *Eur. Respir. J.* **2006**, *27*, 366–373. [CrossRef] [PubMed]
35. Zhang, X.; Yang, L.; Li, Y.; Li, H.; Wang, W.; Ye, B. Impacts of lead/zinc mining and smelting on the environment and human health in China. *Environ. Monit. Assess.* **2011**, *184*, 2261–2273. [CrossRef]
36. Choi, Y.-H.; Kim, H.J.; Sohn, J.R.; Seo, J.H. Occupational exposure to VOCs and carbonyl compounds in beauty salons and health risks associated with it in South Korea. *Ecotoxicol. Environ. Saf.* **2023**, *256*. [CrossRef] [PubMed]
37. Sakunkoo, P.; Thonglua, T.; Sangkham, S.; Jirapornkul, C.; Limmongkon, Y.; Daduang, S.; Tessiri, T.; Rayubkul, J.; Thongtip, S.; Maneenin, N.; et al. Human health risk assessment of PM2.5-bound heavy metal of anthropogenic sources in the Khon Kaen Province of Northeast Thailand. *Heliyon* **2022**, *8*, e09572. [CrossRef]
38. Xue, Q.; Tian, Y.; Liu, X.; Wang, X.; Huang, B.; Zhu, H.; Feng, Y. Potential Risks of PM2.5-Bound Polycyclic Aromatic Hydrocarbons and Heavy Metals from Inland and Marine Directions for a Marine Background Site in North China. *Toxics* **2022**, *10*, 32. [CrossRef]
39. Paatero, P. Least squares formulation of robust non-negative factor analysis. *Chemom. Intell. Lab. Syst.* **1997**, *37*, 23–35. [CrossRef]
40. Paatero, P.; Tapper, U. Positive matrix factorization: A non-negative factor model with optimal utilization of error estimates of data values. *Environmetrics* **1994**, *5*, 111–126. [CrossRef]
41. USEPA. Positive Matrix Factorization Model for Environmental Data Analyses. Available online: https://www.epa.gov/air-research/positive-matrix-factorization-model-environmental-data-analyses (accessed on 1 May 2024).
42. USEPA. Integrated Risk Information System (IRIS). Available online: http://www.epa.gov/iris (accessed on 1 May 2024).
43. Zhang, X.; Eto, Y.; Aikawa, M. Risk assessment and management of PM2.5-bound heavy metals in the urban area of Kitakyushu, Japan. *Sci. Total Environ.* **2021**, *795*, 148748. [CrossRef] [PubMed]
44. Liu, B.; Liang, D.; Yang, J.; Dai, Q.; Bi, X.; Feng, Y.; Yuan, J.; Xiao, Z.; Zhang, Y.; Xu, H. Characterization and source apportionment of volatile organic compounds based on 1-year of observational data in Tianjin, China. *Environ. Pollut.* **2016**, *218*, 757–769. [CrossRef] [PubMed]
45. Li, L.; Xie, S.; Zeng, L.; Wu, R.; Li, J. Characteristics of volatile organic compounds and their role in ground-level ozone formation in the Beijing-Tianjin-Hebei region, China. *Atmos. Environ.* **2015**, *113*, 247–254. [CrossRef]
46. Zhang, X.; Yin, Y.; Wen, J.; Huang, S.; Han, D.; Chen, X.; Cheng, J. Characteristics, reactivity and source apportionment of ambient volatile organic compounds (VOCs) in a typical tourist city. *Atmos. Environ.* **2019**, *215*, 116898. [CrossRef]
47. Yuan, B.; Shao, M.; Lu, S.; Wang, B. Source profiles of volatile organic compounds associated with solvent use in Beijing, China. *Atmos. Environ.* **2010**, *44*, 1919–1926. [CrossRef]
48. Zhou, X.; Li, Z.; Zhang, T.; Wang, F.; Wang, F.; Tao, Y.; Zhang, X.; Wang, F.; Huang, J. Volatile organic compounds in a typical petrochemical industrialized valley city of northwest China based on high-resolution PTR-MS measurements: Characterization, sources and chemical effects. *Sci. Total Environ.* **2019**, *671*, 883–896. [CrossRef]
49. Song, C.; Liu, B.; Dai, Q.; Li, H.; Mao, H. Temperature dependence and source apportionment of volatile organic compounds (VOCs) at an urban site on the north China plain. *Atmos. Environ.* **2019**, *207*, 167–181. [CrossRef]
50. Liu, Y.; Shao, M.; Fu, L.; Lu, S.; Zeng, L.; Tang, D. Source profiles of volatile organic compounds (VOCs) measured in China: Part I. *Atmos. Environ.* **2008**, *42*, 6247–6260. [CrossRef]
51. Jobson, B.T.; Berkowitz, C.M.; Kuster, W.C.; Goldan, P.D.; Williams, E.J.; Fesenfeld, F.C.; Apel, E.C.; Karl, T.; Lonneman, W.A.; Riemer, D. Hydrocarbon source signatures in Houston, Texas: Influence of the petrochemical industry. *J. Geophys. Res. Atmos.* **2004**, *109*. [CrossRef]
52. Zheng, H.; Kong, S.; Xing, X.; Mao, Y.; Hu, T.; Ding, Y.; Li, G.; Liu, D.; Li, S.; Qi, S. Monitoring of volatile organic compounds (VOCs) from an oil and gas station in northwest China for 1 year. *Atmos. Chem. Phys.* **2018**, *18*, 4567–4595. [CrossRef]
53. Barletta, B.; Meinardi, S.; Sherwood Rowland, F.; Chan, C.-Y.; Wang, X.; Zou, S.; Yin Chan, L.; Blake, D.R. Volatile organic compounds in 43 Chinese cities. *Atmos. Environ.* **2005**, *39*, 5979–5990. [CrossRef]
54. Zheng, J.; Shao, M.; Che, W.; Zhang, L.; Zhong, L.; Zhang, Y.; Streets, D. Speciated VOC Emission Inventory and Spatial Patterns of Ozone Formation Potential in the Pearl River Delta, China. *Environ. Sci. Technol.* **2009**, *43*, 8580–8586. [CrossRef] [PubMed]
55. Yuan, Z.; Lau, A.K.H.; Shao, M.; Louie, P.K.K.; Liu, S.C.; Zhu, T. Source analysis of volatile organic compounds by positive matrix factorization in urban and rural environments in Beijing. *J. Geophys. Res. Atmos.* **2009**, *114*. [CrossRef]
56. Zhang, Y.; Yang, W.; Simpson, I.; Huang, X.; Yu, J.; Huang, Z.; Wang, Z.; Zhang, Z.; Liu, D.; Huang, Z.; et al. Decadal changes in emissions of volatile organic compounds (VOCs) from on-road vehicles with intensified automobile pollution control: Case study in a busy urban tunnel in south China. *Environ. Pollut.* **2018**, *233*, 806–819. [CrossRef]
57. Dai, Q.; Li, L.; Yang, J.; Liu, B.; Bi, X.; Wu, J.; Zhang, Y.; Yao, L.; Feng, Y. The fractionation and geochemical characteristics of rare earth elements measured in ambient size-resolved PM in an integrated iron and steelmaking industry zone. *Environ. Sci. Pollut. Res.* **2016**, *23*, 17191–17199. [CrossRef]
58. Bi, X.; Dai, Q.; Wu, J.; Zhang, Q.; Zhang, W.; Luo, R.; Cheng, Y.; Zhang, J.; Wang, L.; Yu, Z.; et al. Characteristics of the main primary source profiles of particulate matter across China from 1987 to 2017. *Atmos. Chem. Phys.* **2019**, *19*, 3223–3243. [CrossRef]

59. Dai, Q.; Bi, X.; Song, W.; Li, T.; Liu, B.; Ding, J.; Xu, J.; Song, C.; Yang, N.; Schulze, B.C.; et al. Residential coal combustion as a source of primary sulfate in Xi'an, China. *Atmos. Environ.* **2019**, *196*, 66–76. [CrossRef]
60. Cui, M.; Chen, Y.; Feng, Y.; Li, C.; Zheng, J.; Tian, C.; Yan, C.; Zheng, M. Measurement of PM and its chemical composition in real-world emissions from non-road and on-road diesel vehicles. *Atmos. Chem. Phys.* **2017**, *17*, 6779–6795. [CrossRef]
61. Wang, M.; Duan, Y.; Zhang, Z.; Huo, J.; Huang, Y.; Fu, Q.; Wang, T.; Cao, J.; Lee, S.-C. Increased contribution to PM2.5 from traffic-influenced road dust in Shanghai over recent years and predictable future. *Environ. Pollut.* **2022**, *313*, 120119. [CrossRef]
62. Hopke, P.K.; Querol, X. Is Improved Vehicular NOx Control Leading to Increased Urban NH_3 Emissions? *Environ. Sci. Technol.* **2022**, *56*, 11926–11927. [CrossRef]
63. Li, X.; Yang, K.; Han, J.; Ying, Q.; Hopke, P.K. Sources of humic-like substances (HULIS) in PM2.5 in Beijing: Receptor modeling approach. *Sci. Total Environ.* **2019**, *671*, 765–775. [CrossRef] [PubMed]
64. Bari, M.A.; Kindzierski, W.B. Characterization of air quality and fine particulate matter sources in the town of Hinton, Alberta. *Atmos. Pollut. Res.* **2018**, *9*, 84–94. [CrossRef]
65. Yao, X.; Zhang, L. Chemical processes in sea-salt chloride depletion observed at a Canadian rural coastal site. *Atmos. Environ.* **2012**, *46*, 189–194. [CrossRef]

toxics

MDPI

Review

Contamination Characterization, Toxicological Properties, and Health Risk Assessment of Bisphenols in Multiple Media: Current Research Status and Future Perspectives

Fangyun Long, Yanqin Ren *, Fang Bi *, Zhenhai Wu, Haijie Zhang, Junling Li, Rui Gao, Zhengyang Liu and Hong Li

State Key Laboratory of Environmental Criteria and Risk Assessment, Chinese Research Academy of Environmental Sciences, Beijing 100012, China
* Correspondence: renyq@craes.org.cn (Y.R.); bifang@craes.org.cn (F.B.)

Abstract: Bisphenols (BPs) are ubiquitous environmental endocrine disruptors that cause various human health hazards and pollute water, soil, and the atmosphere to varying degrees. Although various studies have investigated the pollution characteristics and health hazards of BPs in different media, a systematic review of BPs in the broader environmental context is still lacking. This study highlights the pollution characteristics, detection methods, and risk assessment status of BPs by combining relevant studies from both domestic and international sources, and their environmental distribution characteristics are summarized. The results show that BP pollution is a widespread and complex global phenomenon. Bisphenol A (BPA) remains the predominant component of BPs, which can damage the nervous and reproductive systems. At present, high-performance liquid chromatography–tandem mass spectrometry, high-performance liquid chromatography, and liquid chromatography–tandem mass spectrometry are the main detection methods used for BPs. BPs can also damage the reproductive system, leading to germ cell apoptosis and ovarian damage. Future research should focus on expanding the BP testing repertoire, advancing rapid detection techniques, elucidating toxic mechanisms, conducting comprehensive safety assessments, and developing systematic health risk assessment methods. These efforts will provide a scientific foundation for preventing and controlling emerging pollutants.

Keywords: bisphenols (BPs); detection method; potential hazards; physical properties; health assessment

1. Introduction

Emerging contaminants (ECs), which are a class of natural and synthetic chemicals and biological agents, have recently become a major focus of international environmental attention [1]. As ECs, endocrine-disrupting chemicals (EDCs) have gained increasing public attention and are now a topic of global concern regulated by international conventions. EDCs are defined as exogenous substances that cause adverse health effects in an intact organism [2]. EDCs have been reported to affect endocrine system functions, interfering with hormone action [3]. They have various sources, being commonly found in agricultural supplies, plastics, food contact materials, and personal care products. They have the potential to infiltrate the human body via various routes and affect the endocrine, reproductive, and nervous systems [4–7]. Bisphenols (BPs) are an important group of chemical raw materials commonly used to synthesize polymer materials, including bisphenol A (BPA) and its alternatives. These materials show favourable optical properties and exceptional high-temperature stability [8].

Meanwhile, BPs act as estrogen-like endocrine disruptors and potentially affect human health [9,10]. BPs can affect the reproductive neuroendocrine system via estrogen receptors and the aromatase pathway [10]. Further studies have shown that the immune system is particularly sensitive to BPA, especially mouse Th17 cells, which play a key role in cellular immune mechanisms [11]. Meanwhile, another study also found that some BPs can have a certain impact on the innate immune response; for example, BPA and bisphenol AF (BPAF) can reduce the transcriptional activity of RACK1 promoters, mRNA expression, and protein levels [12]. Today, dendritic cells are recognized at the center stage of adaptive, as well as innate, immune responses and are known to be the most important antigen-presenting cells [9]. Recently, experimental studies have shown that BPA and BPAF can significantly reduce the endocytotic capacity of human monocyte-derived dendritic cells (MDDCs) [9]. This also suggests that exposure to BPs may increase the risk of adaptive and innate immune responses in humans. In addition, BPs, as a class of endocrine disruptors, can further affect the regulation of other pathways, thus triggering tumor transformation [13]. Therefore, it is imperative to conduct more in-depth studies on the effects of BP exposure. BPA is one of the most representative BPs and typical EDCs. It was first synthesized in 1891 [5]. To date, it has been detected in various environmental media, including air, water, soil, dust, and sediment. According to statistical data, the annual production of BPA consistently increased by 4.6% from 2013 to 2019 [5,14–17]. Toxicological studies have demonstrated that BPA exerts considerable endocrine-disrupting effects, impacting human reproductive and nervous systems. Furthermore, evidence suggests that BPA exposure may be associated with autism spectrum disorder (ASD), polycystic ovary syndrome (PCOS), and other health conditions [18,19]. The use of BPA in packaging compounds for bottles, straws, and formulas has been prohibited in China, Canada, the European Union, and the United States (US) by the Food and Drug Administration (FDA) due to its multiple related health risks. These bans were implemented in 2011 (China), 2008 (Canada), 2011 (the European Union), and 2013 (the US). Moreover, France has also prohibited the use of BPA in any food or beverage packaging, with the ban taking effect in 2015 [16,20–22]. Studies have focused on finding alternatives to BPA due to increasing consumer demand and regulatory changes. Reports indicate that studies on BPA alternatives and their analogues have increased annually from 2000 to 2020 [23]. In January 2014, the US Environmental Protection Agency (EPA) Design for the Environment Program assessed chemical alternatives. The assessment included the potential impacts of BPA alternatives in thermal paper on human health and the environment [24]. BPA and its replacements may have multiple sources. For example, studies have discovered that BPA is commonly found in the polycarbonate plastics and epoxy resin industries. It is chemically bonded to polymers and difficult to disperse directly into the air. Therefore, solid waste incineration is one of the most significant sources of BPA in the environment [25]. Other bisphenol compounds are widely used as alternative compounds for BPA due to their similar structure, including bisphenol S (BPS), BPAF, bisphenol F (BPF), and bisphenol Z (BPZ) [8]. BPS has enhanced thermal and optical stability relative to BPA. It is predominantly used as an anti-corrosion agent in epoxy adhesives, a reagent in polymer reactions, and a developer [8,23]. BPAF is used as a crosslinker for fluoropolymers and as a monomer for other polymers, such as polyimides, polycarbonates, and polymers used in food contact applications [26]. Among the BPA alternatives, BPFA and BPS have the highest detection rates in human and environmental samples [27]. BPF is used in the production of thermal paper, in coating materials, and in the plastic processing industry [21]. BPZ is commonly used in the production of cured high-heat-resistant plastic materials [28]. Currently, BPs are widely used in various production and living applications [16]. However, there is a lack of global research investigating the presence of BPs in the environment. Moreover, BPs have a natural resistance to biodegradation, which poses a challenge for future treatment [8]. Thus, it is necessary to systematically summarize and address the previous research on BPs.

This study elucidates the transfer and pollution characteristics of BPs in three main environmental media (water, soil, and the atmosphere) via literature research and analyses. It also preliminarily investigates the transfer mode of BPs in these environments, providing a specific theoretical foundation for further study of BP pollution in the environment. A total of 21 BPs are assessed in terms of their physical properties, and their distribution coefficients are calculated using EPI software https://www.cdc.gov/epiinfo/index.html (accessed on 9 January 2025). The distribution in each phase is further examined. This study serves as a contribution to future targeted detection efforts. Furthermore, the main BP detection methods are summarized, the similarities and differences between the various methods are compared, and the potential risks of BPs in different media are roughly evaluated. The current prevention and control strategies for BPs are also summarized, providing technical and policy support for future BP pollution management efforts. This study provides a scientific basis for future research on the detection and health risk assessment of BPs, and it contributes to the development of scientific approaches for addressing emerging pollutants.

2. Pollution Characteristics

2.1. Physical Properties

In industrial production and application, the selection of chemicals based on analogous physical properties is paralleled by considering alternatives with structurally related compounds [29]. Thus, performing a comparative evaluation between the original chemical and its alternatives is crucial to ensure that the substitutes do not induce the same or more pronounced biological effects while maintaining optimal physical characteristics. This approach aims to reduce the potential risks associated with chemical hazards. BPA alternatives frequently show properties analogous to those of BPA or more superior structural characteristics [8,21,23,28]. To date, approximately 16 bisphenol analogues have been well documented for industrial applications [8]. Based on their structural characteristics, BPA alternatives can be broadly classified into five categories: BPA analogues, compounds with a single phenol group, compounds with aromatic rings, compounds with non-aromatic rings, and acyclic compounds [23]. Among the five categories of BPs, BPA analogues are the most prevalent. They are mostly used to replace BPA as a monomer in epoxy resins and polycarbonate plastics [5]. In this study, 21 types of BPs, including 8 types of BPA analogues, 5 types of monophenol-containing compounds, 3 types of aromatic ring-containing compounds, 3 types of non-aromatic ring compounds, and 3 types of acyclic compounds, were examined. All physical parameters are shown in Table S1. Most BP substances have high melting and boiling points and are insoluble in water. The physical and structural properties of BPA and its replacements are comparable, but some BPA alternatives show distinct physical features from BPA. The water solubility of BPZ is considerably lower than that of BPA (172.7 mg/L), being only 1.5 mg/L. In contrast, gallic acid has a solubility of 1.2 g/L. The boiling point of L-lactic acid is only 122 °C, while BPA has a higher boiling point of 364 °C, which is triple that of L-lactic acid. Newly developed BPA alternatives may differ from BPs and may have different effects. Therefore, it is necessary to pay attention to the physical properties of BPA substitutes. In this study, EPI Suite^{TM} provided by the EPA, Washington, D.C, US was used to calculate several parameters of BPA and its substitutes. EPI is a Windows-based suite of physical/chemical property and environmental fate estimation programmes developed by the EPA and Syracuse Research Corp (SRC), New York, US [30]. EPI was used to calculate the following three parameters of the 21 BPs: the octanol–water partition coefficient (K_{ow}), the soil–water partition coefficient (K_{oc}), and the air–water partition coefficient (K_{oa}). The different distribution coefficients can preliminarily indicate the distribution tendency of each substance in the environmental media [5]. The calculation results are shown in Table S2 and Figure 1. It can be observed that BPA and BPA analogues have a higher $LogK_{ao}$, $LogK_{ow}$, and $LogK_{oc}$, and $LogK_{ao}$ is much higher than $LogK_{ow}$ and $LogK_{oc}$. Most of the BP substances show a high $LogK_{oa}$, which suggests that they are primarily distributed in the air. This is consistent with the low water solubility of BPs in terms of the above physical properties. It can be seen that the proportion of BPA and BPA analogues with a high $LogK_{oa}$ is larger, while the $LogK_{ao}$ of acyclic compounds is relatively small. Compared with BPA and BPA analogues, compounds with one phenol and aromatic ring, acyclic compounds tend to be more hydrophilic. Based on these results, the BP content in the three media is ranked in the order of air > soil > water, indicating that the identification and study of BPs in the atmosphere and soil environment may be enhanced in the future.

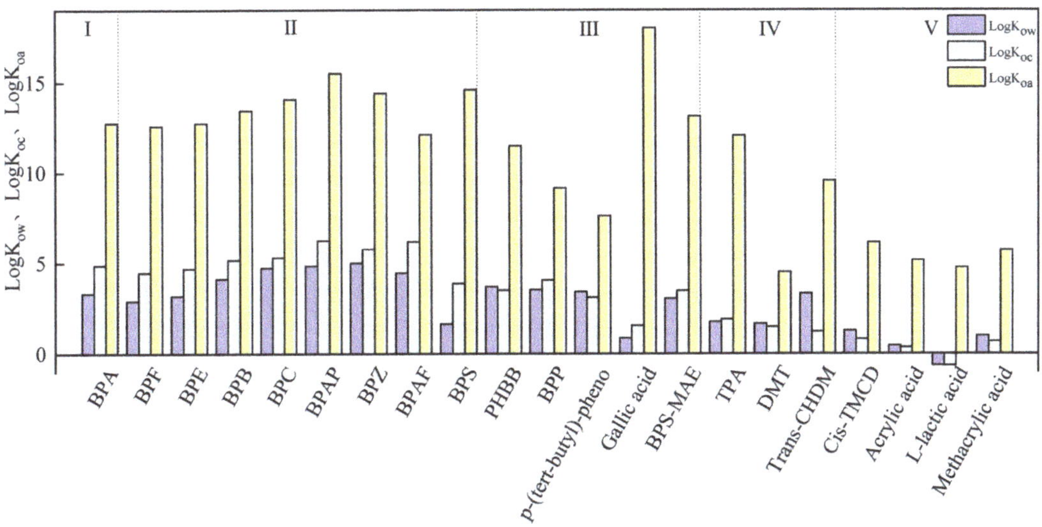

Figure 1. EPI simulation of each BP distribution coefficient (I: BPA; II: BPA analogues; III: BPA replacement compounds with one phenol; IV: BPA replacement compounds with aromatic rings; V: acyclic compounds). EPI Suite^{TM} provided by the EPA was used to calculate several parameters of BPA and its substitutes. The log octanol-water partition coefficient, log K_{ow}, of chemicals is estimated using an atom/fragment contribution method; the Henry's Law constant (air/water partition coefficient) is calculated using both the group contribution and the bond contribution methods; and K_{oc} is estimated using two different models: the Sabljic molecular connectivity method with improved correction factors and the traditional method based on log K_{ow}. After EPI software calculates $logK_{ow}$, $logK_{oa}$, and $logK_{oc}$ parameters of each substance, origin software is used to plot the data.

2.2. Environmental Transmission

In daily life, BPA is mainly used as a plasticizer, and it can be detected in specific medical devices and plastic products, including baby bottles and plastic bags [31]. Factors related to human activities, such as population density, economic development, and sewage discharge, will inevitably have an impact on the BPA content in the environment [32]. However, due to the restrictions placed on BPA by many regulations, its use is decreasing. To meet the needs of BPA use in daily life, BPA alternatives were developed, and these are expected to progressively infiltrate the environment through the production process and daily life activities [8,21,23,28,33]. Pollutants released into the environment can be dispersed through different mechanisms, such as atmospheric transmission [34], bubble rupture [35], infiltration [36,37], and dry and wet deposition [38] across interconnected water, soil, and atmospheric media. These processes contribute to the broader dissemination of pollutants, thereby expanding their diffusion range. Consequently, it is imperative to further investigate BPs' mode of transmission in order to better understand the pollutant's impact on the environment. As demonstrated by regional studies, BPA is found in atmospheric particles of coarse and fine sizes [34]. Soil resuspension is a source of BPA in atmospheric coarse particles; BPA is attracted to these particles, which eventually permeate the atmosphere. A previous study found that a large number of bubbles were generated when waves broke on the ocean, and these bubbles eventually burst and ejected water droplets, which played a key role in the formation of aerosols in the atmosphere [35]. This role also provided a basic framework for the transfer of substances at the water–gas interface [35]. Moreover, this framework provides a basis for the transfer of BPA from water to the atmosphere. However, relevant research data on BPA in marine aerosols are still limited. In the future, the transfer of BPA in sea air can be further studied. Furthermore, a survey found that a flat terrain slowed down the horizontal runoff rate and surface diffuse infiltration caused by atmospheric precipitation and agricultural irrigation; it also found that the vertical and lateral seepage of surface water polluted groundwater with BPA [36]. The infiltration of reclaimed water also polluted underground aquifers [37]. Another study analyzed roof-collected rainwater and found BPA, confirming that atmospheric deposition contributes to BPA pollution by returning it from the atmosphere to soil and surface water [38]. These results indicate that BPA can be transmitted in the environment through interactions at environmental interfaces and various physical mechanisms (Figure 2).

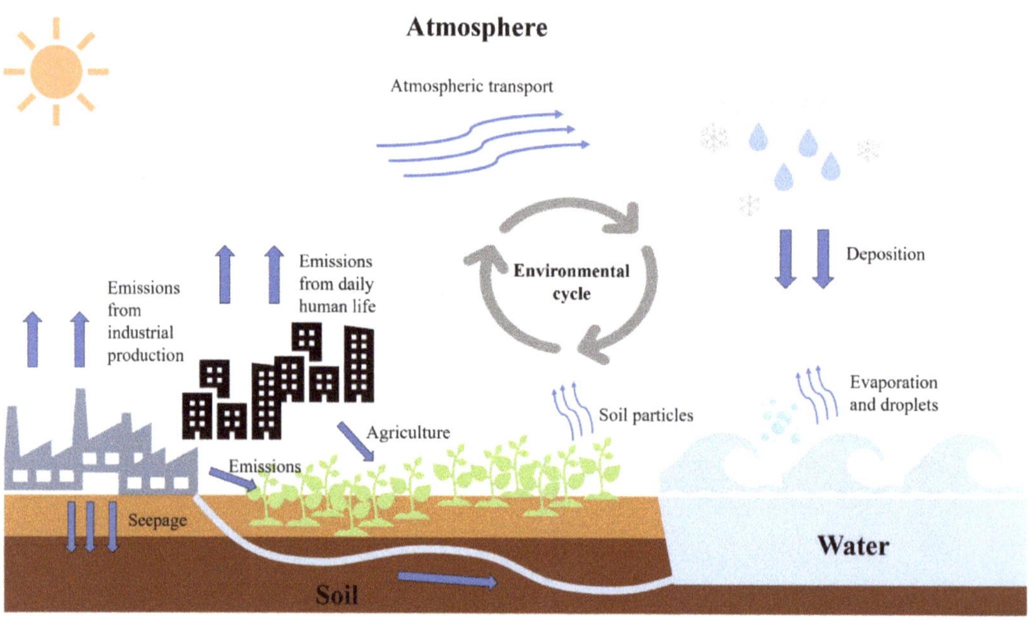

Figure 2. Schematic diagram of BP propagation in various media [32,35–38]. Industrial production, daily life, and agricultural production release BPs to the atmosphere, soil, and surface water. The BPs in the atmosphere diffuse through long-distance atmospheric transport and return to the soil and water through dry and wet settlement, the BPs in the soil surface penetrate into the groundwater and the groundwater flows into the surface water, and the BPs in surface water enter the atmosphere and soil through evaporation and contact so as to achieve environmental circulation.

Most recent test results show that, in the environment, the concentration of BPA is higher than that of other BPs. However, the detection of BP concentration has recently become very limited, particularly in the atmosphere. This review summarized the concentrations of BPA and six BPA analogues in three environments, namely, water, soil, and the atmosphere, as shown in Figure 3 and Table S3.

Human activities and daily necessities are key drivers of water pollution, leading to an increase in BP concentrations in natural water systems due to the discharge of urban domestic wastewater. Moreover, factors such as reduced rainfall, decreased river runoff, a lowered water self-purification potential, and impaired pollutant migration and transformation during the dry season collectively contribute to the accumulation and increased concentration of pollutants in water [39]. The number of sewage outlets and the use of industrial BPs also affect the pollution level of ΣBPs in water. Currently, the concentration of BPA in the global water environment ranges from about 0.09×10^{-3} to 228.04 μg/L [40], being much higher than that of other BPs. For example, in 2021, the Jialing River had the highest concentration of BPA in the wet season at 168.57 ng/L, while the highest concentrations of BPS, BPZ, BPAP, BPAF, and BPP were 1.44 ng/L, 5.7 ng/L, 1.36 ng/L, 3.3 ng/L, and 4.6 ng/L, respectively [39]. Therefore, among the BPs, BPA is still the main pollutant. Several studies have also found seasonal differences in the BPA concentration in water, with the most obvious difference being between winter and summer, showing high seasonal characteristics in summer and low seasonal characteristics in winter [5,22,39,41,42]. A possible reason for this is that BPA is primarily leached in high-temperature, acidic, or alkaline environments [43]. Therefore, seasonal fluctuations in BPA may be related to changes in local water quality, including local water temperature, acidity, and alkalinity. BP pollution is also influenced by waste disposal. A previous study found ΣBPs at 39 ng/L in the leachate of the Longyearbyen landfill site, which was almost 140 times the concentration of ΣBPs in the Longyearbyen River detected during the same period [5]. Therefore, this part of the leachate may enter groundwater and surface water via infiltration and surface runoff. Further, the BPA contamination of food and drinks may result from direct contact with plastic products, and BPA concentrations may marginally increase under heating conditions [44].

BPs can enter soil through direct wastewater discharge, urban runoff, solid waste treatment, etc. Moreover, agricultural activities are also one of the sources of soil BPs [40]. BPA is commonly adsorbed on soil particles due to its low vapour pressure and large octanol–water partition constant [40]. The highest concentration of ΣBPs detected in soils in Zhejiang, China, is 382.5 ng/g [45], which is much higher than the concentration of ΣBP (2.0 ng/g) in soils in Longyearbyen [5]. The results of Mexican soil sample tests indicated that BPA is a major contributor to BPs [46]. In a comprehensive national survey of soil BP contamination in China, BPA was detected in 85% of samples, making it the most prevalent BP [47]. A 2017 soil study in Shenzhen, China, reported that the maximum concentration of BPA was 6783 ng/g, which was much higher than that of other BPs: BPS (333 ng/g), BPF (4913 ng/g), BPZ (225 ng/g), BPAP (26.9 ng/g), BPAF (331 ng/g), and BPP (<5 ng/g) [46]. However, a survey in Zhejiang Province, China, showed that BPs in soil accounted for the main pollution contribution of BPP [48]. This variation in the primary soil components of the local area may be due to differences in industrial discharge conditions and farming methods. Furthermore, BPs are also concentrated in sediments, and their adsorption and desorption by sediments also impact their migration and transformation. The content of BPs in mud bottom varies across basins and regions, and the concentration of ΣBPs in landfill sites is relatively higher than that in other regions [5,8,25,32,49]. Recent studies on the concentration of BPs in soil have been relatively limited. However, survey findings indicate that the concentration of BPs is still detected, even in the Arctic region, where background concentrations are comparatively low. Combined with the above K_{ow} calculation results, these findings underscore the urgent need to focus on BP contamination in soil.

The widespread application of BPs in various products results in their inevitable release into the atmosphere through human activities. These BPs may persist in the atmosphere due to interactions and mutual transmission between different environmental media. However, current research on other BPs in atmospheric environments is particularly limited, with the majority of studies concentrating mainly on the detection of BPA. Due to the strong bonding between BPA and polymers, BPA is not easily released directly from these materials. Thus, the primary source of atmospheric BPA is solid waste combustion. A study showed that BPA in indoor and outdoor air was less affected by temperature fluctuations, while BPA concentration levels were relatively consistent in the cold and warm seasons [48]. Indoor concentrations of BPA may be affected by floor type and furniture [50], which may lead to higher concentrations in indoor air; additionally, the content of BPA differs among indoor environments. In Shanghai, the highest concentration of BPA detected in an office setting was 2.77 μg/g, while in student dormitories, BPA levels were found to be as high as 4.7 μg/g [51]. The BPA concentration found in student dormitories was nearly twice that found in office environments, suggesting that interior decoration may be a source of indoor BPA. Differences in BPA concentration across different indoor settings highlight the influence of specific sources. It has been suggested that paper supplies, including thermal paper and other office supplies, could potentially be a source of indoor BPA due to the presence of BPA components [29]. As per the environmental distribution of BPs, it is predicted that BPs may be distributed in the air in large quantities. Thus, it is imperative to enhance the atmospheric detection of BPs and broaden the application of detected substances in the atmosphere to collect fundamental data for future studies of BPs in the atmosphere.

Based on the above discussion, BPs, as widely used additives, are detected in water, soil, and the atmosphere in various regions, producing extensive global pollution. Environmental conditions (e.g., local water quality acidity), daily requirements (e.g., decoration furniture), and human activities (e.g., agricultural activities and industrial emissions) are likely to have some effect on BPs in the environment. The in-depth development of multimedia BP pollution detection methods is the primary key to solving the problem of BP pollution. It is necessary to establish a more complete BP detection list, expand the detection range and content of BPs, and establish BP standard benchmarks in different environments and production conditions.

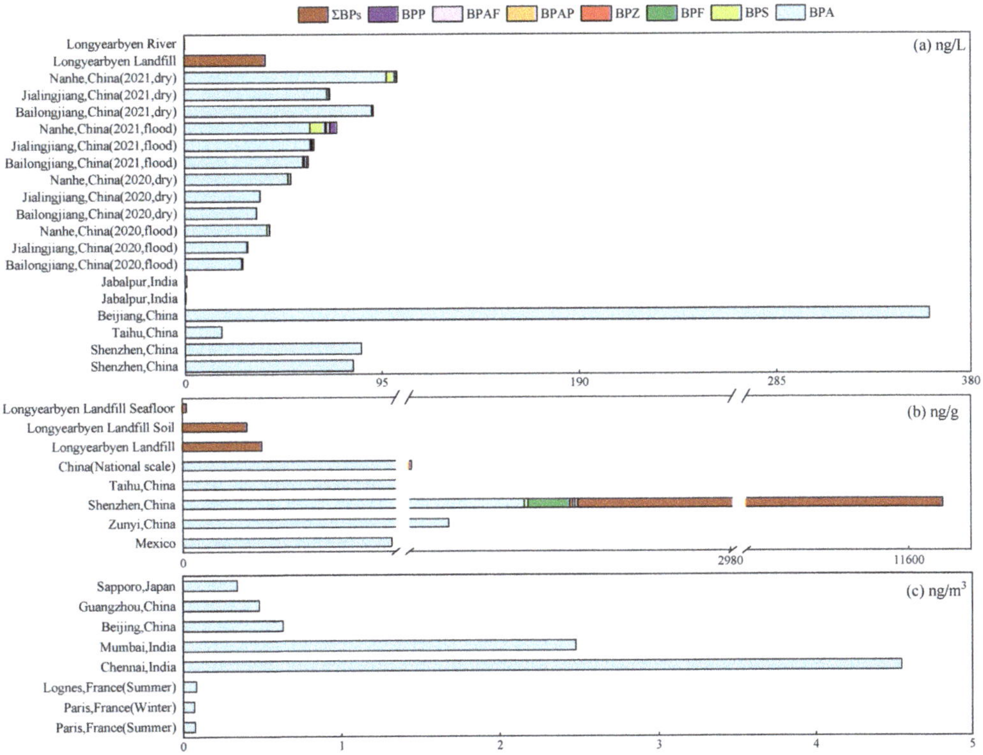

Figure 3. Concentration distribution of BPs in three environmental media: water, soil, and the atmosphere ((**a**) water; (**b**) soil; (**c**) atmosphere) [5,15,22,34,39,41,42,44–49,51]. China's National Knowledge Infrastructure (CNKI) database was selected as the core data source. In CNKI, BPs, soil, atmosphere, and water are used as keywords for search. The detection data are selected in different environments at home and abroad, and Origin software is used for analysis and mapping.

3. Analytical Methods and Detection Techniques

BPs are extensively distributed, being found in various aspects of human daily life, the natural surroundings, and even human waste. To better detect the content of BPs in the environment and human body and to determine their distribution, various efficient and convenient detection methods must be urgently adopted. Table 1 summarizes the commonly used detection methods for BPs and the new detection methods developed for BPA to provide theoretical and technical support for the detection of BPA alternatives.

To date, BPA detection technology has been continuously improved; however, conventional methods, including HPLC, liquid chromatography–tandem mass spectrometry (LC-MS/MS), UPLC-MS/MS, and gas chromatography–mass spectrometry (GC-MS), continue to be used [52]. The detection scheme is optimized by adjusting the pre-treatment methods and instrument settings. For example, acetonitrile/0.1% formic acid aqueous solution was used for gradient elution, the optimal elution volume (7 mL) was extracted thrice, and the BPA content in vegetable oil was determined using HPLC. This method had good linearity, a recovery rate of 90.8%, and a low detection cost [53]. The detection method based on UPLC-MS/MS can optimize the mass spectrum, air enrichment conditions, and elution solvent. A CSH C18 column and an 8:2 (v:v; 5 mL) methanol–water mixture were used for ultrasonic elution, and the recovery rate of this detection method was 68–119% [54]. Zhang et al. [55] used $ZnCl_2$ as a catalyst to synthesize amino-functionalized poly (N-vinylpyrrolidone divinylbenzene) [P(NVP-DVB)]. They then used the prepared P(NVP-DVB) as an adsorbent to detect BPA in the aqueous environment using HPLC and solid-phase extraction. The recovery rate was 99.7%. This method is effective in both simulated and real water samples. Additionally, conventional BPA detection methods based on other composite materials have also been developed. For example, cobalt nanoparticles/nitrogen-doped carbon nanomaterials were modified on the surface of glass carbon electrodes to detect BPA. The enhancement of the device resulted in a 98% recovery rate and excellent electrocatalytic performance for BPA [56]. A self-powered molecule-imprinted photochemical sensor based on a CdSe/ZnS QDs/HOF heterostructure was used to detect BPA, and the highly sensitive and selective detection of BPA was realized [57]. A magnesium phytate-based composite modified electrode was synthesized using the hydrothermal method. Under the action of Mg^{2+}, the electrode has a good response to BPA. The cost of Mg^{2+} phytate-based modified electrodes is low, and the recovery rate can reach 92.5 to 101.5% [58]. The potentiostatic deposition method was used to produce gold nanoparticles on the surface of a glass carbon electrode. These nanoparticles were then used to create a gold–ligand nanoelectrochemical sensor, which was then used to detect the presence of BPA in water. The results showed that the sensor had good anti-interference performance, but the recovery rate was only 87% [59], which is lower than the above magnesium phytate-based modified electrode.

Currently, the overall detection of BPs is relatively limited. HPLC and UPLC-MS/MS are mainly used for detection, and they can detect BP content in drinking water, human urine, biological samples, soil, leachate, and vegetable oil. Besides BPA, HPLC can be used to detect BPS, BPF, BPAF, BPAP, and BPZ. UPLC-MS/MS can also detect nearly 30 types of BPs, including those mentioned above [5,53,60]. Further, the salting-out-assisted liquid–liquid extraction HPLC-MS/MS method can detect nine types of BPs in liquid milk, namely, BPS, BPF, BPE, BPA, BPAP, BPB, BPZ, TBBPA, and TCBPA, after optimizing the traditional method. This method has a detection limit of 0.15 to 0.75 μg/kg [61].

Table 1. BP detection means and various detection parameters.

Detection Settings	Detectable Substances	Optimization Means	Detection Media	Advantage	Disadvantage	Recovery Rate (%)	Detection Limit	Quantification Limit	Linear Range	References
Ultra-high-performance liquid chromatography–tandem mass spectrometry	BPA, BPS, BPF, BPAF, BPAP, BPZ, etc.	-	Air, drinking water, human urine, biological samples, soil, leachate	High sensitivity, good recovery rate, and good repeatability	-	67.6–119	0.007–1.5 μg/L	0.5 μg/kg	0.1–100 μg/mL	[5,52,54,60]
High-performance liquid chromatography	BPA, BPS, BPF, BPAF, and BPAP	-	Vegetable oil, drinking water, human urine	High sensitivity and good repeatability	Difficult to achieve rapid on-site detection	90.8–103.2	0.007 mg/kg	13 μg/kg	0.003–0.7 μg/mL	[52,53,58,60]
Liquid chromatography–mass spectrometry tandem method	BPA, BPS, BPF, BPAP, etc.	Salting-out-assisted liquid–liquid extraction	Milk powder	Rapid test	-	80.8–118.1	0.15–0.75 μg/kg	0.5–2.5 μg/kg	-	[61]
Fourier change infrared spectroscopy	BPA	An amine-functionalized poly (N-vinylpyrrolidone divinylbenzene) adsorbent was prepared for the detection of BPA	Water	Low cost, good functional degree, strong selective adsorption ability, and less environmental pollution	-	99.65	0.5 mg/kg	-	0.5–3.0 μg/mL	[55]
Liquid chromatography–tandem triple quadrupole mass spectrometry	BPA	N-hexane/acetone (4:1) ultrasonic extraction three times	Dust		-	84.5–100	0.002–0.018 mg/kg	-	-	[51]
Photoelectric chemical sensor method without bias voltage	BPA	Self-powered molecularly imprinted photo/photochemical sensing based on CdSe/ZnS QD/HOF heterojunction	-	High sensitivity and high selectivity	-	-	-	-	-	[57]
Electrochemical method of carbon nanotube composites based on cobalt nanoparticles/nitrogen doping	BPA	Electrochemical method of carbon nanotube composites based on cobalt nanoparticles/nitrogen doping	-	High selectivity, good stability, and good reproducibility	-	98.4–104.6	5.0 nmol/L	0.005 μmol/L	0.01–20 (μmol/L)	[56]
Magnesium phytate-modified electrode method	BPA	Magnesium phytate-based modified electrode	-	High sensitivity, wide linear range, good repeatability, low electrode cost, and good stability	-	92.5–101.5	0.1 μmol/L	-	0.8–50 (μmol/L)	[58]
Gold nanoparticle–aptamer electrochemical sensing method	BPA	Gold nanoparticles were prepared on the surface of a glassy carbon electrode via potentiostatic deposition	Water	High sensitivity, wide linear range, good repeatability, low electrode cost, and good stability	-	87.4–110.0	10^{-9} mg/kg	10^{-9} μg/kg	10^{-9}–5×10^{-6} μg/mL	[59]
Solid-phase extraction combined with liquid chromatography–mass spectrometry	BPA	The pH value of the sample is 2 5 mL pure methanol and 5 mL dichloromethane solution	Water		-	74.41–111.2	0.0001–0.0033 μg/L	-	0.1–100 μg/L	[62]

4. Potential Hazards and Risk Assessment

BPs can enter the human body via multiple routes, and their endocrine-disrupting effects may cause irreversible changes in both the nervous and reproductive systems (Figure 4). The hormone-like activity of BPA can disrupt the interactions between sex and thyroid hormones and their specific co-receptors, potentially resulting in neurobehavioral disorders [63]. This disrupting effect of BPA can cause neurobehavioral developmental disorders in children [64], such as attention-deficit hyperactivity disorder, internal and external behaviour problems, social and learning disorders, and ASD, as well as having an impact on the neuroimmune system of children with ASD [18,65,66]. Simultaneously, gender-specific effects of BPA on neurobehavior have been observed, with increased sensitivity in boys [4]. A study also revealed a potential causal association between BPA exposure and the risk of childhood obesity [67], with the risk of childhood obesity increasing with increasing BPA concentrations [67]. Besides direct exposure, maternal BPA exposure during pregnancy may also contribute to developmental abnormalities in offspring. Experiments have demonstrated that perinatal exposure to low doses of BPA can lead to abnormal sexual and social behaviours in offspring rats and disrupt sex-based behavioural differentiation [18]. Prenatal exposure can reduce the relative quality of thymus in male rats, and the concentration of BPA in maternal urine during pregnancy is also positively correlated with the risk of asthma in children [68]. BPA exposure can also impact the normal operation of the mammalian reproductive system. Its exposure can reduce the quality and quantity of male germ cells, indirectly leading to the abnormal development of the male reproductive system and damage to reproductive organs [31,69–71]. Exposure to BPA during lactation can alter the distribution of endogenous spermatocytes in the mouse testis and disrupt spermatogenic cell differentiation [72]. Moreover, the estrogenic effects of BPA may impair the quality of female oocytes, and they could be a potential factor in the development of PCOS [73]. It was found that serum BPA concentrations in infertile women were higher than those in fertile women. Therefore, high BPA concentrations inhibited the development of primordial follicles, primary follicles, and luteum. Long-term low-dose BPA exposure can cause reproductive lesions in female fallopian tubes [19,31,74]. BPA exposure can also result in abnormalities in the estrous cycle and sexual behaviour in females, lead to menstrual disorders, and increase the risk of spontaneous abortion, as well as the incidence of reproductive tract and sexual organ diseases [75]. Besides its endocrine-disrupting effects, BPA can adversely impact the reproductive system by modulating gene expression. For example, it can induce cell proliferation by affecting the cell cycle in the MCF-7 human breast cancer cell line. Moreover, exposure to low doses of BPA may precipitate the early onset of puberty in young children [63,76].

BPA alternatives have similar endocrine-disrupting effects to BPA, especially in terms of reproductive toxicity, and some can even cause stronger reproductive toxicity than BPA [10,26,77,78]. For example, BPS and BPAF are two important BPA analogues that have the potential to cause severe reproductive defects, such as germ cell apoptosis and embryo mortality. High BPS concentrations have been shown to decrease embryo viability [27,29]. The ovarian reserve may be impaired by exposure to BPS compounds, including BPA and BPS [74]. However, their mechanism of action on the toxicity of female reproduction remains unknown. The reproductive toxic effect of BPS has been preliminary investigated in several experiments, suggesting that it has the potential to cause cell cycle arrest and cell proliferation interruption in females [79]. Studies have shown that BPAF also causes reproductive toxicity in males and gonad damage in male rats, and its toxic effect may be stronger than that of BPA [33]. Therefore, future research will prioritize more in-depth and systematic toxicity studies of BPAF. Multiple BPA analogues have also been found to show cytotoxicity, genotoxicity, dioxin-like effects, and neurotoxicity [10,26,77,78]. Studies have also found a significant positive correlation between serum BPs and the risk of Sjogren's disease [80]. Some BPs may have adverse effects on innate and adaptive immune responses. Previous studies have shown that BPA and BPAF can significantly reduce the endocytosis capacity of MDDCs. MDDCs, as the center of adaptive and innate immune response, will cause certain damage to the immune system when MDDCs is affected, and the experimental results show that the effect of BPAF is more significant [9]. The researchers have conducted further research, and the results show that when exposed to BPA and BPAF, the expression of RACK1 is significantly reduced, and the release of related cytokines is also significantly decreased [12]. RACK1 plays an important role in immune activation by regulating the balance of glucocorticoids and androgens [12]. Therefore, BPA and BPAF have significant effects on receptor binding and expression, which will lead to certain damage to the human immune system. It indicates the harm of BPs to the immune system; therefore, it is urgent to further investigate other BPs and confirm the possible harm of other BPs to the human immune system.

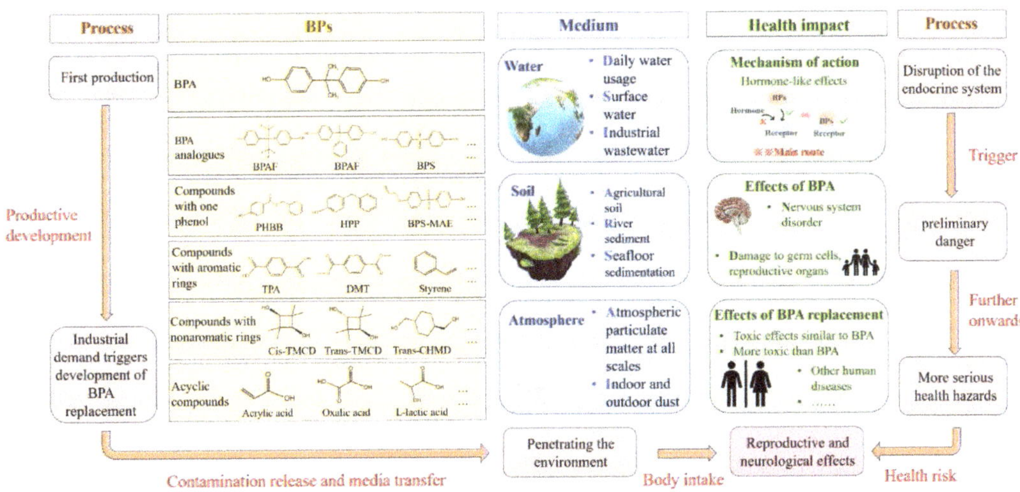

Figure 4. BP production, distribution, and harm to human health. The limited use of BPA has led to the emergence of BPA substitutes; BPA and its substitutes penetrate into the water, soil, and atmospheric environment; and humans are further exposed to BPs in the environment, causing adverse health effects, such as nervous, reproductive, and immune effects.

Although the mechanisms by which BPA affects human health are well documented, there remains a gap in research regarding the toxicity of some BPA substitutes. Considering the structural similarities between BPA and its analogues, as well as the possibility that these substitutes may pose higher toxicity, it is crucial to conduct further safety evaluations of BPA alternatives based on previous studies on BPA [81].

There is a mature health assessment for BPA that can be carried out using a risk assessment model and exposure level. Currently, daily intake (EDI) can be used to evaluate the levels of BPA exposure in different environmental media [47,82]. Simultaneously, the hazard quotient (HI) is used to assess the health risk of BPA, with HI > 1 indicating a carcinogenic risk [82]. The health risk assessment methods of BPA are not special or unique; however, due to the complexity of BPA, the health risks of BPA and its metabolites in vivo are a major concern. After BPA, BPA alternatives will become a new focus due to the continuous expansion of BP substances. However, the majority of BPA alternatives lack certain computational parameters, making it difficult to calculate their health risks. Recently, BPA has been found to be the main chemical substance in BPs, and the proportion of BPA in BPs is higher than that of other BPA alternatives. Therefore, assessing the health risks of BPA has become a priority. In this study, the health risk was evaluated based on BPA data obtained from the literature. The parameters used are shown in Table 2, and the evaluation was carried out using Formulas (1) and (2) [43]:

$$EDI = \frac{(C \times IR)}{BW} \quad (1)$$

$$HI = \sum \frac{EDI}{RfD} \quad (2)$$

Table 2. Relevant parameters.

Parameter	Age	Reference Data
Rfd (μg/g/d)	-	0.05
IR_{Water} (L/d)	Adult	1.85
	Child	0.86
IR_{Soil} (mg/d)	Adult	100
	Child	200
$IR_{Atmosphere}$ (m^3/d)	Adult	12.8
	Child	7.8
BW (kg)	Adult	58.6
	Child	15

As shown in Figure 5, the EDI exposure level shows the phenomenon of water < atmosphere < soil. BPs are not easily soluble in water, and their exposure level in water is relatively low. The EDI value of soil is higher than the average exposure level of the whole environment (1.29×10^{-3}), followed by the atmosphere, suggesting that more attention needs to be paid to BPA pollution in soil and the atmosphere. The average value of the overall environment needs to be lower than the present calculated average, as the EDI value in the water environment is overestimated in the calculation (1.29×10^{-3}). The EDI is consistently higher in children than in adults, which may be attributed to the comparatively low body weight of children. Further, HI < 1, suggesting that the current BPA level poses no carcinogenic risk to children or adults and can be disregarded. Children are at a higher health risk than adults, as evidenced by their health risk being six times greater than that of adults (HI children (0.03) > HI adults (0.005)). This may be attributed to the immaturity of children's bodies and their compromised metabolisms, which can increase their susceptibility to pollutants [83,84]. After analyzing the data, it is evident that the current levels of environmental exposure and associated health risks are not alarmingly high; however, it is important to understand that children are still susceptible to these health risks.

Most BPA alternatives have not undergone comprehensive safety assessments. Current testing indicates that the environmental concentrations of other BPs are much lower than those of BPA. Thus, the current exposure levels and associated health risks of BPA alternatives will likely be considerably lower than those recently estimated for BPA. However, considering the continuous expansion of BP applications and alternatives with a higher toxicity than BPA, the exposure levels and health risks of BPA alternatives should be continuously considered. Potential risk prediction models have been developed for several chemicals to predict the potential risk of BPs without experimental data [85,86]. For example, Hong et al. [86] predicted the activity of 29 types of bio-based platform chemicals in estrogen receptors using computer models, and they found that approximately 15 related species were expected to be estrogen conjugations. Karrer et al. [85] used a model to estimate human BP exposure levels, and they found that diet and tea polyphenols may be the most important sources of BP exposure. However, there is still a considerable lack of experimental data to substantiate the potential health risk assessments for BPA alternatives. Thus, it remains crucial to develop effective health risk assessments for these BPA alternatives, relying on future field observation data.

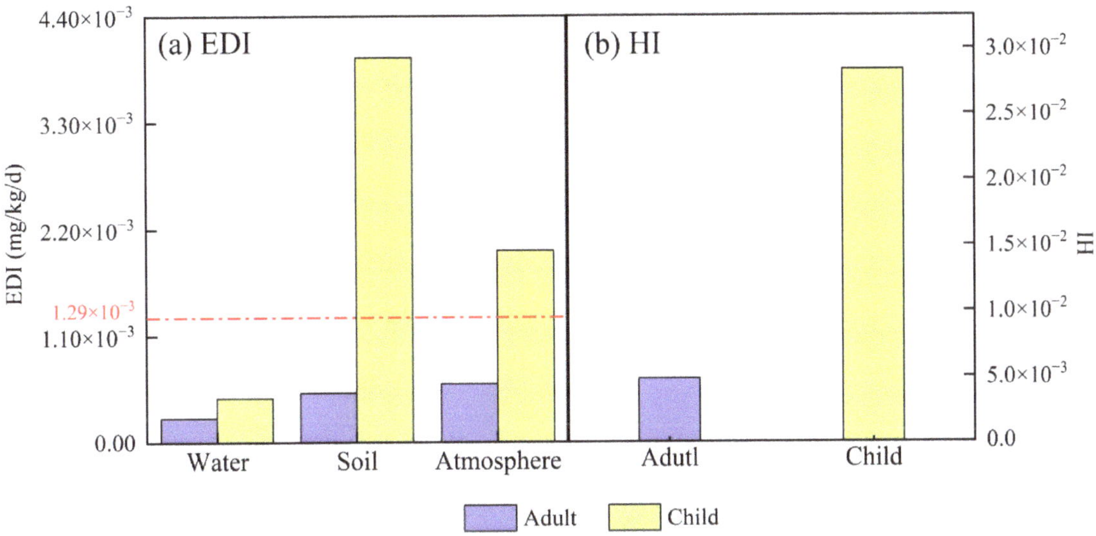

Figure 5. EDI level and HI assessment analyses of BPs in adults and children (notes: $EDI_{water} \times 10^8$) [15,22,39,41,42,44,46–49]. Using CNKI as the search database and BPs, water, soil, and atmosphere as the search keywords, the concentration data of BPs in the three media were retrieved, and the historical concentration data were used to calculate the EDI and HI values according to the empirical formula given by EPA, so as to evaluate the health risks in each environment. The red dotted line is the average exposure level for the whole environment.

5. Suggestions for Future Prevention and Control Countermeasures

In many countries, the prevention and control of the impact of BPs on human health have become issues that require immediate solutions, especially for BPA. Table 3 summarizes the restrictions placed on BPA in several countries and regions; these restrictions focus on limiting the use of BPA in infant products, which is consistent with the above conclusions regarding children's susceptibility to BPs. The Toxin-Free Keiki measure, recently announced by the U.S. state of Hawaii, prevents the use of BPA in reusable food or beverage containers for children as young as three years old. Further, Illinois prohibits the use of BPA in commercial or bank record papers [87]. In 2022, China listed BPA as a pollutant with high pollution and environmental risks [88]. California also contemplated the inclusion of BPA on the list of carcinogens [89].

Table 3. The relevant provisions of international organizations on PAEs.

Region/Organization	Bill	Stipulation	Reference
US. Hawaii	HB139(HD1) Toxin-Free Keiki Bill	BPA is banned from reusable food or drink containers for children as young as three years old	[87]
US. Illinois	HB2076 Bill	BPA is prohibited from being used in commercial or bank record paper	[87]
US. New York	S1076 Bill	No BPA in toys or in cans and other containers containing liquids or beverages intended for children aged three years or younger	[87]
US. New York	S3056 Bill	BPA is prohibited in childcare products for children aged three years or younger	[87]
EU	2011/8/EU	The chemical BPA is prohibited from being used in the production of baby bottles, requiring that all plastic materials that come into contact with food have no more than 0.6 mg/kg of BPA allowed to migrate	[90]
Food and Drug Administration (FDA)	2022-14682	The authorized use of BPA as a food additive is revoked and limited to establish a maximum limit of 0.5 ng/kg in food	[91]
CN, Ministry of Ecology and Environment of the People's Republic of China	Comprehensive List of Environmental Protection (2021 version)	BPA is listed as a "high-pollution, high-environmental-risk" product	[88]

Despite the progress in current BP research, problems remain in terms of pollution prevention efforts. First, a major issue is pollution emissions. Due to the physical and chemical properties of BPA, its main source is solid waste combustion [25]. Therefore, it is necessary to limit the emissions of major emission points, control the pollution emissions of BPA from the source, avoid the direct incineration of plastic waste, and properly handle the recycling and disposal of plastic products. Considering the structural similarities between BPA and its substitutes [29], they may share common emission sources. Therefore, it is crucial to monitor the emissions of other BPs during the combustion of solid waste and to develop comprehensive testing protocols. This approach will enhance the detection of BPA and its replacements in various environments, particularly emphasizing atmospheric and soil conditions. BPs in the atmosphere are mainly affected by direct emissions. Simultaneously, BP pollution in soil is related to agricultural activities, which differ according to the farming method and region, resulting in great differences in soil pollution [46,47]. It is necessary to formulate management measures according to local conditions in order to meet agricultural needs and reduce the pollution of soil by BPs. First, it is crucial to establish more efficient and stable BPA substitutes, enhance the efficacy of BP applications, and reduce their entry into the environment to protect production and life. Second, focus should be placed on advancing detection technologies for BPA and its alternatives and assessing the safety of these compounds. Currently, the primary methods for detecting BPA alternatives include HPLC, LC-MS/MS, and UPLC-MS/MS [5,52–54,60,62]. Recently, new detection methods for BPA have been developed, in addition to traditional detection methods [56–59]. To enhance the efficiency of detecting BPA substitutes, efforts should focus on expanding the detection capabilities of emerging technologies, building on the existing methods for BPA analysis. This approach aims to acquire more comprehensive and accurate data on BP compounds. Moreover, safety assessments of BPs need to be conducted. Currently, safety assessments of BPs are carried out using computer simulations [85,86], but there is still a lack of field observational and experimental data for comparative verification. Therefore, it is necessary to strengthen the combination of theoretical simulation and field observation results in the future. To ensure efficient, low-toxicity, safe, and stable production and to protect life, it is crucial to verify and fully investigate the safety evaluations of BPs, focusing on examining BPA alternatives. Third, it is necessary to formulate relevant policies and regulations. In many regions, regulations for the use of BPA have been established, resulting in the restriction of its use in specific products [87–91]. However, due to the expansion of industrial demand and the continuous development of BPA alternatives, new BPs are emerging in an endless stream [10,26,77,78]. Therefore, along with the safety assessments of BPA, it is essential to broaden restrictions on other toxic BP compounds. Simultaneously, industry-specific standards for BP limits must be established, particularly for sectors such as plastics, including products such as baby pacifiers and paper supplies. This requires increasing regulatory oversight, developing corresponding policies, controlling industrial pollution emissions, and ensuring the proper recycling and disposal of solid waste.

6. Conclusions

BPs have attracted attention due to their potential health impacts. However, despite recent progress in environmental and biological aspects, challenges remain, including rapid on-site detection methods, more refined detection techniques, and comprehensive safety assessments of BPA alternatives. These gaps highlight the necessity for further research and new developments. The second issue is the absence of safety assessments of BPA alternatives. This necessitates in-depth studies of BPs' mechanisms to better identify their health risks and develop appropriate preventative and control approaches. This study offers the following directions for future research. First, BP detection range should be expanded upon, focus should be given to soil and atmosphere pollution, a complete BP detection list should be developed, toxic BPA alternatives should be prevented from entering the environment and biological systems, and an understanding of the pollution and distribution characteristics of related contaminants should be facilitated. Second, future research should prioritize the development of more advanced and efficient detection technologies for BPs to facilitate rapid and accurate assessments. Third, it is crucial to explore the mechanisms underlying BPs' effects and conduct comprehensive safety evaluations to build a rigorous health assessment system. This will ensure that BPs meet industrial requirements without causing detrimental biological health effects. Addressing these research areas is crucial for managing the challenges caused by BP pollution and protecting environmental and human health.

Supplementary Materials: The following supporting information can be downloaded at the following: https://www.mdpi.com/article/10.3390/toxics13020109/s1. Table S1: The molecular formulae of each BP and their associated physical parameters; Table S2: Distribution coefficients for each of the BPs, simulated using EPI; Table S3: Concentration levels of common BPs in water, soil, and atmospheric media worldwide.

Author Contributions: F.L.: conceptualization, methodology, software, and writing—original draft. Y.R.: conceptualization, supervision, and writing—review and editing. F.B.: methodology and investigation. Z.W.: methodology and investigation. H.Z.: investigation. J.L.: investigation. R.G.: investigation. Z.L.: investigation. H.L.: investigation. All authors have read and agreed to the published version of the manuscript.

Funding: This work was supported by the National Key Research and Development Plan (no. 2023YFC3706105).

Informed Consent Statement: Not applicable.

Data Availability Statement: The original contributions presented in this study are included in the article/Supplementary Material. Further inquiries can be directed to the corresponding author(s).

Acknowledgments: This work was supported by the National Research Program for Key Issues in Air Pollution Control (no. DQGG2021301), the Fundamental Research Funds for Central Public Welfare Scientific Research Institutes of China (nos. 2022YSKY-27 and 2019YSKY-018), and the National Natural Science Foundation of China (no. 41907197).

Conflicts of Interest: The authors declare no conflicts of interest.

References

1. Yu, Y.; Wang, S.; Yu, P.; Wang, D.; Hu, B.; Zheng, P.; Zhang, M. A Bibliometric Analysis of Emerging Contaminants (ECs) (2001–2021): Evolution of Hotspots and Research Trends. *Sci. Total Environ.* **2024**, *907*, 168116. [CrossRef] [PubMed]
2. Darbre, P. Overview of Air Pollution and Endocrine Disorders. *Int. J. Gen. Med.* **2018**, *11*, 191–207. [CrossRef]
3. Buoso, E.; Masi, M.; Racchi, M.; Corsini, E. Endocrine-Disrupting Chemicals' (EDCs) Effects on Tumour Microenvironment and Cancer Progression: Emerging Contribution of RACK1. *Int. J. Mol. Sci.* **2020**, *21*, 9229. [CrossRef]
4. Tian, Y.; Zhou, X.; Miao, M.; Li, D.; Wang, Z.; Li, R.; Liang, H.; Yuan, W. Association of Bisphenol A Exposure with LINE-1 Hydroxymethylation in Human Semen. *Int. J. Environ. Res. Public Health* **2018**, *15*, 1770. [CrossRef] [PubMed]
5. Wang, L. The Pollution Characteristics and Bioaccumulation of BPs in Multiple Media at Longyearbyen, Arctic. Master's Thesis, Harbin Institute of Technology, Shenzhen, Guangdong, 2022.
6. Qiao, Y.; Yan, Z.; Feng, C.; Wang, J.; Bai, Y.; Wu, F. Research Focus Analysis of Endocrine Disrupting Chemicals (EDCs) Based on Bibliometrics. *Res. Environ. Sci.* **2022**, *35*, 424–434. [CrossRef]
7. Schneider, M.; Pons, J.-L.; Labesse, G.; Bourguet, W. In Silico Predictions of Endocrine Disruptors Properties. *Endocrinology* **2019**, *160*, 2709–2716. [CrossRef]
8. Yang, J. Endocrine Disrupting Effects of Typical Bisphenols and their Ecological Risks in the Sediments of Taihu Laike. Master's Thesis, Chinese Research Academy of Environmental Sciences, Beijing, China, 2024.
9. Švajger, U.; Dolenc, M.S.; Jeras, M. In Vitro Impact of Bisphenols BPA, BPF, BPAF and 17β-Estradiol (E2) on Human Monocyte-Derived Dendritic Cell Generation, Maturation and Function. *Int. Immunopharmacol.* **2016**, *34*, 146–154. [CrossRef]
10. Qiu, W.; Liu, S.; Chen, H.; Luo, S.; Xiong, Y.; Wang, X.; Xu, B.; Zheng, C.; Wang, K.-J. The Comparative Toxicities of BPA, BPB, BPS, BPF, and BPAF on the Reproductive Neuroendocrine System of Zebrafish Embryos and Its Mechanisms. *J. Hazard. Mater.* **2021**, *406*, 124303. [CrossRef]
11. EFSA Panel on Food Contact Materials, Enzymes and Processing Aids (CEP); Lambré, C.; Barat Baviera, J.M.; Bolognesi, C.; Chesson, A.; Cocconcelli, P.S.; Crebelli, R.; Gott, D.M.; Grob, K.; Lampi, E.; et al. Re-evaluation of the Risks to Public Health Related to the Presence of Bisphenol A (BPA) in Foodstuffs. *EFSA J.* **2023**, *21*, e06857. [CrossRef]
12. Buoso, E.; Kenda, M.; Masi, M.; Linciano, P.; Galbiati, V.; Racchi, M.; Dolenc, M.S.; Corsini, E. Effects of Bisphenols on RACK1 Expression and Their Immunological Implications in THP-1 Cells. *Front. Pharmacol.* **2021**, *12*, 743991. [CrossRef]
13. Masi, M.; Racchi, M.; Travelli, C.; Corsini, E.; Buoso, E. Molecular Characterization of Membrane Steroid Receptors in Hormone-Sensitive Cancers. *Cells* **2021**, *10*, 2999. [CrossRef] [PubMed]
14. Liu, Z.; Zhao, X.; Zhang, Y.; Li, Y.; Qiu, J.; Mu, X.; Jiang, J.; Qian, Y. Advances in potential impacts of bisphenol A and its alternatives on gut-brain regulation and involved mechanisms. *Asian J. Ecotoxicol.* **2023**, *18*, 159–174.
15. Xiong, S.; Wang, X.; Luo, W.; Ma, Y.; Lin, Y.; Wang, M.; Zheng, J. Spatia Distribution, Ecological Risk and Industry-Dependence of Endocrine Disrupting Chemicals in the Beijiang River, South China. *Environ. Chem.* **2021**, *40*, 3803–3814.
16. Wang, Q.; Feng, Q.; Zhu, X. Determination of bisphenols in sediment by accelerated solvent extraction and solid-phase extraction purification coupled with ultra performance liquid chromatography-tandem mass spectrometry. *Chin. J. Chromatogr.* **2023**, *41*, 582–590. [CrossRef] [PubMed]
17. Li, C.; Li, M.; Zhu, A.; Liu, G.; Liu, P. Distribution and ecological risk assessment of bisphenols in Fuhe River and Baiyang Lake. *J. Environ. Health* **2025**, 1–5. Available online: http://kns.cnki.net/kcms/detail/12.1095.r.20240410.1710.004.html (accessed on 9 January 2025).
18. Qin, Y.; Jian, B. Impact of Bisphenol A Pollution on Nervous Immune System of Children with Autism Spectrum Disease. *Environ. Sci. Manag.* **2016**, *41*, 28–31.
19. Chen, W.; Lau, S.-W.; Fan, Y.; Wu, R.S.S.; Ge, W. Juvenile Exposure to Bisphenol A Promotes Ovarian Differentiation but Suppresses Its Growth—Potential Involvement of Pituitary Follicle-Stimulating Hormone. *Aquat. Toxicol.* **2017**, *193*, 111–121. [CrossRef]
20. Usman, A.; Ahmad, M. From BPA to Its Analogues: Is It a Safe Journey? *Chemosphere* **2016**, *158*, 131–142. [CrossRef]

21. Zhang, T.; Xue, J.; Gao, C.; Qiu, R.; Li, Y.; Li, X.; Huang, M.; Kannan, K. Urinary Concentrations of Bisphenols and Their Association with Biomarkers of Oxidative Stress in People Living Near E-Waste Recycling Facilities in China. *Environ. Sci. Technol.* **2016**, *50*, 4045–4053. [CrossRef]
22. Gibson, R.; Durán-Álvarez, J.C.; Estrada, K.L.; Chávez, A.; Jiménez Cisneros, B. Accumulation and Leaching Potential of Some Pharmaceuticals and Potential Endocrine Disruptors in Soils Irrigated with Wastewater in the Tula Valley, Mexico. *Chemosphere* **2010**, *81*, 1437–1445. [CrossRef]
23. Ji, Z.; Liu, J.; Sakkiah, S.; Guo, W.; Hong, H. BPA Replacement Compounds: Current Status and Perspectives. *ACS Sustain. Chem. Eng.* **2021**, *9*, 2433–2446. [CrossRef]
24. US Environmental Protection Agency (EPA). Bisphenol A Alternatives in Thermal Paper. 2014. Available online: https://www.epa.gov/sites/default/files/2015-08/documents/bpa_final.pdf (accessed on 9 January 2025).
25. Li, J.; Wang, G. Airborne Particulate Endocrine Disrupting Compounds in China: Compositions, Size Distributions and Seasonal Variations of Phthalate Esters and Bisphenol A. *Atmos. Res.* **2015**, *154*, 138–145. [CrossRef]
26. Chen, D.; Kannan, K.; Tan, H.; Zheng, Z.; Feng, Y.-L.; Wu, Y.; Widelka, M. Bisphenol Analogues Other Than BPA: Environmental Occurrence, Human Exposure, and Toxicity—A Review. *Environ. Sci. Technol.* **2016**, *50*, 5438–5453. [CrossRef] [PubMed]
27. Shi, Z.; Qin, H.; Gao, J.; Qu, G.; Jiang, G. The effect of bisphenols on the secretion of exosomes. *Environ. Chem.* **2024**, *43*, 1–7.
28. Pastor-Belda, M.; Bastida, D.; Campillo, N.; Pérez-Cárceles, M.D.; Motas, M.; Viñas, P. A Study of the Influence on Diabetes of Free and Conjugated Bisphenol A Concentrations in Urine: Development of a Simple Microextraction Procedure Using Gas Chromatography–Mass Spectrometry. *J. Pharm. Biomed. Anal.* **2016**, *129*, 458–465. [CrossRef]
29. Chen, Y.; Shu, L.; Qiu, Z.; Lee, D.Y.; Settle, S.J.; Que Hee, S.; Telesca, D.; Yang, X.; Allard, P. Exposure to the BPA-Substitute Bisphenol S Causes Unique Alterations of Germline Function. *PLoS Genet* **2016**, *12*, e1006223. [CrossRef]
30. US Environmental Protection Agency (EPA), EPI SuiteTM-Estimation Program Interface. Available online: https://www.epa.gov/tsca-screening-tools/epi-suitetm-estimation-program-interface (accessed on 9 January 2025).
31. Liu, J.; Jin, Y.; Wei, Q.; Jin, Z.; Wei, D.; Jin, Y. Research progress on reproductive toxicity and reproductive system tumors induced by environmental endocrine disrupting chemicals. *J. Environ. Occup. Med.* **2022**, *39*, 833–839.
32. Devi, T.; Saleh, N.M.; Kamarudin, N.H.N.; Roslan, N.J.; Jalil, R.; Hamid, H.A. Efficient Adsorption of Organic Pollutants Phthalates and Bisphenol A (BPA) Utilizing Magnetite Functionalized Covalent Organic Frameworks (MCOFs): A Promising Future Material for Industrial Applications. *Ecotoxicol. Environ. Saf.* **2023**, *268*, 115706. [CrossRef]
33. Zhu, X.; Cao, L.; Liu, Y.; Tang, X.; Miao, Y.; Zhang, J.; Chen, J. Genotoxicity of Bisphenol AF in Rats: Detrimental to Male Reproductive System and Probable Stronger Micronucleus Induction Potency than BPA. *J. Appl. Toxicol.* **2024**, *44*, 428–444. [CrossRef]
34. Fu, P.; Kawamura, K. Ubiquity of Bisphenol A in the Atmosphere. *Environ. Pollut.* **2010**, *158*, 3138–3143. [CrossRef]
35. Jiang, X.; Rotily, L.; Villermaux, E.; Wang, X. Submicron Drops from Flapping Bursting Bubbles. *Proc. Natl. Acad. Sci. USA* **2022**, *119*, e2112924119. [CrossRef] [PubMed]
36. Wang, S.; Rao, Z.; Guo, F.; Liu, C.; Zhan, N.; Wang, Y.; Peng, J.; Yang, H. Occurrence Characteristics and Health Risk Assessment of Endocrine Disrupting Chemicals in Groundwater in Wuxi-Changzhou. *Environ. Sci.* **2021**, *42*, 166–174. [CrossRef]
37. Wang, P. Study on Migration and Transformation of Representative Endocrine Disrupt Chemicals during Reclaimed Water Infltrating into Groundwater. Ph.D. Thesis, Beijing Forestry University, Beijing, China, 2020.
38. Ju, X.; Gao, Z.; Zheng, W.; Zhang, Q. Identification and Derivation of Emerging Contaminants in the Roof Rainwater Confluence. *Environ. Sci.* **2024**, *45*, 4032–4043. [CrossRef]
39. Zhao, B.; Tan, X.; Xue, M.; Lu, J.; Xu, D.; Yang, R.; Zhang, L.; Gou, W. Pollution Status and Distribution Characteristics of Bisphenols in Rivers of Guangyuan City. *Environ. Monit. Forewarning* **2023**, *15*, 17–23.
40. Dueñas-Moreno, J.; Mora, A.; Cervantes-Avilés, P.; Mahlknecht, J. Groundwater Contamination Pathways of Phthalates and Bisphenol A: Origin, Characteristics, Transport, and Fate—A Review. *Environ. Int.* **2022**, *170*, 107550. [CrossRef]
41. Kumawat, M.; Sharma, P.; Pal, N.; James, M.M.; Verma, V.; Tiwari, R.R.; Shubham, S.; Sarma, D.K.; Kumar, M. Occurrence and Seasonal Disparity of Emerging Endocrine Disrupting Chemicals in a Drinking Water Supply System and Associated Health Risk. *Sci Rep* **2022**, *12*, 9252. [CrossRef]
42. Chen, M.; Guo, M.; Liu, D.; Li, J.; Zhang, S.; Shi, L. Occurrence and distribution of typical endocrine disruptors in surface water and sediments from Taihu Lake and its tributaries. *Chin. Environ. Sci.* **2017**, *37*, 4323–4332.
43. Ma, B.; Wang, L.; Tao, W.; Liu, M.; Zhang, P.; Zhang, S.; Li, X.; Lu, X. Phthalate Esters in Atmospheric PM2.5 and PM10 in the Semi-Arid City of Xi'an, Northwest China: Pollution Characteristics, Sources, Health Risks, and Relationships with Meteorological Factors. *Chemosphere* **2020**, *242*, 125226. [CrossRef]
44. Yan, Y.; Huang, Q.; Mou, J.; Luo, W. Bisphenol A secondary pollution in barreled drinking water and its exposure assessment. *Pract. Prev. Med.* **2017**, *24*, 43–45.
45. Xu, Y.; Hu, A.; Li, Y.; He, Y.; Xu, J.; Lu, Z. Determination and Occurrence of Bisphenol A and Thirteen Structural Analogs in Soil. *Chemosphere* **2021**, *277*, 130232. [CrossRef]

46. Ye, Z. Studies on the Pollution Distribution of Typical Bisphenol Compounds in Soil and Earthworms and Toxicity Effects. Master's Thesis, Shenzhen University, Shenzhen, China, 2021.
47. Zhang, Y. Research on Pollution Characteristics and Health Risks of Bisphenols in Soil at National Scale. Master's Thesis, Harbin Institute of Technology, Shenzhen, China, 2022.
48. Teil, M.-J.; Moreau-Guigon, E.; Blanchard, M.; Alliot, F.; Gasperi, J.; Cladière, M.; Mandin, C.; Moukhtar, S.; Chevreuil, M. Endocrine Disrupting Compounds in Gaseous and Particulate Outdoor Air Phases According to Environmental Factors. *Chemosphere* **2016**, *146*, 94–104. [CrossRef] [PubMed]
49. Liu, C.; Yang, Y.; Yu, J.; Xu, J. *Investigation and Environmental Risk Assessment of Octylphenol, Nonylphenol and Bisphenol A in the Sediment of Xiangjiang River in Zunyi City*; Chinese Society for Environmental Sciences: Beijing, China, 2019; pp. 1274–1278.
50. Loganathan, S.N.; Kannan, K. Occurrence of Bisphenol A in Indoor Dust from Two Locations in the Eastern United States and Implications for Human Exposures. *Arch. Env. Contam. Toxicol.* **2011**, *61*, 68–73. [CrossRef] [PubMed]
51. Liu, W.; Wang, Y.; Liu, Y.; Sun, Y.; Wu, M.; Ma, J. Comparative assessment of human exposure to phthalate esters and bisphenol A from different indoor dust. *J. Shanghai Univ. Nat. Sci.* **2019**, *25*, 282–292.
52. Wu, W.; Guan, C.; Wang, G.; Wang, Z.; Chen, Z.; Li, L.; Feng, D. A Study on Determination of 8 Kinds of Bisphenols in Packaged Drinking Water by High Per-Formance Liguid Chromatography-Tandem Mass Spectrometry. *China Meas. Test* **2022**, *48*, 78–82.
53. Liu, Y.; Jiang, W.; Hu, R.; Wu, X. Study of the Method for the Determination of Bisphenol a in Vegetable Oil by High Performance Liquid Chromatography. *Sci. Technol. Cereals Oils Foods* **2022**, *30*, 153–158. [CrossRef]
54. Ling, Y.; Liu, Z. Determination of 14 Environmental Hormones in Workplace Air by Ultra-High Performance Liquid Chromatography-Tandem Mass Spectrometry. *PTCA Part B Chem. Anal.* **2022**, *58*, 1449–1456.
55. Zhang, J.; Cui, Y.; He, L.; Zhang, Y.; Meng, X.; Zhang, S. Modification of P(NVP-DVB) and its application in the detection of BPA in aqueous solution. *Fine Chem.* **2023**, *40*, 75–86. [CrossRef]
56. Tang, J.; Cheng, X.; Cui, X.; Zhu, Y.; Fan, J.; Liu, T.; Zheng, S. Electrochemical Detection of Bisphenol A Based on Co Nanoparticles/N-Doped Carbon Nano-Tubes Composites. *Chin. J. Anal. Lab.* **2023**, *42*, 1–7. [CrossRef]
57. Zhang, X.; Yu, L.; Bai, B.; Guo, C.; Zhang, J.; Yang, Y. A self- powered molecular imprinting photoelectrochemical sensor based on CdSe ZnS QDs/HOFs heterojunction for the detection of bisphenol A. In *Proceedings of the Chinese Society of Food Science and Technology 19th Annual Meeting Paper Abstracts Col-Lection*; Chinese Society of Food Science and Technology: Beijing, China, 2022; pp. 28–29.
58. Zhang, Y.; Li, Y.; Yu, H.; Wang, H. Preparation of magnesium phytate-based modified electrode and its detection for bisphenol A. *Chin. J. Anal. Lab.* **2023**, *42*, 324–330. [CrossRef]
59. Meng, X.; Zhang, Y.; Liu, J.; Tan, F. Gold nanoparticles-aptamer electrochemical sensor for detection of bisphenol A in environmental waters. *Environ. Chem.* **2023**, *42*, 379–387.
60. Wu, W.; Guan, C.; Wang, G.; Wang, Z.; Chen, Z.; Li, L.; Chen, Q.; Cao, J.; Feng, D. Determination of 8 Kinds of Bisphenols in Urineby Solid Phase Extraction Combined with High Performance Liquid Chromatography-Tandem Mass Spectrometry. *China Meas. Test* **2023**, *49*, 75–80.
61. Jiang, K.; Zhang, H.; Cao, H.; Wang, J.; Zhou, X.; Li, X. Salting out Assisted Liquid Liquid Extraction Coupled with HPLC-MS/MS for Determination of 13 Kinds of Bisphenols and Alkyl Phenol Compounds in Liquid Dairy Products. *Food Mach.* **2022**, *225*, 32–36. [CrossRef]
62. Wang, T.; Xu, Q.; Li, J.; Li, J.; Tong, W. Simultaneous determination of 9 alkyl phenols and bisphenol A in water by SPE combined with LC-MS/MS. *Water Wastewater Eng.* **2023**, *59*, 529–534. [CrossRef]
63. Huang, Y.; Zhang, W.; Wang, R.; Su, X. Advances on pollution status and endocrine disrupting effects of bisphenols. *Asian J. Ecotoxicol.* **2022**, *17*, 60–81.
64. Li, Y.; Jiao, Z.; Zhang, X.; Fu, W.; Wang, L.; Gong, H.; Jiang, G. Progress in the treatment technologies toward endocrine disrupter bisphenol A. *Environ. Chem.* **2023**, *42*, 4019–4031.
65. Hong, S.; Hong, Y.; Kim, J.; Park, E.; Shin, M.; Kim, B.; Yoo, H.; Cho, I.; Bhang, S.; Cho, S. Bisphenol A in Relation to Behavior and Learning of School-age Children. *Child Psychol. Psychiatry* **2013**, *54*, 890–899. [CrossRef]
66. Xie, N.; Wang, H.; Wang, S. Research progress on the effects of bisphenol A exposure on children's health. *Matern. Child Health Care China* **2022**, *37*, 1539–1542. [CrossRef]
67. Kim, K.Y.; Lee, E.; Kim, Y. The Association between Bisphenol A Exposure and Obesity in Children—A Systematic Review with Meta-Analysis. *Int. J. Environ. Res. Public Health* **2019**, *16*, 2521. [CrossRef]
68. Tian, H.; Shan, L.; Cui, K.; Ru, S. Pollution status and adverse effects of bisphenols on early life. *Asian J. Ecotoxicol.* **2022**, *17*, 96–111.
69. Wu, Y.; Lu, X.; Ren, Y.; Sun, S. The effect of environmental endocrine disruptoe bisphenol A on female reproductive system. *J. Zunyi Med. Univ.* **2022**, *45*, 399–406. [CrossRef]
70. Xie, Y.; Zhu, Y.; Tao, F. Childhood and adolescent obesity and maternal phthalate exposure during pregnancy. *Chin. J. Sch. Health* **2019**, *40*, 1434–1437. [CrossRef]

71. Tian, J.; Ding, Y.; She, R.; Ma, L.; Du, F.; Xia, K.; Chen, L. Histologic Study of Testis Injury after Bisphenol A Exposure in Mice: Direct Evidence for Impairment of the Genital System by Endocrine Disruptors. *Toxicol. Ind. Health* **2017**, *33*, 36–45. [CrossRef] [PubMed]
72. Huang, X.; Ling, Y.; Zhao, C.; Chen, J.; Zhang, J.; Liu, Y.; Xie, M. Mechanism of meiotic arrest of spermatogenic cells in testis of mice exposed to bisphenol A during lactation. *J. Environ. Occup. Med.* **2021**, *38*, 769–774. [CrossRef]
73. Wang, Y.; Zhu, Q.; Dang, X.; He, Y.; Li, X.; Sun, Y. Local Effect of Bisphenol A on the Estradiol Synthesis of Ovarian Granulosa Cells from PCOS. *Gynecol. Endocrinol.* **2017**, *33*, 21–25. [CrossRef]
74. Zhang, N.; Zhao, Y.; Zhai, L.; Bai, Y.; Wei, W.; Sun, Q.; Jia, L. Urinary Concentrations of Bisphenol A and Its Alternatives: Potential Predictors of and Associations with Antral Follicle Count among Women from an Infertility Clinic in Northern China. *Environ. Res.* **2024**, *249*, 118433. [CrossRef]
75. Zhou, Y.; Lou, R. The effect of bisphenol A on the reproductive system and its mechanism. *Matern. Child Health Care China* **2019**, *34*, 468–474.
76. Li, Y.; Zhang, H.; Kuang, H.; Fan, R.; Cha, C.; Li, G.; Luo, Z.; Pang, Q. Relationship between Bisphenol A Exposure and Attention-Deficit/ Hyperactivity Disorder: A Case-Control Study for Primary School Children in Guangzhou, China. *Environ. Pollut.* **2018**, *235*, 141–149. [CrossRef]
77. Wei, J.; He, Z.; Wang, C.; Hu, Y.; Wang, C.; Cao, H.; Cao, M.; Liang, Y. Progress in Reproductive Toxicity and Human Reproductive Health Risk of Bisphenol A Analogues. *Asian J. Ecotoxicol.* **2022**, *17*, 85–107.
78. Pelch, K.; Wignall, J.A.; Goldstone, A.E.; Ross, P.K.; Blain, R.B.; Shapiro, A.J.; Holmgren, S.D.; Hsieh, J.-H.; Svoboda, D.; Auerbach, S.S.; et al. A Scoping Review of the Health and Toxicological Activity of Bisphenol A (BPA) Structural Analogues and Functional Alternatives. *Toxicology* **2019**, *424*, 152235. [CrossRef]
79. Yu, M.; Yang, Z.; Zhou, Y.; Guo, W.; Tian, L.; Zhang, L.; Li, X.; Chen, J. Mode of Action Exploration of Reproductive Toxicity Induced by Bisphenol S Using Human Normal Ovarian Epithelial Cells through ERβ-MAPK Signaling Pathway. *Ecotoxicol. Environ. Saf.* **2024**, *272*, 116037. [CrossRef]
80. Liao, K.; Zhao, Y.; Qu, J.; Yu, W.; Hu, S.; Fang, S.; Zhao, M.; Jin, H. Association of Serum Bisphenols, Parabens, and Triclosan Concentrations with Sjögren Syndrome in the Hangzhou, China Population. *Sci. Total Environ.* **2024**, *915*, 170031. [CrossRef] [PubMed]
81. Gély, C.A.; Lacroix, M.Z.; Morin, M.; Vayssière, C.; Gayrard, V.; Picard-Hagen, N. Comparison of the Materno-Fetal Transfer of Fifteen Structurally Related Bisphenol Analogues Using an Ex Vivo Human Placental Perfusion Model. *Chemosphere* **2021**, *276*, 130213. [CrossRef] [PubMed]
82. Dehdashti, B.; Nikaeen, M.; Amin, M.M.; Mohammadi, F. Health Risk Assessment of Exposure to Bisphenol A in Polymeric Baby Bottles. *Env. Health Insights* **2023**, *17*, 117863022311515. [CrossRef] [PubMed]
83. Li, H.; Yang, Z.; Xu, X.; Liu, Z.; Xian, J.; Yang, S.; Zhang, C. Pollution Characteristics and Health Risk Assessment of Phthalate Esters in Household Dust in Chengdu, China. *Hum. Ecol. Risk Assess. Int. J.* **2022**, *28*, 958–971. [CrossRef]
84. Zhang, K.; Feng, L.; Zhang, P.; Zhang, D.; Li, X. Pollution Characteristics and Health Risk Assessment of Phthalate Esters (PAEs) in Soils of Typical Agricultural Areas in Qingdao. *Environ. Chem.* **2024**, *43*, 1857–1870.
85. Karrer, C.; Andreassen, M.; Von Goetz, N.; Sonnet, F.; Sakhi, A.K.; Hungerbühler, K.; Dirven, H.; Husøy, T. The EuroMix Human Biomonitoring Study: Source-to-Dose Modeling of Cumulative and Aggregate Exposure for the Bisphenols BPA, BPS, and BPF and Comparison with Measured Urinary Levels. *Environ. Int.* **2020**, *136*, 105397. [CrossRef]
86. Hong, H.; Harvey, B.; Palmese, G.; Stanzione, J.; Ng, H.; Sakkiah, S.; Tong, W.; Sadler, J. Experimental Data Extraction and in Silico Prediction of the Estrogenic Activity of Renewable Replacements for Bisphenol A. *Int. J. Environ. Res. Public Health* **2016**, *13*, 705. [CrossRef]
87. Several States in the United States Have Issued New Regulations to Control Bisphenol A. Available online: https://bz.hgcm.cn/zggmsb/2019-03-25/details.html?edition=2&details=12 (accessed on 9 January 2025).
88. Ministry of Ecology and Environment of the People's Republic of China. Comprehensive List of Environmental Protection (2021 Edition). Available online: https://www.mee.gov.cn/ywdt/hjywnews/202111/t20211103_959023.shtml (accessed on 9 January 2025).
89. Chau, K. Request for Relevant Information on the Carcinogenicity of Bisphenol A (BPA). Available online: https://oehha.ca.gov/proposition-65/crnr/request-relevant-information-carcinogenicity-bisphenol-bpa (accessed on 9 January 2025).
90. Program, H.F. Bisphenol A (BPA). Food and Drug Administration (FDA) 2024. Available online: https://www.fda.gov/food/food-packaging-other-substances-come-contact-food-information-consumers/bisphenol-bpa (accessed on 9 January 2025).
91. Food and Drug Administration, US. Department of Health and Human Services Food and Drug Administration. Available online: https://www.federalregister.gov/documents/2022/07/11/2022-14682/environmental-defense-fund-maricel-maffini-breast-cancer-prevention-partners-clean-water-actionclean (accessed on 9 January 2025).

toxics

MDPI

Article

Impacts of NO_2 on Urban Air Quality and Causes of Its High Ambient Levels: Insights from a Relatively Long-Term Data Analysis in a Typical Petrochemical City in the Bohai Bay Region, China

Xiaoshuai Gao [1,2], Cong An [3], Yongxin Yan [1], Yuanyuan Ji [1], Wei Wei [2], Likun Xue [4], Rui Gao [1], Fanyi Shang [5], Jidong Li [5], Luyao Tan [6] and Hong Li [1,*]

1 State Key Laboratory of Environmental Criteria and Risk Assessment, Chinese Research Academy of Environmental Sciences, Beijing 100012, China; s202265286@emails.bjut.edu.cn (X.G.); yyx_in@163.com (Y.Y.); ji.yuanyuan@craes.org.cn (Y.J.); gaorui@craes.org.cn (R.G.)
2 Department of Environmental Science and Engineering, Beijing University of Technology, Beijing 100124, China; weiwei@bjut.edu.cn
3 Shanghai Key Laboratory of Atmospheric Particle Pollution and Prevention, Department of Environmental Science and Engineering, Fudan University, Shanghai 200438, China; ancong_22@163.com
4 Environment Research Institute, Shandong University, Qingdao 266237, China; xuelikun@sdu.edu.cn
5 Dongying Municipal Ecology and Environment Bureau, Dongying 257000, China; dyhbsfy@126.com (F.S.); shbjzlb@dy.shandong.cn (J.L.)
6 Ltd. of Shandong Environmental Protection Industry Corp., Jinan 250061, China; tanluyyy@163.com
* Correspondence: lihong@craes.org.cn

Abstract: The ambient levels of NO_2 in urban areas in China in recent years have generally shown a downward trend, but high NO_2 concentrations still exist under certain conditions, and the causes for such phenomenon and its impact on air quality remain unclear. Taking Dongying, a typical petrochemical city in the Bohai Bay of China, as an example, this paper analyzed the influence of NO_2 on urban air quality and investigated the causes for the formation of NO_2 with high concentrations. The results indicated that higher daily NO_2 concentrations (>40 $\mu g/m^3$) mainly occurred during January-April and September-December each year, and higher hourly NO_2 concentrations mainly occurred during the nighttime and morning rush hour in Dongying from 2017 to 2023. With the increase in daily NO_2 concentrations, the daily air pollution levels showed a general increasing trend from 2017 to 2023. The occurrence of high NO_2 values in Dongying was affected by the combination of unfavorable meteorological conditions, local emissions and regional transports, and localized atmospheric chemical generation. High-pressure and uniform-pressure weather patterns in 2017–2022, along with land–sea breeze circulation in 2022, contribute to high NO_2 concentrations in Dongying. Boundary layer heights (BLH) in spring (−0.43) and winter (−0.36), wind direction in summer (0.21), and temperature in autumn (−0.46) are the primary meteorological factors driving NO_2-HH (High hourly NO_2 values), while BLH (−0.47) is the main cause for NO_2-HD (High daily NO_2 values). The titration reaction between NO with O_3 is the main cause for NO_2-HH in spring, summer and autumn, and photochemical reactions of aromatics have a significant influence on NO_2-HD. NOx emissions from the thermal power and petrochemical industry in Dongying and air pollution transports from western and southwestern Shandong Province (throughout the year) and from the Bohai Sea (during spring and summer) had serious adverse impact on high NO_2 values in 2022. The results of the study could help to provide a scientific basis for the control of NO_2 and the continuous improvement of air quality in Dongying and similar petrochemical cities.

Keywords: NO_2 pollution characteristics; impacts on air quality; pollution meteorological conditions; causes for high NO_2 values; Bohai Bay region of China

1. Introduction

Atmospheric nitrogen oxides (NOx) include compounds such as nitrous oxide (N_2O), nitric oxide (NO), nitrogen dioxide (NO_2), dinitrogen trioxide (N_2O_3), and dinitrogen tetroxide (N_2O_4). Among these, NO_2 is one of the more stable nitrogen oxides, with higher concentration and longer persistence, playing a significant role in atmospheric chemical reactions) [1]. The formation of NO_2 in the atmosphere is primarily related to the conversion of NO. During the day, the photochemical chain reactions involving volatile organic compounds (VOCs) are key to NO_2 generation, while at night, the reaction of ozone (O_3) with NO is the main pathway for NO_2 generation (Figure S1). The photochemical reactions between NO_2 and VOCs during the day are the only chemical source of O_3 [2]. The reaction between NO_2 and hydroxyl radicals (OH) is the primary source of gaseous nitric acid (HNO_3), and the reactions between NO_2 and alkoxy radicals (RO) are the main sources of organic nitrate compounds in the atmosphere. NO_2 reacts with O_3 to form nitrate radical (NO_3) at night when NO concentration is low. NO_3 can react with VOCs or with NO_2 to form dinitrogen pentoxide (N_2O_5), which then reacts with water or aerosols to produce nitric acid (HNO_3) or nitrate salts (NO_3^-), ultimately leading to the net removal of NO_2 [3]. NO_2 and its related products, such as O_3 and organic nitrate compounds, are key components of photochemical smog, while HNO_3 or NO_3^- are major contributors to acid rain. These pollutants pose potential risks to human health and the eco-environment, they could damage human cardiovascular and pulmonary functions, and other organisms in ecosystems; In addition, photochemical smog could irritate the eye membrane of humans, reduce atmospheric visibility, and contribute to global warming, and acid rain could lead to soil acidification and building corrosion [4,5].

Therefore, studying the pollution characteristics of NO_2, its impact on air quality, and the causes of high NO_2 concentrations is crucial for further understanding atmospheric chemical mechanisms, continuously improving air quality, and reducing its adverse effects on human health and the environment.

The distribution pattern of NO_2 in the atmosphere results from the combined effects of pollution sources, meteorological conditions, and chemical generation. NO_2 sources are mainly divided into natural and anthropogenic sources [6]. Natural sources include lightning, soil emissions, and ammonia oxidation, while anthropogenic sources include industrial facilities and mobile emissions. Ambient NO_2 concentrations are greatly influenced by anthropogenic sources [7,8], which is the primary causes for the "higher in the east, lower in the west" distribution pattern observed in China over the years [9]. The average NO_2 concentration in autumn and winter are relatively higher than in spring and summer [10]. The seasonal differences in anthropogenic NOx emissions and meteorological conditions are important factors contributing to the seasonal variation in NO_2 concentration [11]. The diurnal variation of NO_2 concentration usually shows a "two peaks and one valley" pattern. The morning peak can be attributed to motor vehicle emissions, while the evening peak may be related to unfavorable weather conditions and the conversion of NO with high concentrations during the evening rush period [12]. Additionally, NO_2 exhibits a certain transport capacity [13]. Cheng et al. [14] found that implementing an odd-even license plate policy in the surrounding areas of Beijing significantly reduced NO_2 concentration in the city. Wang et al. [15] found that controlling NOx emissions in upwind cities is crucial for mitigating O_3 pollution in specific cities. Relevant studies have found that meteorological conditions, especially wind speeds (WS) and wind direction (WD), often affect the regional transports of NO_2, leading to high urban NO_2 concentrations [16,17]. High NO_2 concentrations at low WS are influenced by nearby pollution sources, while at high WS are influenced by long-distance transport. Xiao et al. [18] found that under weak large-scale meteorological conditions, local circulations related to terrain can also affect pollutant concentrations. The land–sea breeze characteristics in the Bohai Bay region are distinct [19]. During pollution events, offshore land winds can transport pollutants from Shandong Province to the Bohai Sea, while onshore sea winds can create a convergence zone, inducing upward air currents that cause pollutants to cycle over the bay, further deteriorating air quality in coastal areas [20]. Therefore, to fully understand the pollution characteristics of NO_2 and determine the causes for the formation of NO_2 with high concentrations, it is necessary to investigate from multiple perspectives.

Over the past 20 years, NOx emissions in China have initially increased and then decreased annually, peaking in 2012. Although NOx emissions in China have decreased, the emission structure in many regions has shifted to being dominated by traffic sources [21–23]. However, the overall NOx emissions remain above ten million tons, leading to persistently high ambient NO_2 concentrations in China [24]. The China Ecological Environment Status Bulletin shows that after a sharp decline in 2014 (34.8 $\mu g/m^3$), NO_2 concentrations in China's urban areas fluctuated from 2015 to 2019 (27.5–28.4 $\mu g/m^3$), continued to decline from 2019 (27 $\mu g/m^3$) to 2022 (21 $\mu g/m^3$) but rebounded in 2023 (22 $\mu g/m^3$) with a year-on-year increase of 4.8% [25].Currently, high NO_2 values in China are concentrated in economically developed and densely populated regions of North China, East China, and Central China [26], primarily in autumn and winter. For example, the monthly average NO_2 concentration in Jinan, Shandong Province, exceeded 60 $\mu g/m^3$ in January from 2015 to 2021 [27]. Some cities also experienced peak concentration at night during the summer. In 2018, four prefecture-level cities in Shanxi Province experienced pollution peaks during summer, with the average hourly NO_2 concentration in Lvliang exceeding 80 $\mu g/m^3$ at night [28]. Besides O_3, $PM_{2.5}$, and PM_{10}, NO_2 remains one of the primary pollutants on days with air quality exceedance in urban areas of China [25,29]. For example, the number of days with NO_2 as the primary pollutant and the number of NO_2 exceedance days increased annually from 2018 to 2020 in Chuzhou, Anhui Province, in the Yangtze River Delta region. In 2020, the proportion of days with NO_2 as the primary pollutant reached 5.2–9.0% in coastal cities like Qingdao, Binzhou, Rizhao, and Dongying in Shandong Province [30,31].

Dongying City, located in the northwest of Shandong Province and bordered by Bohai Bay to the north and Laizhou Bay to the east, is one of the representative cities in the Bohai Bay of China. Dongying has a developed petrochemical industry. Transportation primarily relies on road traffic, with extensive motor vehicle use and a high number of trucks in urban areas, leading to substantial NO_X and VOC emissions in Dongying [32], resulting in various environmental issues, such as increased photochemical pollution and frequent smoggy weather [33]. Previous studies on air pollution in Dongying have mainly focused on aerosols, particulate matter, O_3, and their precursors [31,32]. Research on NO_2 has primarily concentrated on pollution characteristics [8,15], with few comprehensive analyses of the impacts and causes of high NO_2 values. This study analyzed the temporal and spatial distribution characteristics of NO_2 concentrations in Dongying based on atmospheric pollutant and meteorological data from 2017 to 2023. The impact of NO_2 on air quality and of high NO_2 values in 2022 on O_3, $PM_{2.5}$, $PM_{2.5}$-related secondary components, and atmospheric oxidation capacity were explored, and the causes of high NO_2 values were investigated. The study results could provide a theoretical basis for NO_2 pollution control and air quality improvement in Dongying and similar petrochemical cities.

2. Materials and Methods

2.1. Study Area and Data Sources

This study focuses on Dongying, a representative city in the Bohai Bay region, China. Dongying is located in the northwest of Shandong Province, bordered by Bohai to the east and north, adjacent to Binzhou City to the west, and neighboring Zibo and Weifang cities to the south (Figure 1a). As a typical resource-based city, Dongying is dominated by heavy industries (petroleum and chemicals), with industrial parks concentrated in the southern region. Consequently, the air quality in Dongying is easily influenced by both maritime and inland pollutant transports. Dongying has jurisdiction over five counties/districts: Dongying District, Hekou District, Kenli District, Lijin County, and Guangrao County. Currently, the city has one Super Atmospheric Observation Station (referred to as "Atmospheric Observatory"), eight state-controlled stations, and four provincial-controlled stations. The Hekou Urban Area and Hekou Development Zone stations are in the northern region, Minfeng Lake and Minfeng Road stations in Kenli District are in the central region, and the remaining stations are in the southern region (Figure 1b).

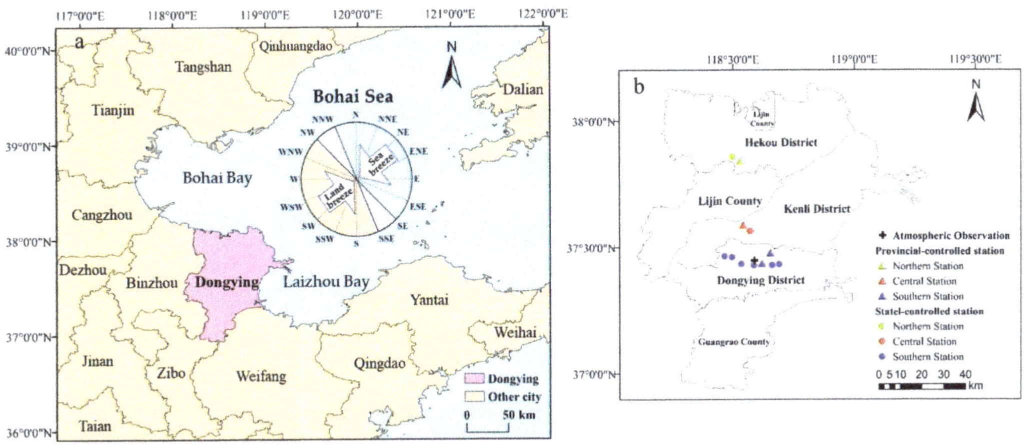

Figure 1. Geographical location of Dongying (**a**), and diagram of observation sites (**b**).

The routine observation items include NO_2 and meteorological parameters—temperature (T), relative humidity (RH), WS, WD, atmospheric pressure (AP), and boundary layer heights (BLH). The enhanced observation items include routine monitoring parameters (NO_2, NO, SO_2, $PM_{2.5}$, O_3), 115 VOCs, OC (organic carbon), EC (elemental carbon), ionic components (NO_3^- and SO_4^{2-}), ultraviolet radiation (UR), and emissions data. All observation items were continuously monitored using automatic monitoring devices, with monitoring station, frequency and duration of data collection, quality assurance, and quality control adhering to various technical specifications. Details regarding data usage, sources, and quality control in this study can be found in Table S1, Text S1, and Text S2, respectively.

2.2. Related Definitions

NO_2 evaluation standards: According to the Technical Regulation for Ambient Air Quality Assessment (on Trial) (HJ 663-2013), the daily NO_2 assessment value is the average NO_2 concentration over a 24 h period, the monthly NO_2 assessment value is the average NO_2 concentration for the month, and the annual NO_2 assessment value is the average NO_2 concentration for the calendar year. According to the Technical Regulation on Ambient Air Quality Index (on Trial) (HJ 633-2012), the daily NO_2 concentration corresponding to the

"Excellent" and "Good" air quality levels are 0–40 μg/m^3 and 40–80 μg/m^3, respectively. The daily assessment limit for NO_2 under the National Ambient Air Quality Secondary Standard is 80 μg/m^3. The World Health Organization (WHO) recommends annual NO_2 Interim Target 2: IT-2 (30 μg/m^3) and IT-3 (20 μg/m^3). For 24 h NO_2, IT-2 is 50 μg/m^3, and the 2021 Air Quality Guideline (AQG) level is 25 μg/m^3 [34].

The significance of NO_2 at different percentiles: Analyzing the arithmetic mean of NO_2 is complemented by evaluating NO_2 at different percentiles to provide a more comprehensive and detailed description of NO_2 pollution characteristics [35,36]. Generally, low percentiles of NO_2 (5th percentile) represent background concentration, while high percentiles (99th and 95th percentiles) indicate concentration during severe NO_2 pollution events [37]. Middle percentiles (75th, 50th, and 25th percentiles) reflect the NO_2 concentration observed during most periods in Dongying.

High daily NO_2 values (NO_2-HD): NO_2-HD is defined as daily average concentration exceeding the 75th percentile for each season in 2022, with seasonal thresholds of >23 μg/m^3 (spring), >18 μg/m^3 (summer), >45 μg/m^3 (autumn), and >44 μg/m^3 (winter) (Figure S2a). NO_2-NHD is defined as the daily average concentration not exceeding the 75th percentile.

High hourly NO_2 values (NO_2-HH): NO_2-HH is defined as hourly concentration exceeding the 75th percentile for each season in 2022, with seasonal thresholds of >25 μg/m^3 (spring), >19 μg/m^3 (summer), >47 μg/m^3 (autumn), and >49 μg/m^3 (winter) (Figure S2b).

NO_2-HH events and NO_2-NHH periods: Based on diurnal variation characteristics of NO_2, NO_2-HH typically occurs during nighttime to morning rush hours. NO_2-HH events are defined as at least five consecutive NO_2-HH occurrences during this period (with spring and summer from 00:00 to 09:00 and 20:00 to 23:00, and autumn and winter from 00:00 to 10:00 and 19:00 to 23:00) (Figure S8). Periods without five consecutive NO_2-HH occurrences are defined as NO_2-NHH periods. In addition, the remaining hours of each season are referred to as the daytime (with spring and summer from 10:00 to 19:00, and autumn and winter from 11:00 to 18:00).

2.3. *Analytical Methods*

2.3.1. Methods for Analyzing Impact of NO_2 on Air Quality

In this study, when analyzing the impact of NO_2 on air pollution levels in Dongying, the number of mildly, moderately, heavily, and severely polluted days refers to the total number of polluted days where various basic pollutants were the primary pollutants, in accordance with the Technical Regulation on Ambient Air Quality Index (on Trial) (HJ 633-2012). The proportion of polluted days was then calculated as the ratio of the number of polluted days corresponding to a specific NO_2 concentration range to the total number of all days in that range. To explore the impact of high NO_2 values on different pollutants in detail, analyses were conducted from the perspectives of NO_2-HH and NO_2-HD in different seasons of Dongying. The analytical framework is detailed in Figure S3, and the main analytical methods are as follows:

Calculation of Secondary Organic Carbon (SOC): The concentration of SOC can be calculated using the OC/EC ratio method. The calculation formula is as follows:

$$SOC = OC - EC \times (OC/EC)_{min} \tag{1}$$

In the formula, SOC represents secondary organic carbon (μg/m^3); OC and EC denote organic carbon and elemental carbon, respectively, and $(OC/EC)_{min}$ represents the minimum value of the OC/EC ratio observed during the study period.

Evaluation indicators for atmospheric oxidation capacity: The atmospheric oxidation capacity is typically represented by O_X concentration, which is generally estimated using the concentration of NO_2 and O_3 ($NO_2 + O_3$). NO_3 is an important nocturnal atmospheric oxidant, and the reaction between NO_3 and NO_2 is the only pathway for the formation of N_2O_5 [38]. The changes in N_2O_5 concentration levels can be indicated by the nighttime O_3 concentration multiplied by the square of the NO_2 concentration ($[NO_2]^2 \times O_3$), which indirectly represents the nighttime atmospheric oxidation capacity.

Transformation Rate of Nitrates and Sulfates: The nitrogen oxidation rate (NOR) and sulfur oxidation rate (SOR) can be used to assess the conversion status of gaseous precursors such as NO_2 and SO_2 into secondary inorganic ions. Higher values of NOR and SOR indicate a greater extent of secondary conversion of NO_2 and SO_2 in the atmosphere. The formulas are as follows:

$$NOR = N_1/(N_1 + N_2) \quad (2)$$

$$SOR = S_1/(S_1 + S_2) \quad (3)$$

In the formula, N_1 and N_2 represent the concentration of NO_3^- and NO_2, respectively (mol/m^3); S_1 and S_2 represent the concentration of SO_4^{2-} and SO_2, respectively (mol/m^3).

2.3.2. Methods for Analyzing the Causes of High NO_2 Values

The NO_2 concentration distribution in 2022 is not significantly different from that in recent years. Additionally, the Atmospheric Observatory and meteorological station data for 2022 are relatively comprehensive. Therefore, this study selects 2022 for analyzing the causes of high NO_2 levels. The analytical framework is detailed in Figure S4, and the main analytical methods are as follows:

Synoptic classification method: This study utilized the synoptic classification software cost733class developed by the European Union COST733 Project (http://www.cost733.org, accessed on 18 March 2024) and employed T-mode principal component Analysis (PCT) method to perform objective synoptic classification and reveal spatial distribution of the 925 hPa geopotential height field and the horizontal full WS (U and V) from 2017 to 2022. Details regarding the synoptic classification method in this study can be found in Text S3.

Criteria for sea–land breeze identification: Based on domestic and international studies on sea–land breeze classification standards [39,40], along with the natural geographic features of Dongying's coastline, the criteria for identifying sea–land breeze events in Dongying are established as follows. Based on the orientation of the coastline, airflow within the SSE to NW range (157.5–315°) is defined as a land breeze, while airflow within the NNW to SE range (0–135° and 337.5–360°) is defined as a sea breeze (Figure 1a). A day is classified as a sea–land breeze day if the following criteria are met: (1) the 24 h average surface wind speed is below 10 $m \cdot s^{-1}$; (2) during the land breeze period (01:00–08:00), the land breeze duration is $\geq$4 h, while the sea breeze duration is $\leq$2 h; and (3) during the sea breeze period (13:00–20:00), the sea breeze duration is $\geq$4 h, while the land breeze duration is $\leq$2 h.

Random forest model: The random forest model, introduced by Breiman in 2001, is a machine learning algorithm based on classification trees. This model is widely used for handling nonlinear relationships, classification, regression, high-order correlations, and variable importance assessment [41,42]. A linear regression model and a random forest model were constructed using Python 3.13. The models randomly selected 90% of the data as the training set and the remaining 10% as the test set. Model performance was evaluated by calculating the coefficient of determination (R^2) and root mean squared error (RMSE) [43]. Furthermore, the mean decrease accuracy (MDA) method was employed to evaluate the importance of each influencing factor. The principle behind MDA is that each variable is randomly assigned new values; if a variable is more important, replacing it randomly will result in a larger prediction error [44].

HYSPLIT mode: The Hybrid Single-Particle Lagrangian Integrated Trajectory Model (HYSPLIT) is primarily used to analyze air mass trajectories and trace the sources of air pollutants. In this study, the Dongying Atmospheric Observatory (37.45° N, 118.59° E) was selected as the starting point for the backward trajectories, with the analysis period covering the four seasons of 2022. The air mass analysis was conducted at an altitude of 300 m, with a backward tracing duration of 24 h. A total of 24 trajectories were calculated per day, resulting in 2208, 2206, 2184, and 2160 effective trajectories for each season. More details can be found in Text S4.

Backward trajectory clustering analysis method: Backward trajectory clustering analysis uses mathematical methods to classify and cluster all air parcel trajectories within a set time based on transport speed and direction. This helps analyze the source and proportion of dominant air masses at the target location, thereby determining the primary pollution source direction [45]. After cluster analysis, this study statistically analyzed the trajectory types and corresponding NO_2 hourly concentration data for each season, and subsequently determined the mean NO_2 concentration and contribution percentage for each trajectory type. More details can be found in Text S5. The formula for calculating the contribution percentage of NO_2 concentration is as follows.

$$P_{NO_2,t} = \frac{\sum C_t}{\sum C} \times 100\% \quad (4)$$

t is the category corresponding to the trajectory, C_t is the hourly mass concentration of NO_2 corresponding to the trajectory with category t, $\sum C_t$ is the sum of hourly mass concentrations of NO_2 corresponding to the trajectory with category t, $\sum C$ is the sum of hourly mass concentrations of NO_2 of all the trajectories, and $P_{NO_2,t}$ is the contribution of NO_2 concentration corresponding to the category of trajectory, t.

Potential source contribution function (PSCF) analysis method: PSCF is a method based on conditional probability functions to identify potential pollution sources [46]. This method relies on backward trajectory results and utilizes a series of functions in the TrajStat plugin to calculate WPSCF values. In this study, NO_2-HH in each season were set as the threshold for the target pollutant. The higher the WPSCF value in an area, the darker the color in the generated graphics, indicating a higher probability that the area contributes to NO_2-HH at the target site. More details can be found in Text S6.

Concentration weight trajectory (CWT) analysis method: The CWT method calculates the weighted concentration of trajectories to quantitatively reflect the pollution severity of different trajectories [47], thus determining the impact of regional transport on NO_2 concentration in Dongying. The higher the WCWT value, the more pollutants transported by the trajectory passing through the grid to the target site, and the greater the contribution to NO_2 concentration at the target site [48]. More details can be found in Text S7.

3. Results and Discussion

3.1. NO_2 Concentration Levels and Variation Characteristics

3.1.1. Concentration Levels

From 2017 to 2023, the annual mean values of NO_2 in Dongying showed a trend of "increase-decrease-stabilization" (Figure 2). The 5th, 25th, 50th, and 75th percentiles of NO_2 in Dongying exhibited a clear downward trend overall, except for a rise in 2019. The 95th and 99th percentiles of NO_2 displayed a fluctuating trend of "increase–decrease–increase," indicating a significant difference in the trends between low and high percentiles of NO_2. Additionally, in 2023, while the low and middle percentiles continued to decline and the high percentiles rebounded, the annual average remained relatively stable. This indicates that the high NO_2 concentration issue in Dongying has not been resolved. The

NO_2 concentration at the 95th percentile in Dongying each year was consistently higher than that at the 75th percentile, further indicating a high NO_2 concentration issue. Notably, in 2019, the annual mean values of NO_2 were the highest, and both the 75th and 95th percentiles were significantly elevated compared to other years. This suggests that the high NO_2 concentration in 2019 may have been driven by increases in the middle and high percentiles. In recent years, the annual NO_2 concentration in Dongying has met IT-2 (30 $\mu g/m^3$), but there is still a significant gap in reaching IT-3 (20 $\mu g/m^3$). In summary, Dongying currently faces a significant issue with high NO_2 concentration, and reducing the high values could notably lower the overall NO_2 concentration levels in the city.

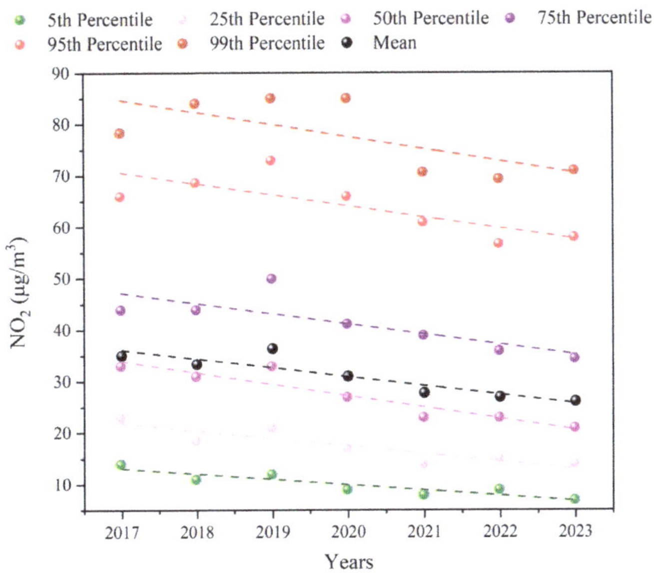

Figure 2. Variations in different percentiles and the annual mean values of the daily average NO_2 concentrations in Dongying from 2017 to 2023.

3.1.2. Temporal Variation Characteristics

The monthly variation of NO_2 in Dongying exhibits a "U" distribution (Figure 3), which is consistent with the findings of most researchers [49,50]. Overall, NO_2 concentration was highest in December, followed by November, October, and January. A significant decline in NO_2 concentration was observed each year from January to February. The period from March to August marked a rapid decrease in NO_2 levels, reaching the lowest values in July and August, while the period from August to October shows a rapid increase. The calendar of NO_2 in Dongying from 2017 to 2023 (Figure 4) shows that NO_2 with high concentrations (>40 $\mu g/m^3$) was concentrated from January to mid-March and from September to late December in 2017, with exceedances (>80 $\mu g/m^3$) occurring in late October. Compared to 2017, NO_2 with high concentrations in 2018 was concentrated from October to December and March. The days in January and February were more dispersed, but days where exceedances occurred in mid-January and were concentrated. In 2019, compared to 2018, the days of NO_2 with high concentrations increased and became more concentrated, with exceedance days mainly occurring in January and being more dispersed. From 2020 to 2023, the days of NO_2 with high concentrations significantly decreased and became more dispersed. In 2023, Dongying still experienced 46 days with NO_2 concentrations exceeding IT-2 (50 $\mu g/m^3$) and 159 days not meeting the 2021 AQG level (25 μg/m^3), indicating a relatively high daily average NO_2 level in the city. Overall, the days of NO_2 with high concentrations in Dongying exhibited a distribution pattern of "few and dispersed days—increasing and concentrated days—gradually decreasing and more dispersed days" from 2017 to 2023. High NO_2 concentrations mainly occurred from September to December and January to April, with exceedances usually happening in late October, November, December, and January.

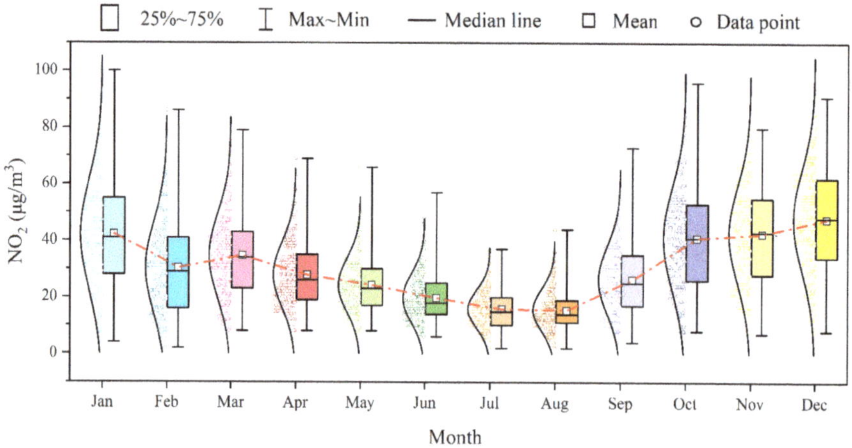

Figure 3. Monthly variation in the daily average NO2 concentrations in Dongying from 2017 to 2023.

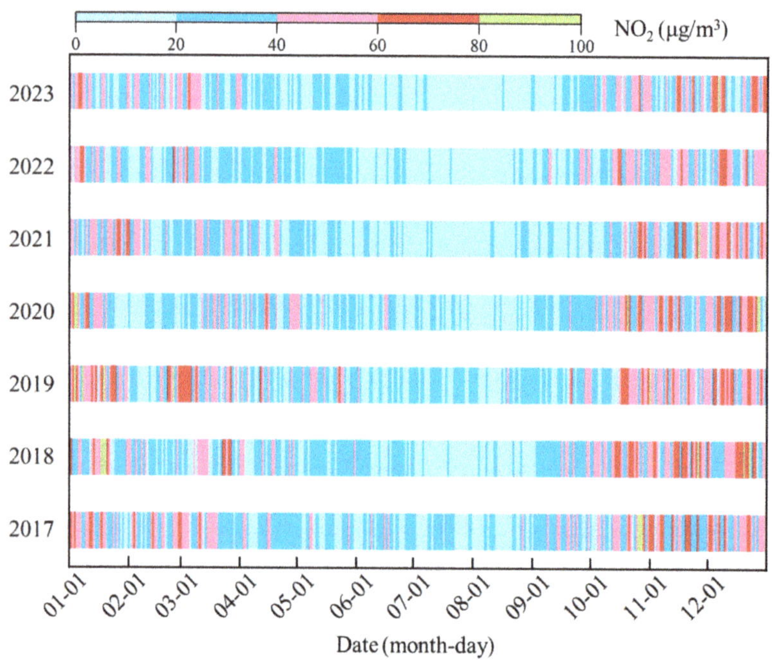

Figure 4. NO_2 concentrations calendar in Dongying from 2017 to 2023.

From 2017 to 2023, the concentrations of NO_2 during high NO_2 concentrations periods (from September to December and January to April) and non-high NO_2 concentrations periods, and the annual mean value of NO_2 concentration in Dongying all exhibited a "single peak and single valley" pattern (Figure 5 and Figure S5). The peak occurred around 8:00 a.m., and the valley appeared around 2:00 PM. During the morning traffic rush (6–9 a.m.), NO_2 concentration slightly increased, then slowly decreased after 9 a.m. After 4:00 p.m., NO_2 concentration rose significantly, remained at high levels, and exhibited a slow upward trend during nighttime to morning rush hours. Most research findings indicate that daily NO_2 variations exhibit a "double peak and single valley" pattern, with distinct morning and evening peaks [23,51]. In comparison, the hourly NO_2 concentration in Dongying did not show a significant peak at night, and the morning peak was smaller, indicating that in addition to vehicle emissions, other pollution sources also influence NO_2 concentration in Dongying. Overall, from 2017 to 2023, the concentrations of NO_2 during high NO_2 concentrations periods and non-high NO_2 concentrations periods, and the annual mean value of NO_2 concentration all showed a downward trend. However, changes in NO_2 concentration mainly occurred during nighttime to morning rush hours, with minimal changes during daytime hours.

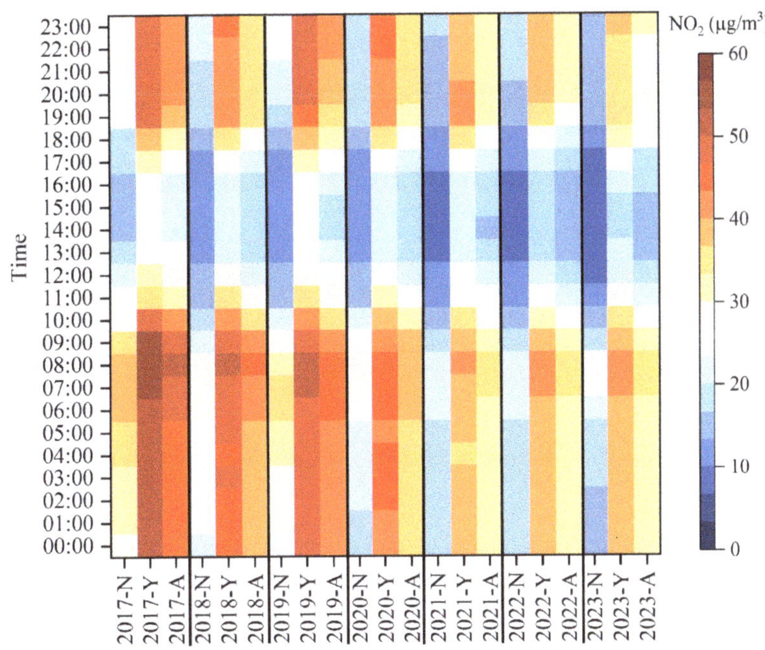

Figure 5. Diurnal variation of NO_2 concentrations in Dongying from 2017 to 2023. ("Year-N" represents the NO_2 concentration during non-high NO_2 concentrations periods of a specific year, "Year-Y" represents the NO_2 concentration during high NO_2 concentrations periods of a specific year, and "Year-A" represents the annual mean NO_2 concentration of a specific year).

3.1.3. Spatial Variation Characteristics

The interannual variation of NO_2 concentration in different regions of Dongying from 2017 to 2023 is shown in Figure 6. From different monitoring sites, during non-high NO_2 concentrations periods, the highest average NO_2 concentration was at the Gengjiangcun site, and the lowest was at the Minfeng Lake site in Kenli District. During periods of high NO_2 concentrations, the highest average NO_2 concentration was also at the Gengjiangcun site, while the lowest was at the Hekou urban site. From the annual variation at different sites, the Gengjiangcun site consistently showed higher average NO_2 concentration, while the Hekou urban site had lower averages. From different regions, during non-high NO_2 concentrations periods, the southern region had the highest average NO_2 concentration (19 μg/m^3), and the central region had the lowest (15 μg/m^3). During high NO_2 concentrations periods, the southern region had the highest average NO_2 concentration (37 μg/m^3), while the northern region had the lowest (34 μg/m^3). From the annual variation at different regions, the southern region had the highest average NO_2 concentration (19 μg/m^3), while the central region had the lowest (15 μg/m^3). Overall, from the perspectives mentioned above, the NO_2 concentration levels in the southern region of Dongying are relatively higher compared to other regions.

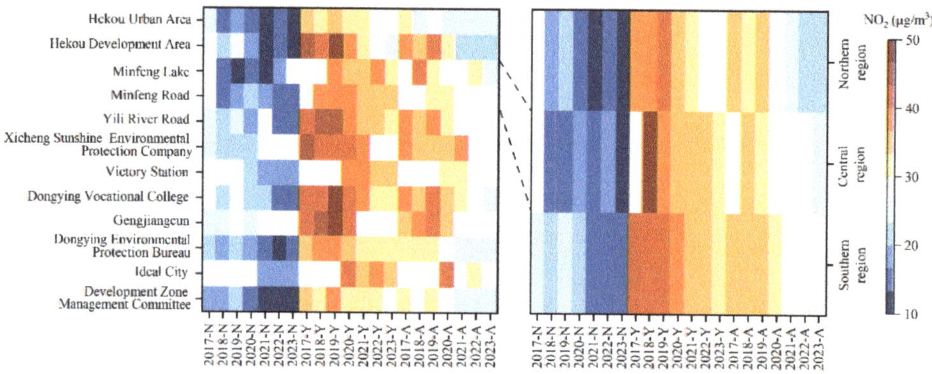

Figure 6. Interannual variation of NO_2 concentrations in different regions of Dongying from 2017 to 2023. ("Year-N" represents the NO_2 concentration during non-high NO_2 concentrations periods of a specific year, "Year-Y" represents the NO_2 concentration during high NO_2 concentrations periods of a specific year, and "Year-A" represents the annual mean NO_2 concentration of a specific year. Blank sections indicate non-operational monitoring sites or missing data).

3.2. *Impact of NO_2 on Air Quality*

3.2.1. Impact of NO_2 on Air Quality Levels

From the perspective of the primary pollutant (Figure S6), the number of days with NO_2 as the primary pollutant in Dongying was the lowest in 2017 (1 d), higher from 2018 to 2020 (14–20 days), and relatively lower from 2021 to 2023 (10–13 days) with a fluctuating trend. The proportion of days with NO_2 daily assessment value of "excellent" showed an overall increasing trend, reaching its lowest point in 2019 (64%) and experiencing a slight rebound in 2023. Exceedance days for NO_2 occurred between 2017 and 2020 (3–5 days) and in 2023 (1 day). Overall, as a pollutant in air quality assessment, NO_2 itself currently has a reduced impact on the air quality in Dongying.

According to Figure 7, from 2017 to 2019, the number of mildly polluted days generally increased and then gradually decreased with rising NO_2 concentration. From 2020 to 2023, the number of mildly polluted days steadily decreased with rising NO_2 concentration. The number of moderately polluted days showed an increase, followed by a decrease and then another increase with rising NO_2 concentration. Severe pollution days mainly occurred in 2017 (1 day) and 2021 (4 days), with corresponding NO_2 concentration ranges of 70–80 and 30–40 μg/m^3, respectively. Unlike other polluted days, the number of severe pollution days is relatively small, and there is no obvious pattern of variation. From 2017 to 2023, the number of mildly, moderately, and heavily polluted days was higher when the NO_2 concentration were 10–30 μg/m^3, 20–30 μg/m^3, and 50–80 μg/m^3, respectively. When the NO_2 concentration is below 20 μg/m^3, the proportion of polluted days is relatively low (<25%). Within the 20–50 μg/m^3 range, the proportion remained stable (32–35%). However, when the NO_2 concentration exceeded 60 μg/m^3, the proportion of polluted days increased significantly (>60%). In summary, the overall air quality in Dongying showed that both the air pollution level and the proportion of polluted days increased with rising daily NO_2 concentrations.

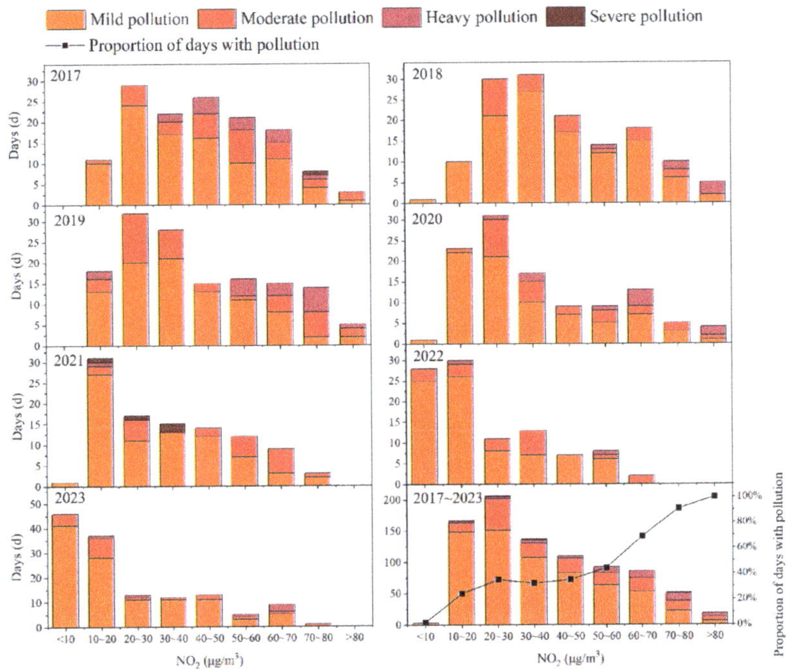

Figure 7. Distribution of air pollution levels corresponding to NO_2 daily average concentration ranges in Dongying from 2017 to 2023.

3.2.2. Impact of High NO_2 Values on O_3, $PM_{2.5}$, and Atmospheric Oxidation Capacity in 2022

According to Figure S7, the number of NO_2-HD days in Dongying in 2022 showed no significant seasonal variation, with similar counts in spring (21 days), summer (21 days), autumn (19 days), and winter (22 days). On a monthly scale, December and March had the highest number of NO_2-HD days (11 days), followed by June and November (9 days), while February, April, and September had the fewest (4 days), indicating some inter-monthly variability. As shown in Figure S8, the proportion of NO_2-HH in spring and summer peaked around 00:00–01:00 (8% and 7%) and again around 07:00 (8% and 10%). In contrast, NO_2-HH in autumn and winter remained relatively stable between 19:00 and 10:00, the following day, ranging from 5% to 7% and 4% to 6%, respectively. Overall, NO_2-HH in Dongying primarily occurred from nighttime to the morning peak hours (with spring and summer being 00:00–09:00 and 20:00–23:00, and autumn and winter 00:00–10:00 and 19:00–23:00).

As shown in Figure 8, compared to NO_2-NHH periods, NO_2-HH events in all seasons lead to significant increases in $PM_{2.5}$ and $PM_{2.5}$-bounded NO_3^-, SO_4^{2-}, and SOC concentrations, and a notable decrease in O_3 concentration. This indicates that high NO_2 values can enhance the concentrations of O_3, $PM_{2.5}$ and $PM_{2.5}$-bounded secondary components during nighttime to morning rush hours, with the most pronounced effects in autumn and winter. The higher the NO_2 concentration in a season, the lower the nighttime O_3 concentration, which is related to the NOx titration reaction [2]. As shown in Figure 9, during NO_2-HH events across all seasons, the concentrations of O_X and N_2O_5 are significantly higher compared to NO_2-NHH periods, indicating that high NO_2 values can enhance the atmospheric oxidation capacity at night, consistent with the findings of Alberto Notario et al. [52]. In spring, summer, and winter, the NOR values decrease during NO_2-HH events, whereas the opposite is observed in autumn. Similarly, SOR values decrease during NO_2-HH events in spring, autumn, and winter, but increase in summer. This may be because the formation of NO_3^- and SO_4^{2-} at night is significantly influenced by nighttime O_3 concentration [53]. During NO_2-HH events, high NO concentration results in the titration of O_3 by NO and the concentrations of NO_2 and SO_2 are relatively high, leading to lower NOR and SOR values. In summary, NO_2-HH events can increase O_3, $PM_{2.5}$ and $PM_{2.5}$-bounded secondary components concentrations, as well as O_X and N_2O_5 concentrations during nighttime to morning rush hours. They also lead to lower O_3 concentration, NOR, and SOR values. Moreover, the analysis of impact of high NO_2 values on daytime O_3, $PM_{2.5}$, and atmospheric oxidation capacity in Dongying in 2022 is detailed in Figures S9 and S10 and Text S8.

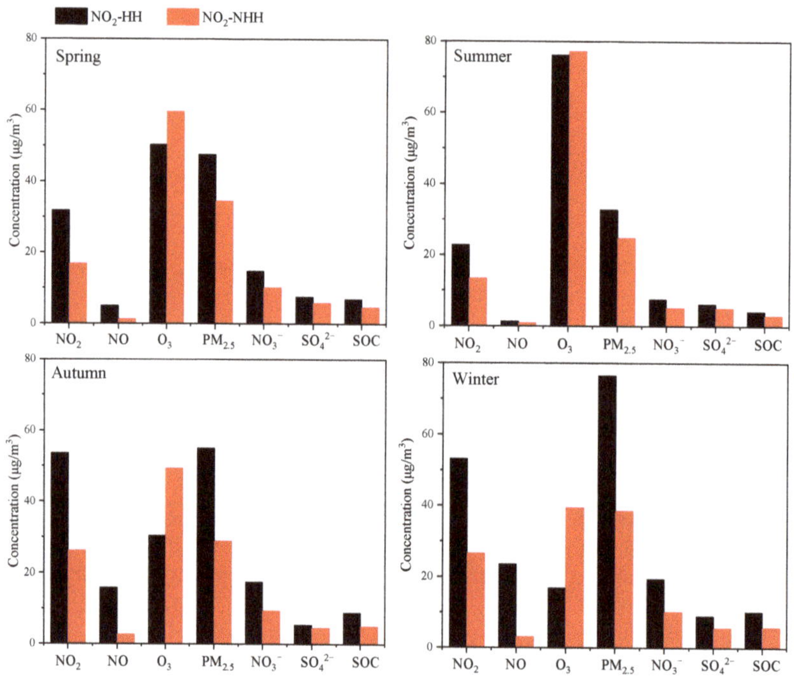

Figure 8. Differences in the concentrations of O_3, $PM_{2.5}$ and $PM_{2.5}$-bounded secondary components during nighttime to morning rush hours between NO_2-HH events and NO_2-NHH periods in Dongying in 2022.

As shown in Figure 10, compared to NO_2-NHD days, $PM_{2.5}$ and $PM_{2.5}$-bounded NO_3^-, SO_4^{2-}, and SOC concentrations are significantly higher on NO_2-HD days across all seasons, indicating that NO_2-HD can promote increases in concentrations of $PM_{2.5}$ and $PM_{2.5}$-bounded secondary components on the same day, with the most pronounced effect in autumn and winter. In spring, autumn, and winter, O_3 concentration on NO_2-HD days is lower than on NO_2-NHD days, whereas the opposite is true in summer. In spring, summer, and autumn, MDA8-O_3 concentration on NO_2-HD days is higher than on NO_2-NHD days, with the opposite observed in winter, likely due to higher daytime NO emissions in winter. As shown in Figure 11, O_X concentration is significantly higher on NO_2-HD days across all seasons, indicating that NO_2-HD can enhance the atmospheric oxidation capacity. In spring and summer, NOR values on NO_2-HD days are lower than on NO_2-NHD days, whereas the opposite is observed in autumn and winter. Conversely, SOR values are higher on NO_2-HD days in summer, autumn, and winter, but lower in spring, indicating a noticeable seasonal difference in the impact of NO_2-HD on NOR and SOR. In summary, NO_2-HD can increase MDA8-O_3, $PM_{2.5}$ and $PM_{2.5}$-bounded secondary component concentrations, as well as O_X concentration, while reducing the average O_3 concentration on the same day.

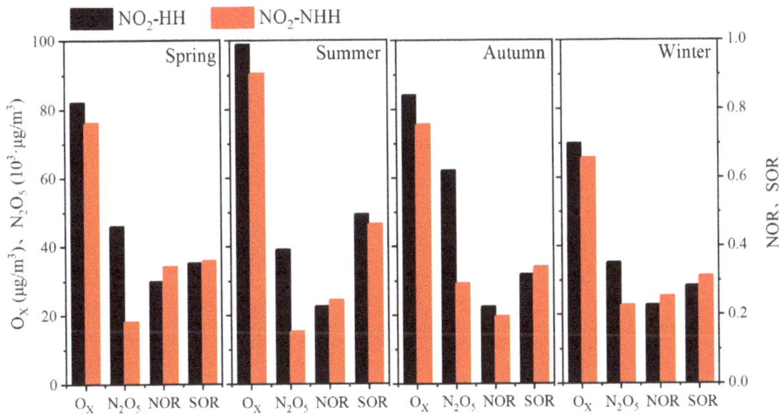

Figure 9. Differences in the concentrations of O_X and N_2O_5, and values of NOR and SOR during nighttime to morning rush hours between NO_2-HH events and NO_2-NHH periods in Dongying in 2022.

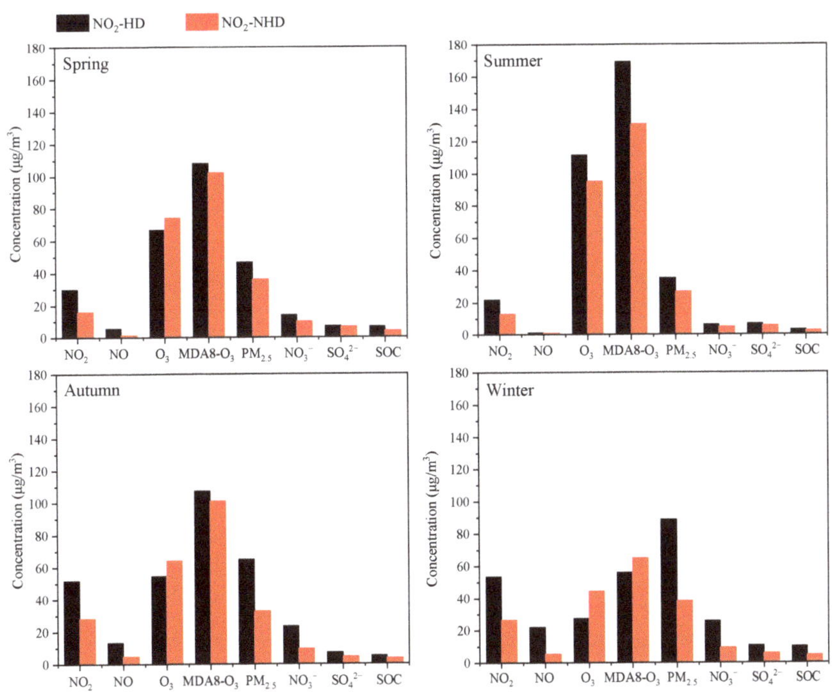

Figure 10. Differences in the concentrations of O_3, $PM_{2.5}$ and $PM_{2.5}$-bounded secondary components on the same day between NO_2-HD days and NO_2-NHD days in Dongying in 2022.

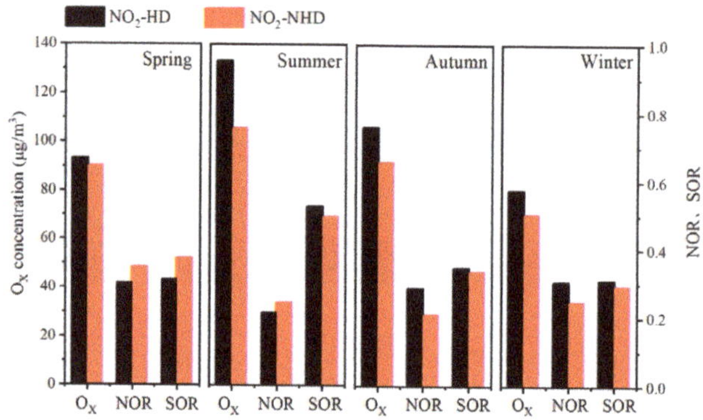

Figure 11. Differences in the concentrations of O_X, and values of NOR and SOR on the same day between NO_2-HD days and NO_2-NHD days in Dongying in 2022.

3.3. *Causes of High NO_2 Values*

3.3.1. Meteorological Conditions

Using the PCT method, the circulations of the 925-hPa geopotential height field from 2017 to 2022 were classified into nine patterns (Figure S11)—namely, "offshore high-pressure rear" (Type 1), "high-pressure front" (Type 2), "high-pressure inside" (Type 3), "high-pressure top front" (Type 4), "uniform pressure" (Type 5), "west-high and east-low" (Type 6), "subtropical high" (Type 7), "inverted trough" (Type 8), and "low-pressure field" (Type 9).

From 2017 to 2022, Type 1 occurs most frequently (20.5%), with large variations in the daily average maximum temperature (T_{max}-mean) and daily average relative humidity (RH-mean). Type 4 (16.6%), Type 2 (14.1%), and Type 3 (12.1%) followed, with these patterns generally characterized by low T_{max}-mean and high RH-mean. Type 6 (8.4%) and Type 5 (6.7%) exhibited large ranges in T_{max}-mean, with the former having sparse isobars and lower RH-mean, and the latter having dense isobars and higher RH-mean. Type 7 (9.5%), Type 8 (7.6%), and Type 9 (4.4%) mainly occur in summer, characterized by higher T_{max}-mean, occasional precipitation, and higher RH-mean (Table 1).

Table 1. Frequency and major characteristics of nine synoptic patterns of the 925-hPa geopotential height field in Dongying from 2017 to 2022.

Synoptic Pattern	Frequency	Major Characteristics
Type1	20.5%	Dongying near the offshore high-pressure rear, with dominant southeasterly wind, T_{max}-mean of 21.8 °C, RH-mean of 56.6%, and Pre-mean of 1.4 mm.
Type2	14.1%	High-pressure center over the Mongolian Plateau. Dongying under the front of the high-pressure, with dominant southeasterly wind, T_{max}-mean of 10.4 °C, RH-mean of 53.9%, and Pre-mean of 0.8 mm.
Type3	12.1%	High-pressure center over the North China Plain. Dongying inside the high-pressure, with dominant easterly wind, T_{max}-mean of 16.9 °C, RH-mean of 64.0%, and Pre-mean of 1.4 mm.
Type4	16.6%	High-pressure center over the North China Plain. Dongying under the top-front of the high-pressure, with dominant southwesterly wind, T_{max}-mean of 14.5 °C, RH-mean of 57.0%, and Pre-mean of 0.9 mm.
Type5	6.7%	Dongying under the control of a weak pressure field with sparse isobars and stable synoptic condition, with dominant southerly wind, T_{max}-mean of 17.3 °C, RH-mean of 64.0%, and Pre-mean of 1.9 mm.
Type6	8.4%	Dongying between a high- and low-pressure systems with dense isobars, with dominant southeasterly and southwesterly winds, T_{max}-mean of 19.2 °C, RH-mean of 56.8%, and Pre-mean of 0.6 mm.
Type7	9.5%	Dongying in the northwest of the western Pacific subtropical high, with dominant southeasterly wind, T_{max}-mean of 29.2 °C, RH-mean of 72.4%, and Pre-mean of 8.0 mm.
Type8	7.6%	Dongying near the inverted trough, with dominant southeasterly wind, T_{max}-mean of 25.8 °C, RH-mean of 73.2%, and Pre-mean of 6.6 mm.
Type9	4.4%	Dongying inside the low-pressure field, with dominant southeasterly wind, T_{max}-mean of 29.3 °C, RH-mean of 69.1%, and Pre-mean of 2.1 mm.

Notes: T_{max}-mean refers to the daily average maximum temperature; RH-mean refers to the daily average relative humidity; and Pre-mean refers to the daily average precipitation.

As shown in Figure 12, when the proportion of "subtropical high" (Type 7), "inverted trough" (Type 8), and "low-pressure field" (Type 9) was higher, the NO_2 concentration in Dongying was relatively low. Conversely, when the proportion of these types decreased and the proportion of high-pressure synoptic patterns (Type 1, Type 2, Type 3, and Type 4) increased, the NO_2 concentration rose. This is because Type 7, Type 8, and Type 9 usually occur in summer, with higher T_{max}-mean, occasional precipitation, and higher RH-mean, which facilitate the removal of pollutants. When Dongying was under "offshore high-pressure rear" (Type 1), the impact of the cold high-pressure was waning, resulting in stable synoptic conditions and low RH, which can lead to a higher concentration of accumulated pollutants and higher NO_2 levels. When Dongying was under "high-pressure front" (Type 2) and "high-pressure top front" (Type 4), the passage of cold air masses resulted in a significant drop in T and RH. Additionally, the dominant wind was typically a low-speed southwest wind, which was conducive to the increase in NO_2 concentration. When Dongying was under "high-pressure inside" (Type 3), the activity of the cold high-pressure was weak, leading to conditions such as temperature inversion, which can result in NO_2 exceedances. Moreover, when Dongying was under "uniform pressure" (Type 5), horizontal atmospheric movement is weak, often accompanied by strong radiative inversions or low-level inversions, making high NO_2 concentration more likely [54]. The above analysis indicates that Type 7, Type 8, and Type 9 help to reduce NO_2 concentration, whereas high-pressure synoptic patterns (Type 1, Type 2, Type 3, and Type 4) and "uniform pressure" (Type 5) tend to trigger NO_2 pollution.

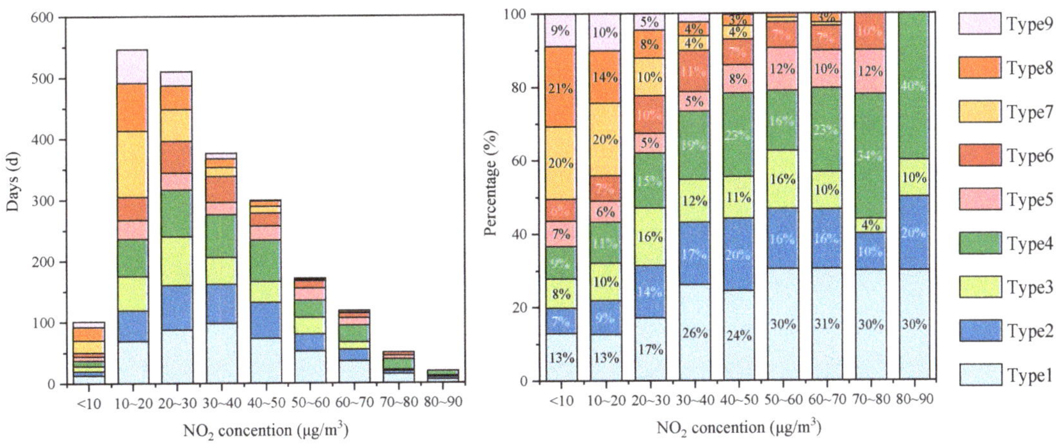

Figure 12. Frequencies and percentages of different synoptic patterns corresponding to NO_2 daily average concentration ranges in Dongying from 2017 to 2022.

The monthly variation in NO_2 concentrations on land–sea breeze and non-land–sea breeze days in Dongying City in 2022 is shown in Figure 13a. Overall, the average NO_2 concentration on land–sea breeze days in Dongying (29.1 $\mu g/m^3$) was higher than that on non-land–sea breeze days (26.5 $\mu g/m^3$). In January, March, and September to October, the NO_2 concentration on land–sea breeze days in Dongying was higher than that on non-land–sea breeze days. From June to August and November to December, the NO_2 concentrations on land–sea breeze days were similar to those on non-land–sea breeze days. In February and from April to May, the NO_2 concentration on land–sea breeze days was lower than that on non-land–sea breeze days. The above analysis indicates that during high NO_2 concentrations periods, the land–sea breeze has a significant contribution to the elevated daily average NO_2 concentrations in Dongying City.

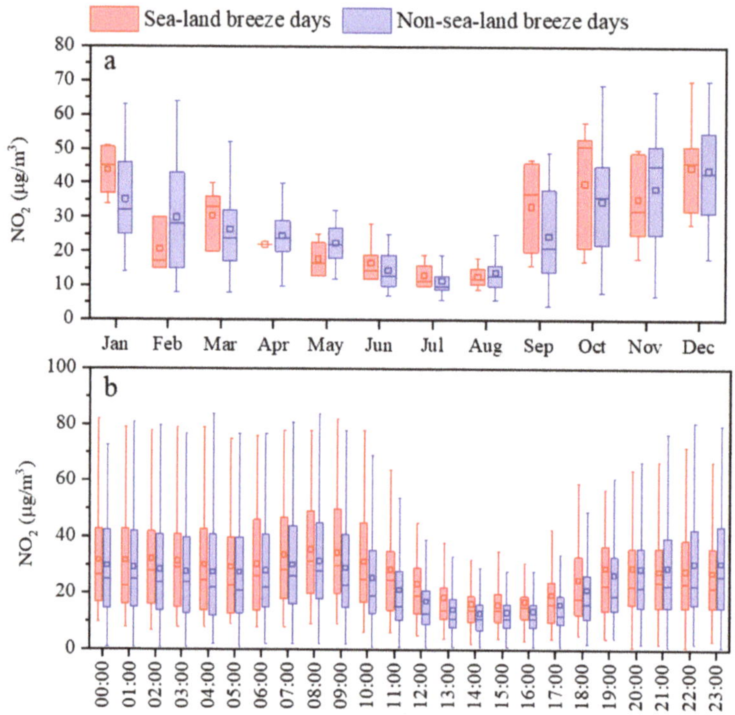

Figure 13. The monthly (**a**) and daily (**b**) variation in NO_2 concentrations on land–sea breeze and non-land–sea breeze days in Dongying in 2022.

The daily variation in NO_2 concentrations on land–sea breeze and non-land–sea breeze days in Dongying City in 2022 is shown in Figure 13b. Overall, the hourly NO_2 concentrations on land–sea breeze days were significantly higher than those on non-land–sea breeze days. From 00:00 to 20:00, the NO_2 concentrations on land–sea breeze days were higher than those on non-land–sea breeze days. From 00:00 to 04:00, the difference between the two was small, ranging from 1.9 to 3.6 $\mu g/m^3$. From 05:00 to 15:00, the difference gradually increased and then decreased, with the greatest difference observed at 11:00 (7.3 $\mu g/m^3$). From 16:00 to 18:00, the concentration difference between the two remained between 3.3 and 3.5 $\mu g/m^3$. From 19:00 to 20:00, the difference gradually decreased, and by 21:00, the NO_2 concentration on land–sea breeze days was lower than on non-land–sea breeze days. The possible explanation for this phenomenon is as follows: during the early phase of land breeze dominance, pollutants from the continent are transported to the air over the Bohai Sea. Meanwhile, ship emissions in the Bohai region also contribute to air pollution [55], and currently, NOx and VOC emissions from ships in China remain relatively high [56]. Consequently, NO_2 gradually accumulates over the Bohai Sea. During the transition from land breeze to sea breeze, NO_2 pollution plume over the Bohai Sea are gradually transported back to Dongying City. Under the combined influence of transport and local emissions, the hourly NO_2 concentration is higher than that on non-land–sea breeze days [20]. During the period of sea breeze dominance, NO_2 is continuously transported from the Bohai Sea to Dongying, resulting in a large difference between the two. During the transition from sea breeze to land breeze, the NO_2 concentration in the sea area decreases and is gradually carried back to the air over the Bohai Sea by the dominant land breeze, resulting in a smaller difference in NO_2 concentrations between land–sea breeze and non-land–sea breeze days, with the concentration on land–sea breeze days at certain hours being lower.

The variation of NO_2 concentrations with T, RH, WS, and WD in Dongying from 2017 to 2022 is shown in Figure 14. High NO_2 concentration (>40 μg/m^3) are notably concentrated when T from −5 to 15 °C and RH from 30 to 80%. Daily average NO_2 concentration exceeding standards is more likely when T are below 10 °C and RH ranges from 50 to 70%. It is noteworthy that NO_2 concentration does not show a strong linear relationship with T and RH. This can be explained by the findings of Huang et al. [57], who, through extensive data analysis, discovered that NO_2 concentrations are primarily influenced by BLH and AP, especially by BLH. When WS in Dongying are below 1 m/s, both the level of daily average NO_2 concentration and the frequency of high NO_2 concentration are relatively high. When WD is N to NW and WS range from 2 to 3 m/s, high NO_2 concentration are more frequent. When WD is W to S, the level of daily average NO_2 concentration and the frequency of high NO_2 concentration decrease with increasing WS. Overall, high NO_2 concentration are more likely to occur when WS are below 1 m/s. While when WS exceeds 1 m/s, southwestern winds facilitate a higher NO_2 concentration. This may be due to local pollutant accumulation at low WS and regional pollutant transport at high WS [58,59].

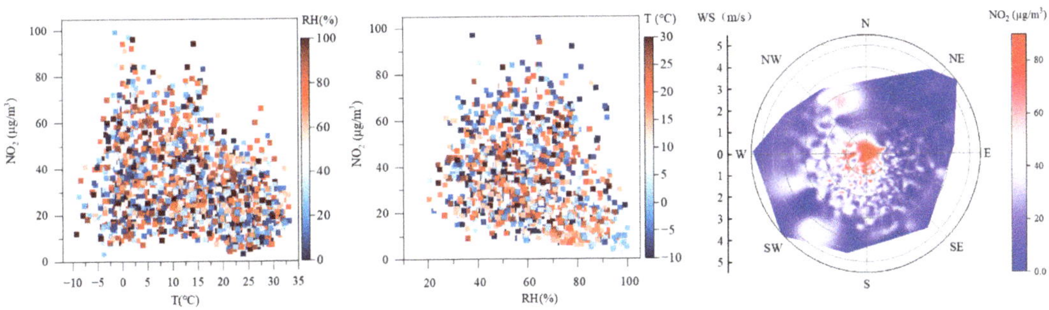

Figure 14. Variation of NO_2 concentrations with T, RH, WS, and WD in Dongying from 2017 to 2022.

In addition, this study analyzed the correlation between the daily average NO_2 concentrations and meteorological factors from 2017 to 2022 (Table S2). From a seasonal perspective, WS (−0.18 to −0.57) and RH (−0.30 to −0.60, except in winter) exhibited relatively consistent correlations across all seasons, whereas T and AP showed seasonal variations. Previous studies have indicated that the strength and correlation of meteorological factors affecting NO_2 concentrations vary depending on the season [60] or the study period [61]. From an annual perspective, NO_2 concentrations showed a significant negative correlation with WS (−0.43), T (−0.45), and RH (−0.38), while exhibiting a positive correlation with AP (0.46). These results are consistent with findings from Handan [62], as well as studies in Shijiazhuang, Foshan and Ya'an regarding WS, T and AP, although no significant correlation with RH was observed in these regions [60,63,64]. The possible reasons for these findings include higher NO_2 emissions in winter, while elevated T in summer enhances photochemical consumption; low WS hinders pollutant dispersion; high RH may promote the heterogeneous reaction of NO_2 [65]; and high AP leads to descending airflows and suppresses pollutant dispersion.

In summary, high NO_2 concentrations in Dongying tend to occur under meteorological conditions characterized by relatively low T (−5 to 15 °C), moderate RH (30–80%), southwesterly WD, and WS below 1 m/s.

The factor importance obtained after building a random forest model based on meteorological data is shown in Figure 15. Overall, the main factors influencing the hourly NO_2 concentrations in Dongying City in 2022 were BLH, while the dominant factors for the daily average NO_2 concentrations were BLH, WS, and UR. Studies have shown that a decrease in BLH is a key reason for the significant increase in NO_2 concentrations at night and across seasons [66,67]. From the difference between NO_2-HH and NO_2-NHH, the importance of BLH significantly increased during the spring and winter for NO_2-HH (with correlations of −0.43 and −0.36, and importance contribution differences of 10% and 12%, respectively). This is because BLH is generally lower in spring and winter, which facilitates the occurrence of high NO_2 values, thus contributing to the main cause of NO_2-HH formation. In summer, the importance of WD increased significantly (correlation of 0.21, importance contribution difference of 10%), likely due to the transport of NO_2 air masses in Dongying City, which is a major factor in the formation of NO_2-HH. In autumn, the importance of T increased significantly (correlation of −0.46, importance contribution difference of 4%), which is related to Dongying City's low-temperature environment, making it conducive to the formation of NO_2-HH. From the difference between NO_2-HD and NO_2-NHD, the importance of BLH significantly increased, indicating that BLH is the primary factor for the occurrence of NO_2-HD.

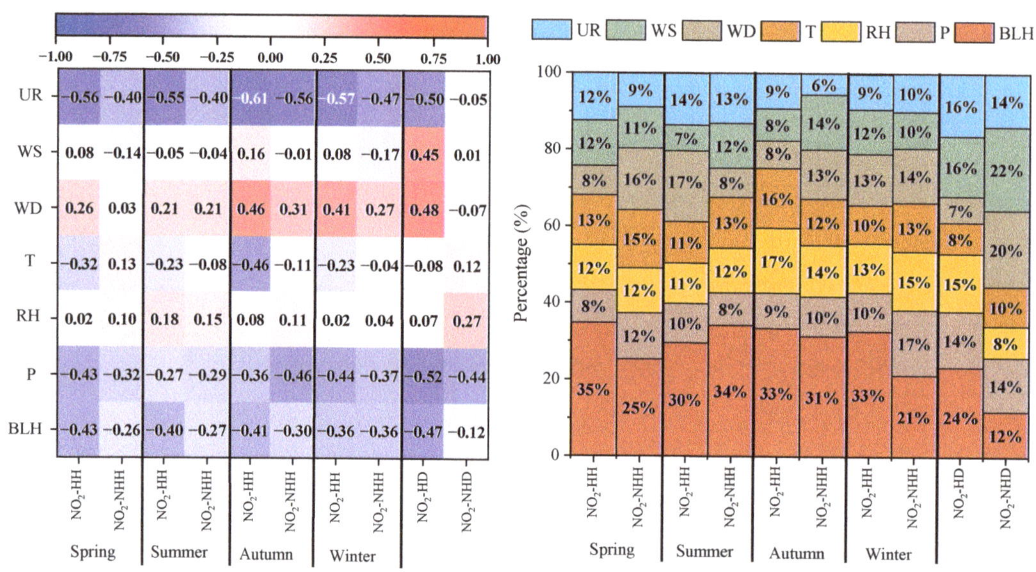

Figure 15. The correlation between NO_2 and meteorological factors (**a**), and the importance share of each factor (**b**) on days with NO_2-HH events and NO_2-NHH periods in each season of Dongying in 2022.

3.3.2. Chemical Generation

The chemical reactions between NO, NO_2, VOCs, and O_3 affect the concentration of NO_2. As shown in Figure 16, compared to NO_2-NHH periods, all types of VOCs increase during NO_2-HH events, with alkanes and halogenated hydrocarbons showing a more pronounced increase, while OVOCs increase to a lesser extent. This suggests that the significant increase in VOCs during NO_2-HH events, particularly alkynes, alkanes, and alkenes, may be associated with the formation of NO_2-HH. In terms of changes in pollutant concentrations, except for a marked decrease in O_3 concentration during NO_2-HH events, concentrations of other pollutants are higher, with NO showing the greatest increase. This is related to NO_2-related chemical reactions, particularly the titration reaction between NO and O_3 and the series of nighttime chemical reactions involving NO_2 [3]. The balance of these reactions significantly influences hourly NO_2 concentration levels. Overall, NO_2-HH events are more likely to occur when VOCs and NO concentrations are high, and O_3 concentration is low.

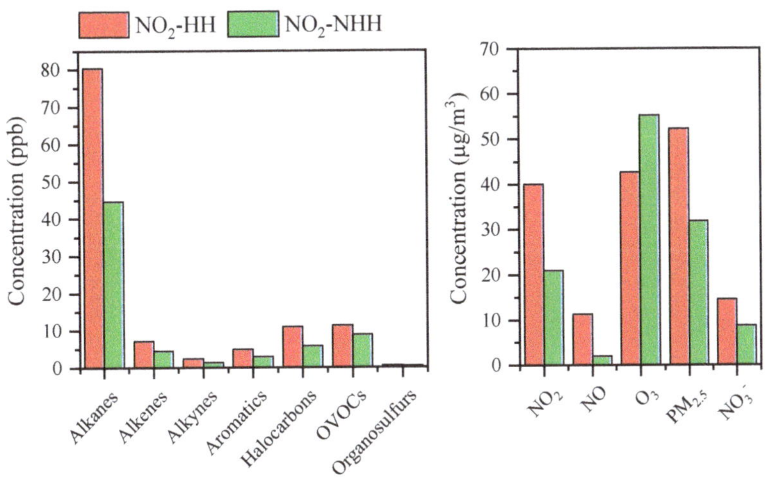

Figure 16. Differences in concentrations of different types of VOCs and related pollutants during nighttime to morning rush hours between NO_2-HH events and NO_2-NHH periods in Dongying in 2022.

As shown in Figure 17, on NO_2-HH event days, VOCs and NO concentrations increase significantly during nighttime and morning rush hours, exhibiting a unimodal pattern, with more pronounced photochemical consumption at noon. In contrast, on NO_2-NHH days, VOCs and NO concentrations remain relatively stable at night, with a smaller increase during the early morning rush and less photochemical consumption at noon. The high concentrations of VOCs and NO in Dongying contribute to intense photochemical reactions during the day, leading to higher NO_2 levels. On NO_2-HH event days, the peak NO concentration in the evening is higher, which rapidly reduces nighttime O_3 levels through titration reactions, while the NO_2 concentration remains stable after a rapid increase. Xia et al. [2] found that an increase in the diurnal variation of NO_2 is accompanied by a larger diurnal variation in O_3, which is consistent with the variations in NO_2 and O_3 on NO_2-HH event days in Dongying. Additionally, studies have shown that when NO concentration is high, NO_3 is primarily removed through reaction with NO to form NO_2, which suppresses the nighttime NO_2 net consumption rate about NO_3. Therefore, the NO_3^- formation rate is lower during the evening peak of NO_2-HH event days. The stability of NO_2 in the early morning indicates that on NO_2-HH event days, there is ample NO, and NO_2 is in a dynamic equilibrium between generation and consumption, with O_3 continuously depleted and NO_3^- increasing. Conversely, on NO_2-NHH days, after the NO_2 concentration rises during the evening peak, it shows a decreasing trend along with VOCs in the early morning. This is due to insufficient NO at night and higher O_3 concentration, which dominate the NO_2 net consumption reactions. Overall, daytime NO_2

concentration are primarily influenced by intense photochemical reactions, with higher VOCs and NO concentrations leading to higher daytime NO_2 levels. During nighttime and morning rush hours, NO and O_3 significantly affect NO_2 concentration.

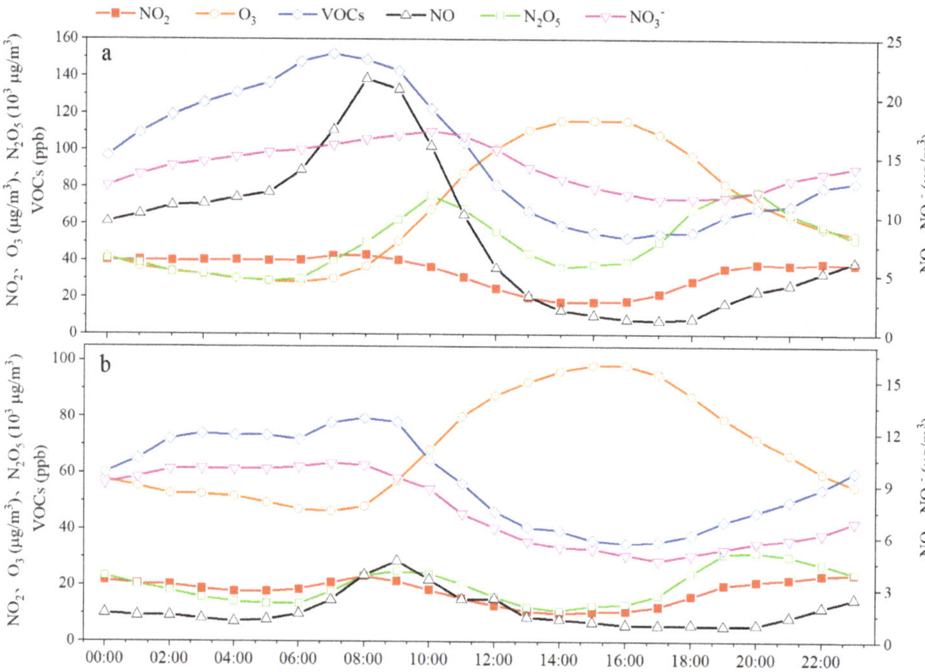

Figure 17. Diurnal variations in different air pollutants on days with NO_2-HH events (**a**) and NO_2-NHH periods (**b**) in Dongying in 2022.

The factor importance obtained after building a random forest model based on pollutant data is shown in Figure 18. Overall, the main pollutants influencing the hourly NO_2 concentrations in Dongying City were NO, O_3, and Aromatics, while the dominant factors for the daily NO_2 concentrations were NO, Aromatics, Alkenes, and O_3. From the difference between NO_2-HH and NO_2-NHH, the importance of O_3 and NO increased for NO_2-HH, except in winter, while the importance of Aromatics decreased across all seasons. Other pollutants showed no consistent pattern. This suggests that NO_2-HH in spring, summer, and autumn is primarily driven by the titration reaction between NO and O_3, while the day–night variation in Aromatics is related to the concentrations of oxidants OH and NO_3 [68]. The weakening of reactions with NO_3 reduces the net consumption of NO_2. From the difference between NO_2-HD and NO_2-NHD, the importance of Alkenes decreased, while that of Aromatics increased. This is because Alkenes are key participants in the nighttime chemical reaction chain, consuming NO_3 and leading to net consumption of NO_2 [69]. Both Alkenes and Aromatics exhibit strong photochemical activity during the day, participating in photochemical chain reactions to produce NO_2 [70].

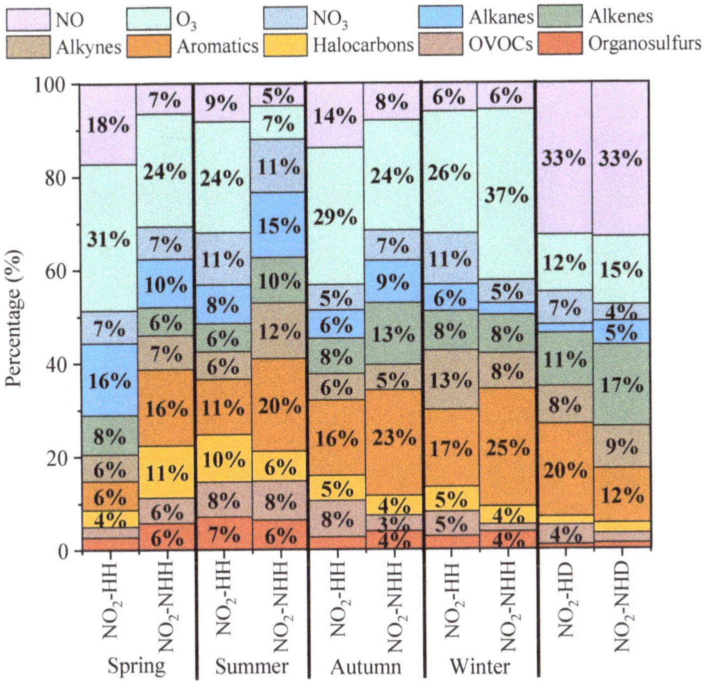

Figure 18. The importance share of each pollutant's influence on NO_2 concentrations on days with NO_2-HH events and NO_2-NHH periods in each season of Dongying in 2022.

Overall, daytime NO_2 concentration are primarily influenced by intense photochemical reactions, with higher VOCs and NO concentrations leading to higher daytime NO_2 levels. During nighttime and morning rush hours, NO and O_3 significantly affect NO_2 concentration. The titration reaction between NO with O_3 is the main cause for NO_2-HH in spring, summer and autumn, and photochemical reactions of Aromatics have a significant influence on NO_2-HD.

3.3.3. Sources Analysis

As shown in Figure 19, NOx emissions in Dongying are generally higher on NO_2-HD days compared to NO_2-NHD days across all seasons, with the greatest difference in winter and the smallest in summer. This indicates that local NOx emissions have a significant impact on NO_2-HD in Dongying. Seasonal variations in NOx emissions are highest in winter and lowest in spring, which differs from the seasonal variations in NO_2 concentration, suggesting that NO_2 distribution may be influenced by additional factors. Industry-specific NOx emissions data reveal that thermal power (51–57%) and petrochemical industry (26–32%) are the largest contributors to NOx emissions across all seasons. Compared to NO_2-NHD days, on NO_2-HD days, the proportion of emissions from thermal power increases in spring, summer, and autumn, while the proportion from the petrochemical industry increases in winter. Therefore, thermal power and petrochemical industry are likely major sources affecting NO_2-HD. In summary, local emissions (thermal power and petrochemical industry) have a certain impact on NO_2-HD in Dongying.

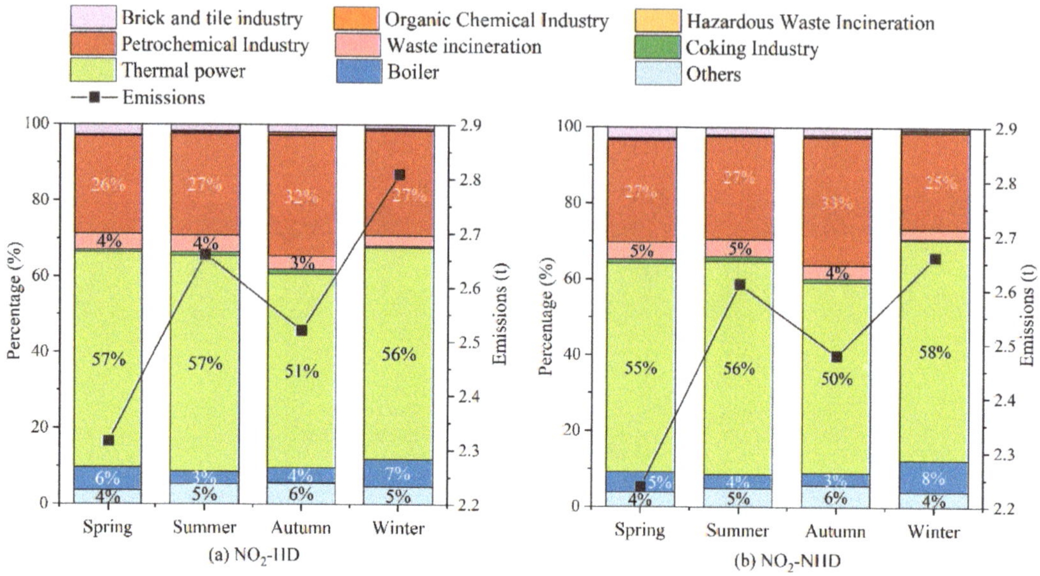

Figure 19. Differences in emissions and percentages of NOx from different industry between NO_2-HD days and NO_2-NHD days for each season in Dongying in 2022.

Cluster analysis (Figure 20 and Figure S12) reveals significant seasonal differences in the direction, speed, and altitude of air masses at Dongying sites. In spring, northwest trajectory 3 accounts for 43.1%, with the shortest trajectory, indicating slower air mass movement and a higher potential for local pollutant accumulation. This trajectory also operates at a lower altitude, facilitating local horizontal pollutant transport. The corresponding average NO_2 concentration for this trajectory was 21.1 μg/m^3, contributing 48.1%. Southwest trajectory 2, which accounts for 34.9%, has a relatively short trajectory and the lowest altitude, leading to pollutant transport from southern cities in Shandong Province. Additionally, trajectory 1, with a lower proportion, has a higher altitude and exerts minimal impact on Dongying in spring. In summer, the southeastward trajectory had the highest share (44.9%) and contribution (45.7%). The average concentration for this trajectory matched the summer average NO_2 concentration, indicating that this trajectory is the primary source of air masses maintaining summer NO_2 levels. Trajectory 3 had the lowest share (23.1%) but the highest average concentration (18.4 μg/m^3), which facilitates the formation of high NO_2 values. Additionally, Dongying City in summer may be affected by the transport of air masses from the Bohai Bay (32%). Similarly, in autumn and winter, the local influence on Dongying City is significantly stronger, primarily from air masses coming from the south and southwest. This is related to air mass transport and pollutant accumulation [71]. The lifespan of NO_2 is relatively short (1–2 days), primarily influenced by local transport rather than long-distance transport [72].

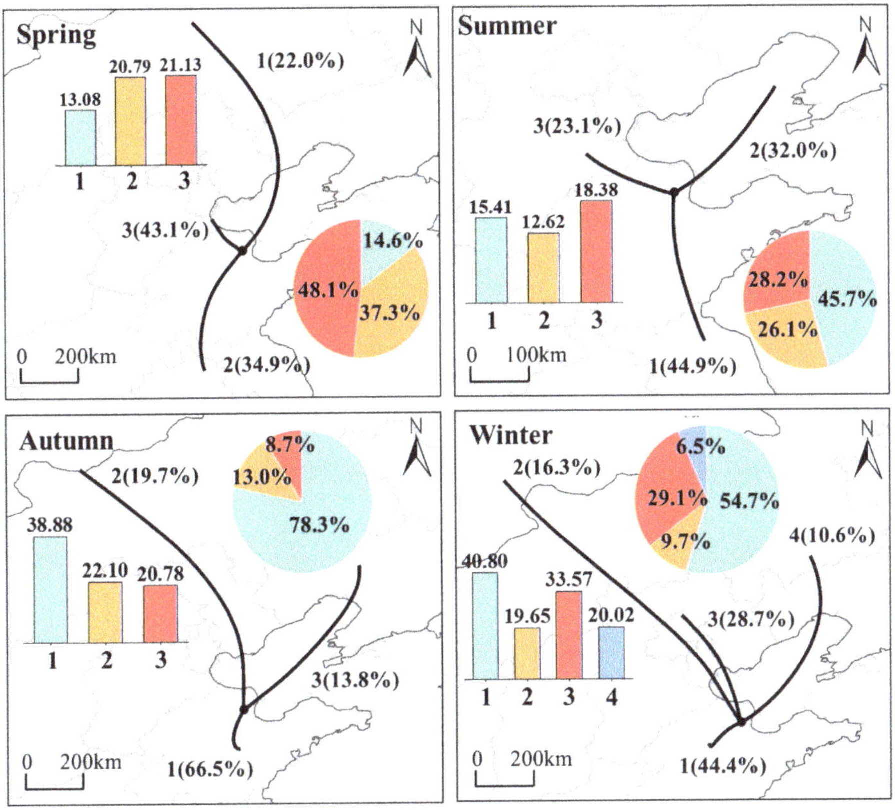

Figure 20. Cluster analysis of backward trajectories in Dongying in 2022. (The bar chart represents the average NO_2 concentration ($\mu g/m^3$) corresponding to each type of trajectory, and the pie chart represents the percentage contribution of NO_2 concentration corresponding to each type of trajectory).

The distribution of WPSCF and WCWT in Dongying in 2022 is shown in Figure 21. From the WPSCF distribution, in spring, the WPSCF values for the southwestern regions of Dongying and the small area around the Bohai are relatively high, with a focus on cities in western Shandong Province, such as Jinan, Binzhou, and Taian, with western Taian having the highest potential impact on NO_2-HH. In summer, the WPSCF values for nearby cities of Dongying, particularly Binzhou and Zibo, are higher, indicating a significant influence from nearby cities. In autumn, potential source regions affecting NO_2-HH in Dongying are more widely distributed, primarily in central Dezhou, Jinan, Taian, and parts of Linyi. In winter, the main potential source regions affecting NO_2-HH in Dongying are located in northeastern Handan, southwestern Dezhou, and central areas of Jinan and Taian. From the WCWT distribution, the distribution of WCWT values is quite similar to that of WPSCF values, with the areas of high concentration contribution aligning well with the potential source regions that significantly impact NO_2-HH at the observation site. In spring, autumn, and winter, areas with WPSCF values greater than 0.5 generally correspond to areas with high WCWT values, while in summer, areas with WPSCF values greater than 0.25 align with areas of high WCWT values. This correspondence is related to the NO_2 threshold settings in the seasonal PSCF analysis. In addition to the regions corresponding to the high WPSCF values, there are some areas with high WCWT values in the Bohai Sea of China during spring and summer and in the northwest of Dongying during winter.

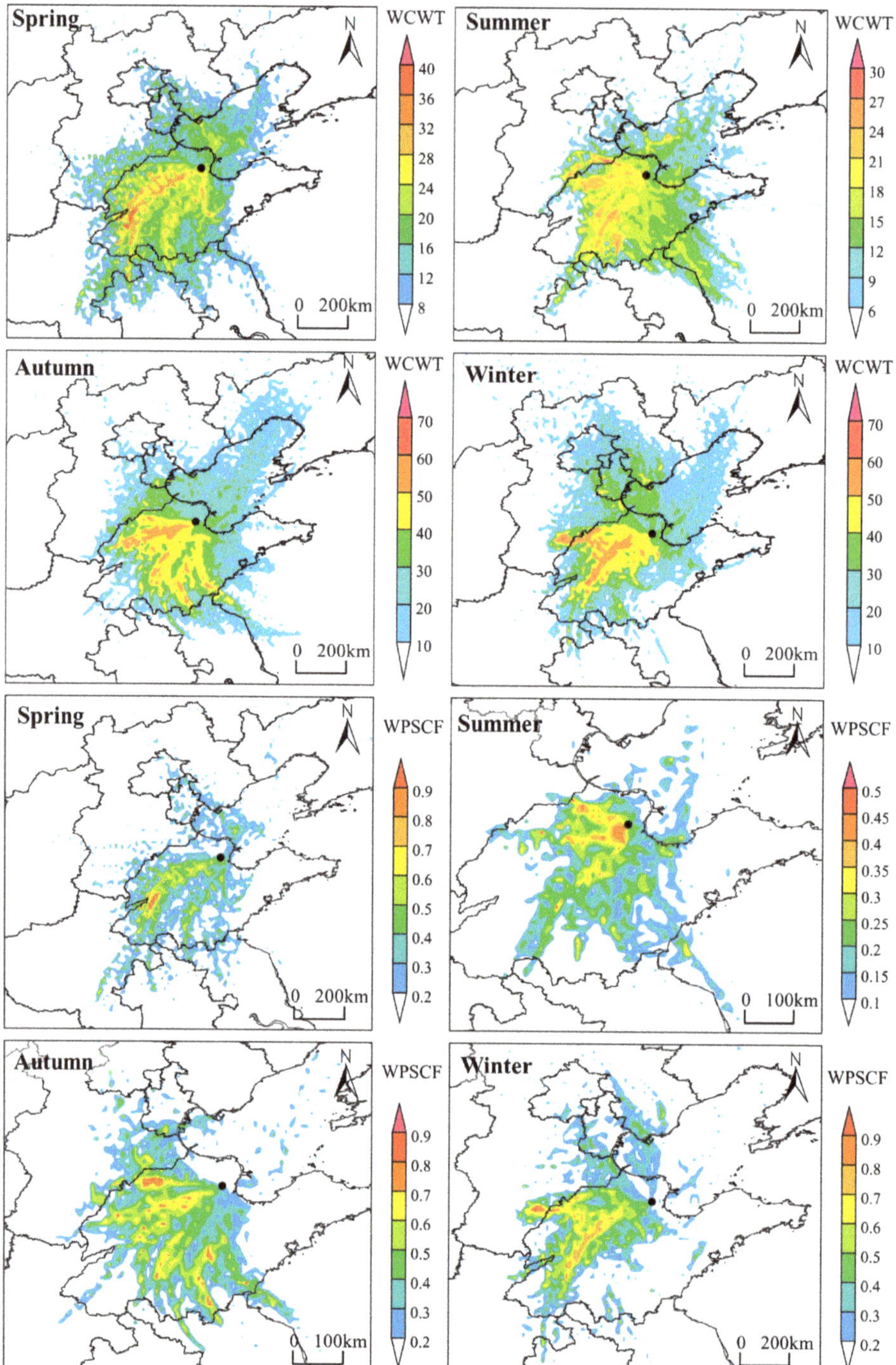

Figure 21. Distribution of WPSCF and WCWT in Dongying in 2022 (The black dot represents Dongying Atmospheric Observatory).

In summary, the distribution of high WCWT and WPSCF values is quite similar and aligns closely with the main air mass transport pathways. NO_2 regional transport in Dongying primarily occurs within Shandong Province. The potential source regions that contribute most significantly to the occurrence probability and concentration of NO_2-HH at the Dongying site are mainly located in the western and southwestern parts of Shandong Province (with wider distribution in autumn and winter), including Jinan, Binzhou, Taian, and Zibo. Additionally, during spring and summer, there is a certain influence from the Bohai Sea of China.

Dongying City is located in the Bohai Bay region, which has a unique geographical location and meteorological conditions [19,55]. In the future, building on outstanding international research, it is highly necessary to conduct long-term and large-scale studies on NO_2 to provide scientific support for regional NO_2 control and air quality improvement.

4. Conclusions

In this study, we explored the effects of NO_2 on the ambient air quality in Dongying, a typical petrochemical city in Bohai Bay of China, from 2017 to 2023 and the impacts of high NO_2 values on the concentrations of O_3, $PM_{2.5}$ and $PM_{2.5}$-bounded secondary components, and atmospheric oxidation capacity in 2022, and investigated the reasons for the formation of high NO_2 values in Dongying.

(1) From 2017 to 2023, the annual assessment values of NO_2 concentrations in Dongying exhibited a trend of initially decreasing and then stabilizing, with monthly variation characterized by a "U" distribution, diurnal variation characterized by a "single-peak, single-valley" distribution, and spatial variation characterized by a "high in the south-central part and low in the north" distribution. Higher daily NO_2 concentrations mainly occurred during January-April and September-December each year, and higher hourly NO_2 concentrations mainly occurred during the nighttime and morning rush hours.

(2) NO_2 is a contributing factor to the increase in various pollutant concentrations, thereby affecting air quality levels. High hourly NO_2 values (NO_2-HH) events promoted the increase in the concentrations of $PM_{2.5}$ and $PM_{2.5}$-bounded NO_3^-, SO_4^{2-} and SOC, and O_X during the nighttime and morning rush hours, and they also facilitated the increase in the concentrations of O_3, $PM_{2.5}$ and $PM_{2.5}$-bounded NO_3^-, SO_4^{2-} and SOC, and O_X, as well as NOR and SOR values during the daytime. High daily NO_2 values (NO_2-HD) could promote the increase in the concentrations of O_3 (MDA8-O_3), $PM_{2.5}$ and $PM_{2.5}$-bounded NO_3^-, SO_4^{2-} and SOC, and O_X on the same day.

(3) The occurrence of high NO_2 values was affected by the combination of unfavorable meteorological conditions, local emissions and regional air pollution transports, and localized atmospheric chemical generation. High-pressure and uniform-pressure weather patterns, along with low temperatures (−5~15 °C), moderate humidity (30~80%), wind speeds below 1 m/s, and southwesterly winds, are conducive to high daily NO_2 values. In 2022, land–sea breeze circulation significantly contributed to high NO_2 concentrations. Random forest analysis indicates that in 2022, BLH in spring (−0.43) and winter (−0.36), WD in summer (0.21), and T (−0.46) in autumn were the primary factors contributing to NO_2-HH, while BLH (−0.47) is the main cause for NO_2-HD. The titration reaction between NO with O_3 is the main cause for NO_2-HH in spring, summer and fall, and photochemical reactions of Aromatics have a significant influence on NO_2-HD. NOx emissions from the thermal power and petrochemical industry in Dongying and air pollution transports from western and southwestern Shandong Province (throughout the year) and from the Bohai Sea (during spring and summer) had serious adverse impact on high NO_2 values.

It is recommended that Dongying strengthens its control over NO_X emissions from local thermal power and petrochemical industries, and actively promotes the establishment of an air pollution prevention and control mechanism in collaboration with cities in western and southwestern Shandong Province and the Bohai Bay region of China to effectively reduce NO_2 concentrations and prevent high NO_2 values, thereby continuously improving air quality.

Supplementary Materials: The following supporting information can be downloaded at: https://www.mdpi.com/article/10.3390/toxics13030208/s1, Figure S1: Conceptual diagram of NO_2 sources, sinks, and impacts; Figure S2: Box plots of daily average concentrations (a) and hourly concentrations (b) of NO_2 in Dongying for each season in 2022; Figure S3: Technical roadmap of analyzing the impact of NO_2 on air quality in Dongying; Figure S4: Technical roadmap of analyzing the causes of high NO_2 values in Dongying in 2022; Figure S5: Comparison of diurnal variations of NO_2 concentrations in Dongying from 2017 to 2023; Figure S6: Percentages of days with different NO_2 air quality levels and number of days with NO_2 as the primary pollutant in Dongying from 2017 to 2023; Figure S7: Comparison of monthly distribution of NO_2-HD days in Dongying in 2022; Figure S8: Comparison of seasonal distribution of NO_2-HH in Dongying in 2022; Figure S9: Differences in the daytime concentrations of O_3, $PM_{2.5}$ and $PM_{2.5}$-bounded secondary components between NO_2-HH events and NO_2-NHH periods in Dongying in 2022; Figure S10: Differences in the daytime concentrations of O_X, and values of NOR and SOR between NO_2-HH events and NO_2-NHH periods in Dongying in 2022; Figure S11: Classification of 925-hPa synoptic patterns in Dongying from 2017 to 2022; Figure S12: Mean barometric pressure variation curves in the vertical direction during the movement of clustered trajectories in each season in 2022; Text S1: Data sources; Text S2: Data quality control; Text S3: Synoptic classification method; Text S4: HYSPLIT mode; Text S5: Backward trajectory clustering analysis; Text S6: PSCF analysis; Text S7: CWT analysis; Text S8: Analysis of impact of high NO_2 values on daytime O_3, $PM_{2.5}$, and atmospheric oxidation capacity in Dongying in 2022; Table S1: Data types and data sources for each part of analysis in this study; Table S2: Correlation between NO_2 concentration and meteorological factors in different seasons in Dongying from 2017 to 2022 [73–86].

Author Contributions: X.G.: Formal analysis, Investigation, Methodology, Visualization and Writing—original draft. C.A.: Validation and Writing—review and editing. Y.Y.: Software. Y.J.: Methodology. W.W.: Methodology. L.X.: Methodology. F.S.: Data curation. J.L.: Data curation. R.G.: Project administration, Supervision, and Methodology. L.T.: Resources. H.L.: Conceptualization, Methodology, Supervision, Validation, Funding acquisition, Writing—review and editing. All authors have read and agreed to the published version of the manuscript.

Funding: This work was supported by the Ministry of Ecology and Environment of China under the project (DQGG202121) and the Dongying Ecological and Environmental Bureau (2021DFKY-0779).

Institutional Review Board Statement: Not applicable.

Informed Consent Statement: Not applicable.

Data Availability Statement: The raw data supporting the conclusions of this article will be made available by the authors on request.

Conflicts of Interest: Author Luyao Tan was employed by the company Ltd. of Shandong Environmental Protection Industry Corp. The remaining authors declare that the research was conducted in the absence of any commercial or financial relationships that could be construed as a potential conflict of interest.

References

1. Tang, X.Y.; Zhang, Y.H.; Shao, M. *Atmospheric Environmental Chemistry*; Higher Education Press: Beijing, China, 2024.
2. Xia, N.; Du, E.Z.; Guo, Z.D.; Vries, W.D. The diurnal cycle of summer tropospheric ozone concentrations across Chinese cities: Spatial patterns and main drivers. *Environ. Pollut.* **2021**, *286*, 117547. [CrossRef] [PubMed]

3. Yan, Y.H.; Wang, S.S.; Zhu, J.; Guo, Y.L.; Tang, G.Q.; Liu, B.X.; AN, X.X.; Wang, Y.S.; Zhou, B. Vertically increased NO_3 radical in the nocturnal boundary layer. *Sci. Total Environ.* **2021**, *763*, 142969. [CrossRef]
4. Lai, H.K.; Hedley, A.J.; Thach, T.Q.; Wong, C.M. A method to derive the relationship between the annual and short-term air quality limits-Analysis using the WHO Air Quality Guidelines for health protection. *Environ. Int.* **2013**, *59*, 86–91. [CrossRef]
5. Zong, Z.; Tian, C.; Li, J.; Syed, J.H.; Zhang, W.; Fang, Y.; Jiang, Y.; Nasir, J.; Mansha, M.; Rizvi, S.H.H.; et al. Isotopic Interpretation of Particulate Nitrate in the Metropolitan City of Karachi, Pakistan: Insight into the Oceanic Contribution to NOx. *Environ. Sci. Technol.* **2020**, *54*, 7787–7797. [CrossRef] [PubMed]
6. Richter, A.; Burrows, J.P.; Nuss, H.; Granier, C.; Niemeier, U. Increase in tropospheric nitrogen dioxide over China observed from space. *Nature* **2005**, *437*, 129–132. [CrossRef]
7. Zhu, Y.J.; Zhan, Y.; Wang, B.; Li, Z.; Qin, Y.Q.; Zhang, K.S. Spatiotemporally mapping of the relationship between NO_2 pollution and urbanization for a megacity in Southwest China during 2005–2016. *Chemosphere* **2019**, *220*, 155–162. [CrossRef]
8. Lalitaporn, P.; Kurata, G.; Matsuoka, Y.; Thongboonchoo, N.; Surapipith, V. Long-term analysis of NO_2, CO, and AOD seasonal variability using satellite observations over Asia and intercomparison with emission inventories and model. *Air Qual. Atmos. Health.* **2013**, *6*, 655–672. [CrossRef]
9. Guan, Q.D.; Zuo, X.Q.; Li, S.H. Analysis of Distribution Characteristics and Driving Factors of Nitrogen Dioxide Concentration in China. *Geomat. Spat. Inf. Technol.* **2021**, *44*, 85–89+97.
10. Nishanth, T.; Praseed, K.M.; Kumar, M.K.S.; Valsaraj, K.T. Observational Study of Surface O_3, NOx, CH_4 and Total NMHCs at Kannur, India. *Aerosol Air Qual. Res.* **2014**, *14*, 1074–1088. [CrossRef]
11. Gao, J.H.; Zhu, B.; Wang, Y.Z.; Kang, H.Q. Distribution and long-term variation of tropospheric NO_2 over China during 2005 to 2013. *China Environ. Sci. (Chin. Ed.).* **2015**, *35*, 2307–2318.
12. Yang, G.F.; Li, S.Y.; Li, W.L. Spatial and temporal distribution characteristics of nitrogen dioxide concentration in Chinese cities. *Environ. Sci. Technol.* **2019**, *42*, 200–206.
13. Wang, Y.J.; Yaluk, E.A.; Chen, H.; Jiang, S.; Huang, L.; Zhu, A.S.; Xiao, S.L.; Xue, J.; Lu, G.B.; Bian, J.T.; et al. The Importance of NOx Control for Peak Ozone Mitigation Based on a Sensitivity Study Using CMAQ-HDDM-3D Model During a Typical Episode Over the Yangtze River Delta Region, China. *J. Geophys. Res. Atmos.* **2022**, *127*, e2022JD036555. [CrossRef]
14. Cheng, N.; Li, Y.; Sun, F.; Chen, C.; Wang, B.; Li, Q.; Wei, P.; Cheng, B. Ground-Level NO_2 in Urban Beijing: Trends, Distribution, and Effects of Emission Reduction Measures. *Aerosol Air Qual. Res.* **2018**, *18*, 343–356. [CrossRef]
15. Wang, Y.J.; Jiang, S.; Huang, L.; Lu, G.B.; Kasemsan, M.; Yaluk, E.A.; Liu, H.Q.; Liao, J.Q.; Bian, J.T.; Zhang, K.; et al. Differences between VOCs and NOx transport contributions, their impacts on O_3, and implications for O_3 pollution mitigation based on CMAQ simulation over the Yangtze River Delta, China. *Sci. Total Environ.* **2023**, *872*, 162118. [CrossRef]
16. Yu, H.L.; Lin, Y.C.; Kuo, Y.M. A time series analysis of multiple ambient pollutants to investigate the underlying air pollution dynamics and interactions. *Chemosphere* **2015**, *134*, 571–580. [CrossRef]
17. Tang, Z.Y.; Xu, J.C.; Zhang, Y.B.; Luo, H.; Yang, X.C. Analysis of Vehicle Exhaust Pollutants and Meteorological Conditions in Xi'an. *Meteorol. Environ. Sci.* **2020**, *43*, 70–78.
18. Xiao, B.; Jia, H.W.; Xu, J.J.; Kang, Y.M.; Zhong, K. Effects of land and sea breeze on pollutant diffusion in northern Shanghai. *China Environ. Sci.* **2022**, *42*, 1552–1561.
19. Long, Q.; Wang, F.; Wang, C.; Mi, X.Y. Characteristics of sea-land breeze and turbulence intensity on the north shore of Bohai Bay. *J. Appl. Oceanogr.* **2020**, *39*, 303–311.
20. Bei, N.; Zhao, L.; Wu, J.; Li, X.; Feng, T.; Li, G. Impacts of sea-land and mountain-valley circulations on the air pollution in Beijing-Tianjin-Hebei (BTH): A case study. *Environ. Pollut.* **2018**, *234*, 429–438. [CrossRef]
21. Guo, Y.H.; Wang, Z.F.; Kang, H.; Gu, C.; Wang, G.; Zhang, X.X.; Ji, Y.; Li, M. Impact of automobile vehicles exhaust emissions on metropolitan air quality: Analysis study on the air pollution change before and after the Spring Festival in Urumqi City, China. *Acta Sci. Circumstantiae* **2014**, *34*, 1109–1117.
22. Zhang, H.F.; Wang, S.X.; Hao, J.M.; Wang, X.M.; Wang, S.L.; Chai, F.H.; Li, M. Air pollution and control action in Beijing. *J. Clean. Prod.* **2016**, *112*, 1519–1527. [CrossRef]
23. Pan, Y.J.; Li, Y.; Chen, J.H.; Shi, J.C.; Tian, H.; Zhang, J.; Zhou, J.; Chen, X.; Liu, Z.; Qian, J. Method for High-resolution Emission Inventory for Road Vehicles in Chengdu Based on Traffic Flow Monitoring Data. *Environ. Sci.* **2020**, *41*, 3581–3590.
24. OPCPC-CSES. *Blue Book on the Prevention and Control of Ozone Pollution in China*; Science publishing Company: Beijing, China, 2023.
25. MEE. *China Ecological Environment Status Bulletin*; China Ministry of Ecology and Environment: Beijing, China, 2023.
26. Li, Y.; Shi, G.; Chen, Z. Spatial and temporal distribution characteristics of ground-level nitrogen dioxide and ozone across China during 2015–2020. *Environ. Res. Lett.* **2021**, *16*, 124031. [CrossRef]
27. Xu, H. Temporal and spatial distribution characteristics and pollution sources of NO_2 pollution in Yangtze River Delta. Master's Thesis, Anhui University, Anhui, China, 2021.
28. Li, H.; Zhang, J.; Wen, B.; Huang, S.; Gao, S.; Li, H.; Zhao, Z.; Zhang, Y.; Fu, G.; Bai, J.; et al. Spatial-Temporal Distribution and Variation of NO_2 and Its Sources and Chemical Sinks in Shanxi Province, China. *Atmosphere* **2022**, *13*, 1096. [CrossRef]

29. MEE. *China Ecological Environment Status Bulletin*; China Ministry of Ecology and Environment: Beijing, China, 2022.
30. Wang, N.N.; Zhu, C.Y.; Li, W.; Qiu, M.Y.; Wang, B.L.; Li, X.Y.; Jiang, B.D.; Qu, X.Y.; Li, Z.S.; Cheng, H.C. Air quality improvement assessment and exposure risk of Shandong Province in China during 2014 to 2020. *Int. J. Environ. Sci. Technol.* **2023**, *20*, 9495–9504. [CrossRef]
31. Xing, H.; Zhu, L.; Chen, B.; Niu, J.; Li, X.; Feng, Y.; Fang, W. Spatial and temporal changes analysis of air quality before and after the COVID-19 in Shandong Province, China. *Earth Sci. Inform.* **2022**, *15*, 863–876. [CrossRef]
32. Wu, L. Volatile Organic Compound Emissions from the Oilfield in the Yellow River Delta Region, Northern China and Their Impact on Regional Ozone Pollution. Master's Thesis, Shandong University, Jinan, China, 2019.
33. Fan, X.Y.; Lian, L.S.; Wang, M. Characteristics of air pollution and its relationship with meteorological factors in Bohai Rim urban agglomerations. *J. Atmos. Environ. Opt.* **2022**, *17*, 506–520.
34. World Health Organization. WHO Global Air Quality Guidelines: Particulate Matter ($PM_{2.5}$ and PM_{10}). In *Ozone, Nitrogen Dioxide, Sulfur Dioxide Carbon Monoxide*; World Health Organization: Geneva, Switzerland, 2021; Volume 1302.
35. Wei, Z.Z.; Huuang, B.Z.; Jiang, Y.D.; Guo, Y.; Wang, M. Comparative analysis of air quality evaluation by arithmetic method, percentile method and moving average method and the information it reveals. *Environ. Sustain. Dev.* **2020**, *45*, 123–133.
36. Huang, L.; Liu, D.; Cai, D.J.; Chen, S.J.; Dong, H.; Lin, G.Z.; Wang, B.G.; Yang, J. Health risk assessment of exposure to multiple pollutants in Guangzhou. *China Environ. Sci. (Chin. Ed.).* **2022**, *42*, 5418–5426.
37. Thomas, M.A.; Devasthale, A. Typical meteorological conditions associated with extreme nitrogen dioxide (NO_2) pollution events over Scandinavia. *Atmos. Chem. Phys.* **2017**, *17*, 12071–12080. [CrossRef]
38. Wangberg, I.; Etzkorn, T.; Barnes, I.; Platt, U.; Becker, K.H. Absolute determination of the temperature behavior of the NO_2 + NO_3 + (M)$\leftrightarrow$$N_2O_5$ + (M) equilibrium. *J. Phys. Chem. A.* **1997**, *101*, 9694–9698. [CrossRef]
39. Zhao, D.; Xin, J.; Wang, W.; Jia, D.; Wang, Z.; Xiao, H.; Liu, C.; Zhou, J.; Tong, L.; Ma, Y.; et al. Effects of the sea-land breeze on coastal ozone pollution in the Yangtze River Delta, China. *Sci. Total Environ.* **2022**, *807*, 150306. [CrossRef] [PubMed]
40. Chen, Y.; Yang, C.; Xu, L.; Fan, X.; Shi, J.; Zheng, R.; Hong, Y.; Li, M.; Liu, T.; Chen, G.; et al. Variations of chemical composition of NR-PM1 under the influence of sea land breeze in a coastal city of Southeast China. *Atmos. Res.* **2023**, *285*, 106626. [CrossRef]
41. Breiman, L. Random Forests. *Mach. Learn.* **2001**, *45*, 5–32. [CrossRef]
42. Lam, K.-L.; Cheng, W.-Y.; Su, Y.; Li, X.; Wu, X.; Wong, K.-H.; Kwan, H.-S.; Cheung, P.C.-K. Use of random forest analysis to quantify the importance of the structural characteristics of beta-glucans for prebiotic development. *Food Hydrocoll.* **2020**, *108*, 106001. [CrossRef]
43. Xu, J.; Huang, X.; Wang, N.; Li, Y.; Ding, A. Understanding ozone pollution in the Yangtze River Delta of eastern China from the perspective of diurnal cycles. *Sci. Total Environ.* **2021**, *752*, 141928. [CrossRef]
44. Bentéjac, C.; Csörgő, A.; Martínez-Muñoz, G. A comparative analysis of gradient boosting algorithms. *Artif. Intell. Rev.* **2021**, *54*, 1937–1967. [CrossRef]
45. Ma, X.F.; Wang, Y.; Yan, S.M.; Feng, F. Transport Characteristics and Potential Source Areas of $PM_{2.5}$ in Yangquan City Under Different Pollution Levels. *Environ. Sci.* **2024**, *45*, 3858–3869.
46. Begum, B.A.; Kim, E.; Jeong, C.H.; Lee, D.W.; Hopke, P.K. Evaluation of the potential source contribution function using the 2002 Quebec forest fire episode. *Atmos. Environ.* **2005**, *39*, 3719–3724. [CrossRef]
47. Hsu, Y.K.; Holsen, T.M.; Hopke, P.K. Comparison of hybrid receptor models to locate PCB sources in Chicago. *Atmos. Environ.* **2003**, *37*, 545–562. [CrossRef]
48. Wang, Z.F.; Liu, W.; Fu, H.X.; Meng, H. Characteristics of Ozone and its Relationship with Meteorological Factors during 2014-2021 in Jinan and Qingdao, China. *Res. Environ. Sci.* **2023**, *36*, 673–683.
49. Xiao, K.; Wang, Y.; Wu, G.; Fu, B.; Zhu, Y. Spatiotemporal Characteristics of Air Pollutants (PM_{10}, $PM_{2.5}$, SO_2, NO_2, O_3, and CO) in the Inland Basin City of Chengdu, Southwest China. *Atmosphere* **2018**, *9*, 74. [CrossRef]
50. Cai, K.; Li, S.S.; Zheng, F.B.; Yu, C.; Zhang, X.Y.; Liu, Y.; Li, Y.J. Spatio-temporal Variations in NO_2 and $PM_{2.5}$ over the Central Plains Economic Region of China during 2005-2015 Based on Satellite Observations. *Aerosol Air Qual. Res.* **2018**, *18*, 1221–1235. [CrossRef]
51. Shareef, M.M.; Husain, T.; Alharbi, B. Analysis of Relationship between O_3, NO, and NO_2 in Riyadh, Saudi Arabia. *Asian J. Atmos. Environ.* **2018**, *12*, 17–29. [CrossRef]
52. Notario, A.; Bravo, I.; Adame, J.A.; Díaz-de-Mera, Y.; Aranda, A.; Rodríguez, A.; Rodríguez, D. Analysis of NO, NO_2, NOx, O_3 and oxidant ($O_X = O_3 + NO_2$) levels measured in a metropolitan area in the southwest of Iberian Peninsula. *Atmos. Res.* **2012**, *104*, 217–226. [CrossRef]
53. An, C.; Yan, Y.X.; Gao, X.S.; Yan, X.Y.; Ji, Y.Y.; Shang, F.Y.; Li, J.D.; Tan, L.Y.; Gao, R.; Bi, F.; et al. A Comparative Investigation of the Characteristics of Nocturnal Ozone Enhancement Events and Their Effects on Ground-Level Ozone and $PM_{2.5}$ in the Central City of the Yellow River Delta, China, in 2022 and 2023. *Atmosphere* **2024**, *15*, 475. [CrossRef]
54. Gao, X.R.; Deng, X.J.; Tan, H.B.; Wang, C.L.; Wang, N.; Yue, D.L. Characteristics and analysis on regional pollution process and circulation weather types over Guangdong Province. *Acta Sci. Circumstantiae* **2018**, *38*, 1708–1716.
55. Zheng, S.C.; Jiang, F.; Feng, S.Z.; Liu, H.; Wang, X.Y.; Tian, X.D.; Ying, C.Y.; Jia, M.W.; Shen, Y.; Lyu, X.P.; et al. Impact of Marine Shipping Emissions on Ozone Pollution During the Warm Seasons in China. *J. Geophys. Res. Atmos.* **2024**, *12*, e2024JD040864. [CrossRef]

56. Luo, Z.Y.; Lv, Z.F.; Zhao, J.C.; Sun, H.T.; He, T.K.; Yi, W.; Zhang, Z.N.; He, K.B.; Liu, H. Shipping-related pollution decreased but mortality increased in Chinese port cities. *Nat. Cities.* **2024**, *1*, 295–304. [CrossRef]
57. Huang, Y.X.; Guo, B.; Sun, H.X.; Liu, H.J.; Chen, S.X. Relative importance of meteorological variables on air quality and role of boundary layer height. *Atmos. Environ.* **2021**, *267*, 118737. [CrossRef]
58. Deng, Y.Y.; Li, J.; Li, Y.Q.; Wu, R.R.; Xie, S.D. Characteristics of volatile organic compounds, NO_2, and effects on ozone formation at a site with high ozone level in Chengdu. *J. Environ. Sci.* **2019**, *75*, 334–345. [CrossRef] [PubMed]
59. Li, A.; Zhang, J.; Xie, P.H.; Hu, Z.K.; Xu, J.; Mou, F.S.; Wu, F.C.; Liu, J.G.; Liu, W.Q. Variation of temporal and spatial patterns of NO_2 in Beijing using OMI and mobile DOAS. *Sci. China Chem.* **2015**, *58*, 1367–1376. [CrossRef]
60. Ling, A.P. Correlation Analysis between Concentrations of SO_2, NO_2, CO and Meteorological Conditions in Ya'an. *Geogr. Sci. Res.* **2019**, *8*, 341–350.
61. Yang, Y.; Wang, K.; Cui, C.; Liu, Y.H.; Huang, L.K. Analysis of relationship between air pollution and meteorological conditions of Harbin. *Chin. J. Environ. Eng.* **2015**, *9*, 5945–5950.
62. Wang, Q.; Han, J.D.; Sun, Y.; Luo, J.H.; Sun, S.J.; Li, X.L.; Qi, T.Y. Spatio-temporal characteristics and source apportionment of NO_2 in Handan. *Environ. Chem.* **2025**, *44*, 234–242.
63. Wang, S.; Nie, S.S.; Feng, Y.P.; Cui, J.S.; Chen, J.; Liu, D.X.; Shi, W.Y. Spatio-Temporal Evolution Characteristics and Source Apportionment of and N in Shijiazhuang. *Environ. Sci.* **2021**, *42*, 2679–2690.
64. Wang, M.Y.; Ding, H.; Xu, R.; Liu, Y.H. The correlation factors of the change of ambient air NO_2 concentration: An analysis based on improved Apriori algorithm. *J. Trop. Meteorol.* **2022**, *38*, 890–900.
65. Khoder, M.I. Atmospheric conversion of sulfur dioxide to particulate sulfate and nitrogen dioxide to particulate nitrate and gaseous nitric acid in an urban area. *Chemosphere* **2002**, *49*, 675–684. [CrossRef]
66. Zheng, C.W.; Zhao, C.F.; Li, Y.P.; Wu, X.L.; Zhang, K.Y.; Gao, J.; Qiao, Q.; Ren, Y.Z.; Zhang, X.; Chai, F.H. Spatial and temporal distribution of NO2 and SO2 in Inner Mongolia urban agglomeration obtained from satellite remote sensing and ground observations. *Atmos. Environ.* **2018**, *188*, 50–59. [CrossRef]
67. Zhang, L.S.; Lee, C.S.; Zhang, R.Q.; Chen, L.F. Spatial and temporal evaluation of long term trend (2005–2014) of OMI retrieved NO_2 and SO_2 concentration in Henan Province, China. *Atmos. Environ.* **2017**, *154*, 151–166. [CrossRef]
68. Warneke, C.; De Gouw, J.A.; Goldan, P.D.; Kuster, W.C.; Williams, E.J.; Lerner, B.M.; Jakoubek, R.; Brown, S.S.; Stark, H.; Aldener, M.; et al. Comparison of daytime and nighttime oxidation of biogenic and anthropogenic VOCs along the New England coast in summer during New England Air Quality Study 2002. *J. Geophys. Res.-Atmos.* **2004**, *109*, D10309. [CrossRef]
69. Salisbury, G.; Rickard, A.R.; Monks, P.S.; Allan, B.J.; Bauguitte, S.; Penkett, S.A.; Carslaw, N.; Lewis, A.C.; Creasey, D.J.; Heard, D.E.; et al. Production of peroxy radicals at night via reactions of ozone and the nitrate radical in the marine boundary layer. *J. Geophys. Res.-Atmos.* **2001**, *106*, 12669–12687. [CrossRef]
70. Zhang, Y.; Xue, L.; Carter, W.P.L.; Pei, C.; Chen, T.; Mu, J.; Wang, Y.; Zhang, Q.; Wang, W. Development of ozone reactivity scales for volatile organic compounds in a Chinese megacity. *Atmos. Chem. Phys.* **2021**, *21*, 11053–11068. [CrossRef]
71. Wang, Z.J.; Huo, J.; Du, H.Y.; Wang, D.W.; Li, J.; Zhang, C.B.; Zhang, T.; Wang, W.; Wang, H.B.; Yang, W.Y. Long term characteristics and potential sources of $PM_{2.5}$ in Rizhao City from 2015 to 2019. *China Environ. Sci. (Chin. Ed.)* **2021**, *41*, 3969–3980.
72. Cheng, M.M.; Jiang, H.; Guo, Z. Evaluation of long-term tropospheric NO_2 columns and the effect of different ecosystem in Yangtze River Delta. *Procedia Environ. Sci.* **2012**, *13*, 1045–1056. [CrossRef]
73. MEP. *Ambient Air Quality Standards (GB 3095-2012)*; China Environmental Science Press: Beijing, China, 2012.
74. MEP. *Technical Regulation on Ambient Air Quality Index (on Trial) (HJ 633-2012)*; China Environmental Science Press: Beijing, China, 2012.
75. MEP. *Technical Regulation for Ambient Air Quality Assessment (on Trial) (HJ 663-2013)*; China Environmental Science Press: Beijing, China, 2013.
76. Huth, R. A circulation classification scheme applicable in GCM studies. *Theor. Appl. Climatol.* **2000**, *67*, 1–18. [CrossRef]
77. Tang, G.Q.; LI, X.; Wang, X.K.; Xin, J.Y.; Hu, B.; Wang, L.L.; Ren, Y.F.; Wang, Y.S. Effects of Synoptic Type on Surface Ozone Pollution in Beijing. *Environ. Sci.* **2010**, *31*, 573–578.
78. Liu, Y.; Li, Y.h.; Hou, X.G.; Ma, W. Research on Density and Diffusion Trajectory of NO_2 during Heavy Pollution Period in Urumqi. *Environ. Sci. Technol.* **2017**, *40*, 33–39.
79. Li, Y.J.; An, X.Q.; Fan, G.Z. Transport pathway and potential source area of atmospheric particulates in Beijing. *China Environ. Sci.* **2019**, *39*, 915–927.
80. Wang, Y.; Chai, F.H.; Wang, Y.H.; Liu, M. Transport Characteristics of Air Pollutants over the Yangtze Delta. *Environ. Sci.* **2008**, *29*, 1430–1435.
81. Wang, Y.Q.; Zhang, X.Y.; Draxler, R.R. TrajStat: GIS-based software that uses various trajectory statistical analysis methods to identify potential sources from long-term air pollution measurement data. *Environ. Model. Softw.* **2009**, *24*, 938–939. [CrossRef]
82. Wang, Z.F.; Zhang, W.J.; Li, M.; Lv, B.; Fu, H.X.; Sun, F.J.; Lv, C.; Bian, M. Analysis of Heavy Air Pollution Episode with Combined Sand Storm and High PM2.5 Occurred in Jinan in 2018. *Res. Environ. Sci.* **2021**, *34*, 2588–2598.
83. Liu, N.; Yu, Y.; He, J.J.; Zhao, S.P. Analysis of Air Pollutant Transport in Winter in Lanzhou. *Res. Environ. Sci.* **2015**, *28*, 509–516.

84. Kulshrestha, U.C.; Raman, R.S.; Kulshrestha, M.J.; Rao, T.N.; Hazarika, P.J. Secondary aerosol formation and identification of regional source locations by PSCF analysis in the Indo-Gangetic region of India. *J. Atmos. Chem.* **2009**, *63*, 33–47. [CrossRef]
85. Hong, Q.Q.; Liu, C.; Hu, Q.H.; Xing, C.Z.; Tan, W.; Liu, H.R.; Huang, Y.; Zhu, Y.; Zhang, J.S.; Geng, T.Z.; et al. Evolution of the vertical structure of air pollutants during winter heavy pollution episodes: The role of regional transport and potential sources. *Atmos. Res.* **2019**, *228*, 206–222. [CrossRef]
86. Xu, X.; Akhtar, U.S. Identification of potential regional sources of atmospheric total gaseous mercury in Windsor, Ontario, Canada using hybrid receptor modeling. *Atmos. Chem. Phys.* **2010**, *10*, 7073–7083. [CrossRef]

toxics

MDPI

Review

Atmospheric Heavy Metal Pollution Characteristics and Health Risk Assessment Across Various Type of Cities in China

Zhichun Cha [1], Xi Zhang [1,2], Kai Zhang [1,*], Guanhua Zhou [3], Jian Gao [1,*], Sichu Sun [1], Yuanguan Gao [1] and Haiyan Liu [4]

1 State Key Laboratory of Environmental Criteria and Risk Assessment, Chinese Research Academy of Environmental Sciences, Beijing 100012, China; chazhichun0509@163.com (Z.C.); x-zhang@kitakyu-u.ac.jp (X.Z.); scsun1117@163.com (S.S.); gaoyg@craes.org.cn (Y.G.)

2 Faculty of Environmental Engineering, The University of Kitakyushu, 1-1 Hibikino, Wakamatsu, Kitakyushu 808-0135, Fukuoka, Japan

3 School of Instrumentation and Optoelectronic Engineering, Beihang University, Beijing 100191, China; zhouguanhua@buaa.edu.cn

4 Yinchuan Ecological Environment Monitoring Centre, Yinchuan 640100, China; 15804665746@163.com

* Correspondence: zhangkai@craes.org.cn (K.Z.); gaojian@craes.org.cn (J.G.)

Abstract: This study investigates the spatiotemporal trends and health risks of nine atmospheric heavy metals (Pb, As, Mn, Ni, Cr, Cd, Zn, Cu, Fe) in $PM_{2.5}$ across 50 Chinse cities, comparing resource-industrial cities (RICs) and general cities (GCs) before (2014–2018) and after (2019–2021) China's 2018 Air Pollution Prevention and Control Action Plan. Post-2018, concentrations of all metals except Fe declined significantly (33–77%), surpassing $PM_{2.5}$ reductions (25%). Geospatial analysis revealed elevated heavy metal levels in northern and southern regions in China, aligning with industrial and mining hotspots. While RICs exhibited persistently higher metal concentrations than GCs, the inter-city gap narrowed post-2018, with RICs achieving greater reduction. Pre-2018, the combined non-carcinogenic hazard index (HI < 1) remained below safety thresholds, but the combined carcinogenic risk total (CRT) for children exceeded 10^{-4}, driven primarily by As and Cr(VI). HIs were 1.5–2.0 times higher in RICs than in GCs. Post-2018, the CRT declined by 69.0–71.1%, aligning with reduced heavy metal levels. Despite improvements, CRTs necessitate targeted mitigation for As (contributing 81.1–86.2% to CRT) and Cr(VI) (11.7–14.0%). These findings validate the policy's effectiveness in curbing industrial and vehicular emissions but underscore the need for metal-specific controls in resource-intensive regions to safeguard child health.

Keywords: $PM_{2.5}$; resource-industrial cities (RICs); general cities (GCs); atmospheric heavy metal pollution; health risk assessment

1. Introduction

In recent years, atmospheric particulate matter, particularly $PM_{2.5}$, has emerged as a pollutant in China's ambient air, linked to significant public health risks, including respiratory and cardiovascular diseases [1–3]. Globally, particulate matter pollution accounts for 3% of deaths from cardiopulmonary diseases and 5% of lung cancer fatalities [4,5]. Heavy metals, including arsenic (As), cadmium (Cd), chromium (Cr), copper (Cu), manganese (Mn), nickel (Ni), lead (Pb), and zinc (Zn), are critical components of $PM_{2.5}$ that can induce various toxic effects through inhalation [6]. Metals like Mn and Ni can catalyze hydroxyl radical production, inducing cellular damage through lipid peroxidation and protein oxidation [7], while Zn, Cr, Cu, and Pb exhibit synergistic toxicity in lung epithelial cells [8]. Prenatal exposure to As, Cd, Mn, and Pb has been linked to childhood asthma [9]. The World Health Organization (WHO) identifies As, Ni, Cr(VI), Cd, and other heavy metals as carcinogens, establishing stringent air quality guidelines to mitigate exposure [10].

Atmospheric heavy metals primarily originate from industrial emissions, mining activities, vehicular exhaust, fossil fuel combustion, waste incineration, construction, and crustal dust [11,12]. Urban pollution profiles vary significantly due to the differences in energy structures and industrial composition, [13]. Current studies predominantly focus on specific cities or regions. For instance, Pb and Cu levels in urban and traffic areas of Navarra, Spain, significantly exceed the rural background, although remaining below WHO thresholds [14]. Similarly, petrochemical operations emit substantial Pb, Hg, Ni, and Cr in the surrounding environment [15], while coal mining areas show the dominance of Zn, Mn, and Pb, constituting over 82% of total metal content [16]. In China, a meta-analysis of 14 cities identified Foshan, Wuhan, Xi'an, Jinan, and Shenzhen as hotspots for Pb, As, Ni, Cr, and Cd pollution [17]. Regional studies have further characterized heavy metals trends: Duan et al. analyzed pre-2013 pollution patterns in Chin [18], while Li et al. revealed a south to north enrichment gradual linked industrial structure distribution [19]. Yu et al. noted declining $PM_{2.5}$ -bound metals (Cd, Cr, Ni, Pb, Zn, Hg, and As), except for the element Cu, since 2017 [20].

Despite these insights, systematic comparisons of heavy metal pollution between resource-industrial cities (RICs) and general cities (GCs) remain limited, particularly in assessing how industrial emissions, vehicle density, and coal combustion drive spatial disparities. Mining and smelting activities—such as crushing, grinding, and refining—generate particulate matter laden with hazardous metal concentrations [21], while industries like steel manufacturing and cement production exacerbate air quality deterioration [22]. Existing studies often sample in mixed-use areas of residential, commercial, and transportation to mitigate the influence of dominant industrial sources. However, this approach may introduce errors if classification cities rely solely on the national list of resource cities (2013), thus compromising the accuracy and representativeness of the results.

The Three-Year Action Plan to Win the Blue Sky Defense War (2018–2020) launched by China's Central Government prioritized regional air quality governance. This policy calls for vigorously adjusting and optimizing the industrial structure, energy structure, transportation structure, and land use structure and strengthening regional joint prevention and control. Comprehensively treating industrial pollution, building a clean, low-carbon, and efficient energy system, and implementing major special actions significantly reduce pollutant emissions. To address this, local governments enhanced their efforts to understand and control heavy air pollution and tightened the prevention and control measures of heavy air pollution, leading to a significant reduction in atmospheric heavy metals pollution levels across various cities [20].

This study addresses these gaps by analyzing $PM_{2.5}$-bound heavy metal (Pb, As, Mn, Ni, Cr, Cd, Zn, Cu, Fe) in 50 Chinese cities pre- (2014–2018) and post-policy (2019–2021). We categorized the collected cities into resource-industrial cities (RICs) and general cities (GCs) based on functional zoning and sampling site representativeness, avoiding biases from oversimplified classifications. The primary objective is to (1) compare the spatiotemporal distribution changes in $PM_{2.5}$ and its associated heavy metal concentrations before and after the 2018 policy implementation. Additionally, it analyzed the differences in $PM_{2.5}$ and heavy metal concentrations between the RICs and GCs areas. Aecondary objectives are to (2) quantify health risks (combined non-carcinogenic hazard index, HI; combined carcinogenic risk total, CRT) for adults and children and (3) evaluate the efficacy of the

2018 policy in mitigating disparities and residual risks. These findings may provide policymakers with actionable strategies for refining emission controls in high-risk regions and safeguarding public health.

2. Materials and Methods

2.1. Data Sources and Processing

The data were obtained from the literature from 2014 to 2021, as well as from existing national air quality monitoring data. The literature inclusion followed strict criteria about data quality, temporal coverage, and analytical methods: (1) Studies must report the $PM_{2.5}$-bound heavy metals concentration (Pb, As, Mn, Ni, Cr, Cd, Zn, Cu, and Fe) with a documented quality control procedure; (2) The sampling period was between 2014 and 2021, with ≥20 samples collected cumulatively and the sampling period exceeding one month (>30 days); (3) Analytical methods for concentration quantification via inductively Coupled Plasma Mass Spectrometry (ICP-MS), Inductively Coupled Plasma Emission Spectrometry (ICP-AES), or X-ray fluorescence (XRF). A total of 63 datasets spanning 50 cities met these criteria. The geometric means and standard deviations of $PM_{2.5}$ and heavy metals concentrations for each city were used to account for log-normal distribution patterns in environmental data.

2.2. Methodology for Classifying Urban Typologies

Cities are classified as RICs and GCs were based on the National Sustainable Development Plan for Resource Cities (2013–2020) and the functional zoning of sampling sites. RICs were defined as cities where natural resources extraction (e.g., minerals and forestry) dominants the economy, with sampling sites located in or near industrial zones (e.g., smelters, coral-fired plants). GCs included urban areas with sampling sites in residential or commercial zones, lacking proximate industrial activity. For example, cities with sampling points within industrial areas were categorized as RICs, while others were designated GCs. This dual classification—combining policy definitions with a spatial context—ensures the representative characterization of industrial emission impacts on atmospheric heavy metal concentrations.

2.3. Health Risk Assessment

This study assessed health risk associated with $PM_{2.5}$-bound heavy metal using the U.S. Environmental Protection Agency (EPA) risk assessment model [23–25]. The participants were categorized into adult and child to account for differences in physical characteristics and respiratory systems. Non-carcinogenic risk (hazard quotient, HQ) and carcinogenic risk (CR) were evaluated across three exposure pathways: ingestion, inhalation, and dermal contact. For chromium, only hexavalent chromium (Cr(VI)), calculated as 1/7 of the total Cr concentration, was considered a carcinogen [26].

The exposure formulas of the three pathways are as follows [24]:

$$CDI_{ing} = C_{ing} \times \frac{IngR \times EF \times ED \times CF}{BW \times AT} \tag{1}$$

$$EC_{inh} = C_{inh} \times \frac{ET \times EF \times ED}{AT_n} \tag{2}$$

$$DAD_{der} = C_{der} \times \frac{SA \times AF \times ABS \times EF \times ED \times CF}{BW \times AT} \tag{3}$$

where CDI_{ing} is the daily intake via ingestion (mg·(kg·d)$^{-1}$); EC_{inh} is the inhalation exposure concentration (μg·m^{-3}); DAD_{der} is the dermal absorbed dose (mg·(kg·d)$^{-1}$); C_{inh} is the concentration of heavy metals in particulate matter (μg·m^{-3}); and C_{ing} and C_{der} are the contents of heavy metals in particulate matter (mg·kg^{-1}). The definitions and values of other parameters in the formulas are shown in Table 1.

Table 1. Parameters and their values entered in the health risk assessment model.

Argument	Definition	Unit	Numerical Value	
			Children	Adult
ABS	Skin absorption factor		0.03 (As), 0.1 (Pb), 0.001 (Cd), 0.01 (else)	
AF	Skin adhesion factor	mg·cm^{-2}	0.2	0.07
AT	Mean life	d	ED × 365 (noncarcinogenic effect) 70 × 365 (carcinogenesis)	ED × 365 (noncarcinogenic effect) 70 × 365 (carcinogenesis)
AT_n	Mean life	h	ED × 365 × 24 (noncarcinogenic effect) 70 × 365 × 24 (carcinogenesis)	ED × 365 × 24 (noncarcinogenic effect) 70 × 365 × 24 (carcinogenesis)
BW	Per capita weight	kg	15	70
CF	Conversion factor	kg·mg^{-1}	1.0×10^{-6}	1.0×10^{-6}
ED	Exposure period	a	6	24
EF	Exposure frequency	d·a^{-1}	180	180
ET	Exposure time	h·d^{-1}	24	24
IngR	Hand–mouth intake	mg·d^{-1}	200	100
SA	Skin surface area	cm^2	2800	5700

Equations (4)–(9) were used to calculate the HQ and CR of a single element to the human body:

$$HQ_{ing} = \frac{CDI_{ing}}{RfD_0} \tag{4}$$

$$HQ_{inh} = \frac{EC_{inh}}{RfC_i \times 1000} \tag{5}$$

$$HQ_{der} = \frac{DAD_{der}}{RfD_0 \times GIABS} \tag{6}$$

$$CR_{ing} = CDI_{ing} \times SF_0 \tag{7}$$

$$CR_{inh} = EC_{inh} \times IUR \tag{8}$$

$$CR_{der} = DAD_{der} \times \frac{SF_0}{GIABS} \tag{9}$$

To assess the overall potential for non-carcinogenic risks and carcinogenic risks posed by multi-element exposure, the combined non-carcinogenic HI and combined CRT were estimated as the sum of HQi and CRi, assuming additive effects [23,24,27]. The HI and CRT are calculated as follows:

$$\mathrm{HI} = \sum HQ_i \tag{10}$$

$$\mathrm{CRT} = \sum CR_i \tag{11}$$

In the formulas, RfD_0, RfCi, GIABS, SF_0, and IUR represent the oral reference dose, inhalation reference concentration, gastrointestinal absorption factor, oral slope factor, and inhalation unit risk, respectively. The parameter values are detailed in Table 2 [24]. When $HI \leq 1$, there is no non-carcinogenic risk; when $HI > 1$, a non-carcinogenic risk is indicated.

When CRT < 10^{-6}, there is a negligible risk; when $10^{-6} \leq$ CRT $\leq 10^{-4}$, there is a tolerable risk; and if CRT > 10^{-4}, there is a significant cancer risk.

Table 2. Reference doses of heavy metals.

Types of Heavy Metals	RfD_0	RfC_i	GIABS	SF_0	IUR
Pb	3.50×10^{-3}	3.52×10^{-3}	1	0.0085	0.000012
As	3.00×10^{-4}	1.50×10^{-5}	1	1.5	0.0043
Mn	1.43×10^{-5}	0.50×10^{-4}	1	—	—
Ni	2.00×10^{-2}	9.00×10^{-5}	0.04	—	0.00026
Cr	3.00×10^{-3}	1.00×10^{-4}	0.025	0.5	0.084
Cd	1.00×10^{-3}	1.50×10^{-5}	0.025	—	0.0018
Zn	3.00×10^{-1}	3.01×10^{-1}	1	—	—
Cu	4.00×10^{-2}	4.02×10^{-2}	1	—	—

3. Results and Discussion

3.1. Changes in $PM_{2.5}$ and Heavy Metal Concentrations Before and After 2018

The specific content and proportions of heavy metal elements in $PM_{2.5}$ across cities and regions in China before and after 2018 are detailed in Table A1. The average concentration of $PM_{2.5}$ decreased by 25%, from 76.3 $\mu g \cdot m^{-3}$ before 2018 to 57.3 $\mu g \cdot m^{-3}$ after 2018. This decline indicates the effectiveness of the Three-Year Action Plan to Win the Blue Sky Defense War promulgated and implemented after 2018. This policy-driven improvement aligns with increased air quality compliance rates and reduced heavily polluted days post-2018 [28].

Pre-2018, the dominant heavy metal in $PM_{2.5}$ were Fe (741.2 $ng \cdot m^{-3}$), Zn (434.7 $ng \cdot m^{-3}$), and Pb (127.0 $ng \cdot m^{-3}$), followed by Cu, Mn, Cr, As, Ni, and Cd (Table A1). Post-2018, all metals except Fe exhibited significant reductions: Cu (−77.1%), Cd (−73.7%), Ni (−73.1%), and Cr (−70.6%) showed the steepest declines, while Fe concentrations decreased marginally (−5.7%), with its proportional contribution to $PM_{2.5}$ increasing from 0.42% to 1.02%. The persistence of Fe, Mn, and Pb is liked to China's steel industry, the world's largest since 1996 [29]. Despite the annual fluctuation in crude steel production (0.9%, −2.3%, 1.2%, 5.7%, 6.6%, 8.3%, 5.2%, and −3% from 2014 to 2021), sustained industrial activity limited reductions in these metals. Steel manufacturing relies on iron ore, manganese ore, and recycled metals, generating Fe-rich particulate emissions [5,30]. Consequently, soils near the mining, smelting, and metallurgical industries through sedimentation typically exhibit elevated levels of Fe, Pb, Zn, and Mn [3,31].

The consistent growth in steel production has led to a relatively smaller decrease in the concentration of Fe, Mn, Zn, and Pb compared with other heavy metals in $PM_{2.5}$. Furthermore, there are significant correlations among these four elements, as evidenced by Spearman correlation analysis (Tables 3 and 4). Although the correlation analysis for Fe was less satisfactory after 2018 due to the smaller dataset, a significant correlation ($r > 0.5$) was observed among the top four elements before 2018 and the remaining three elements afterward, suggesting similar sources for Pb, Mn, Zn, and Fe. Studies indicated that the correlation between Fe, Zn, Pb, and Mn is considered a major marker of motor vehicle emissions [14]. Therefore, the small percentage decrease in these heavy metal concentrations may be attributed to the growth in steel production and the number of motor vehicles in China.

To compare the spatial distribution of heavy metals in $PM_{2.5}$ across China before and after the implementation of the policy, 36 major cities and regions before 2018 and 26 major cities and regions after 2018 were divided into six regions (North, Northeast, Northwest,

South, East, Southwest China) based on the environmental inspection jurisdiction [32]. The concentrations of heavy metals in these six regions are shown in Figures 1 and 2. Before 2018, the high concentrations of heavy metals clustered in North China, South China, and parts of Northwest China (e.g., Xining and Lanzhou). After 2018, high concentrations persisted in North China and South China, with new hotspots in Southwest China (e.g., Chengdu, Zunyi and Kunming). Before 2018, Cr, Cd, and Cu hotspots occurred in Northwest (e.g., Xining and Lanzhou) and South China (e.g., Hengyang and Changsha). However, after 2018, Cr and Fe became concentrated in Southwest cities (e.g., Chengdu, Panzhihua, Guiyang). A persistent north–south disparity in heavy metals concentrations exists, with northern regions exhibiting higher levels due to industrial density and energy reliance on coal [32,33].

Table 3. Correlation analysis results of heavy metal elements before 2018.

Item	Pb	As	Mn	Ni	Cr	Cd	Zn	Cu	Fe
Pb	1								
As	0.532 **	1							
Mn	**0.812** **	0.372	1						
Ni	0.742 **	0.359	0.625 **	1					
Cr	0.804 **	0.487 *	0.652 **	0.842 **	1				
Cd	0.783 **	0.602 **	0.767 **	0.678 **	0.633 **	1			
Zn	**0.767** **	0.663 **	**0.749** **	0.728 **	0.761 **	0.597 **	1		
Cu	0.835 **	0.262	0.758 **	0.917 **	0.729 **	0.633 *	0.701 **	1	
Fe	**0.669** *	0.595	**0.866** **	0.689 *	0.845 **	0.300	**0.508**	0.761 **	1

Note: The black bold letters in the table show the correlation coefficients between Fe, Mn, Zn, and Pb. * indicates $p < 0.05$; ** indicates $p < 0.01$; data with * and ** indicate significant correlation.

Table 4. Correlation analysis results of heavy metal elements after 2018.

Item	Pb	As	Mn	Ni	Cr	Cd	Zn	Cu	Fe
Pb	1								
As	0.412 *	1							
Mn	**0.630** **	0.636 **	1						
Ni	0.446 *	0.482 *	0.498 *	1					
Cr	0.476 *	0.351	0.700 **	0.706 **	1				
Cd	0.678 **	0.421	0.676 **	0.347	0.396	1			
Zn	**0.657** **	0.450	**0.852** **	0.414	0.627 **	0.739 **	1		
Cu	0.379	−0.011	0.340	0.315	0.168	0.543 *	0.338	1	
Fe	**0.524**	0.607	**0.611**	0.333	0.167	0.468	0.286	0.071	1

Note: The black bold letters in the table show the correlation coefficients between Fe, Mn, Zn, and Pb. * indicates $p < 0.05$; ** indicates $p < 0.01$; data with * and ** indicate significant correlation.

3.2. *Comparison of $PM_{2.5}$ and Heavy Metal Concentrations Between RICs and GCs Before and After 2018*

In RICs, the total $PM_{2.5}$ concentration decreased by 18.7%, from 86.7 $\mu g \cdot m^{-3}$ before 2018 to 70.5 $\mu g \cdot m^{-3}$ after 2018 (Table A1). Among the heavy metals in $PM_{2.5}$, Fe exhibited the highest concentration, with average levels of 1279.7 before 2018 and 941.7 $ng \cdot m^{-3}$ after 2018, followed by Zn (595.1 and 264.4 $ng \cdot m^{-3}$), Pb (205.9 and 79.3 $ng \cdot m^{-3}$), Cu (179.8 and 26.6 $ng \cdot m^{-3}$), Mn (95.3 and 47.5 $ng \cdot m^{-3}$), Cr (47.4 and 12.9 $ng \cdot m^{-3}$), As (19.7 and 5.5 $ng \cdot m^{-3}$), Ni (19.9 and 5.0 $ng \cdot m^{-3}$), and Cd (10.1 and 2.1 $ng \cdot m^{-3}$). All nine heavy metal elements showed significant decreasing trends: Cu decreased by 85.2%, Cd by 79.2%, Ni by 74.9%, Cr by 72.8%, As by 72.1%, Pb by 61.5%, Zn by 55.6%, Mn by 50.2%, and Fe by 26.4%. Notably, the proportion of Fe in $PM_{2.5}$ increased from 0.19% before 2018 to 1.40% after 2018, while the proportion of other heavy metals declined. This trend is likely attributed to the stringent regulatory restriction on industrial pollution emissions, particularly in sectors linked to natural resource extraction and processing. However, Fe exhibited a comparatively smaller decline than other metals, potentially due to persistent industrial activity or unique emission sources [34].

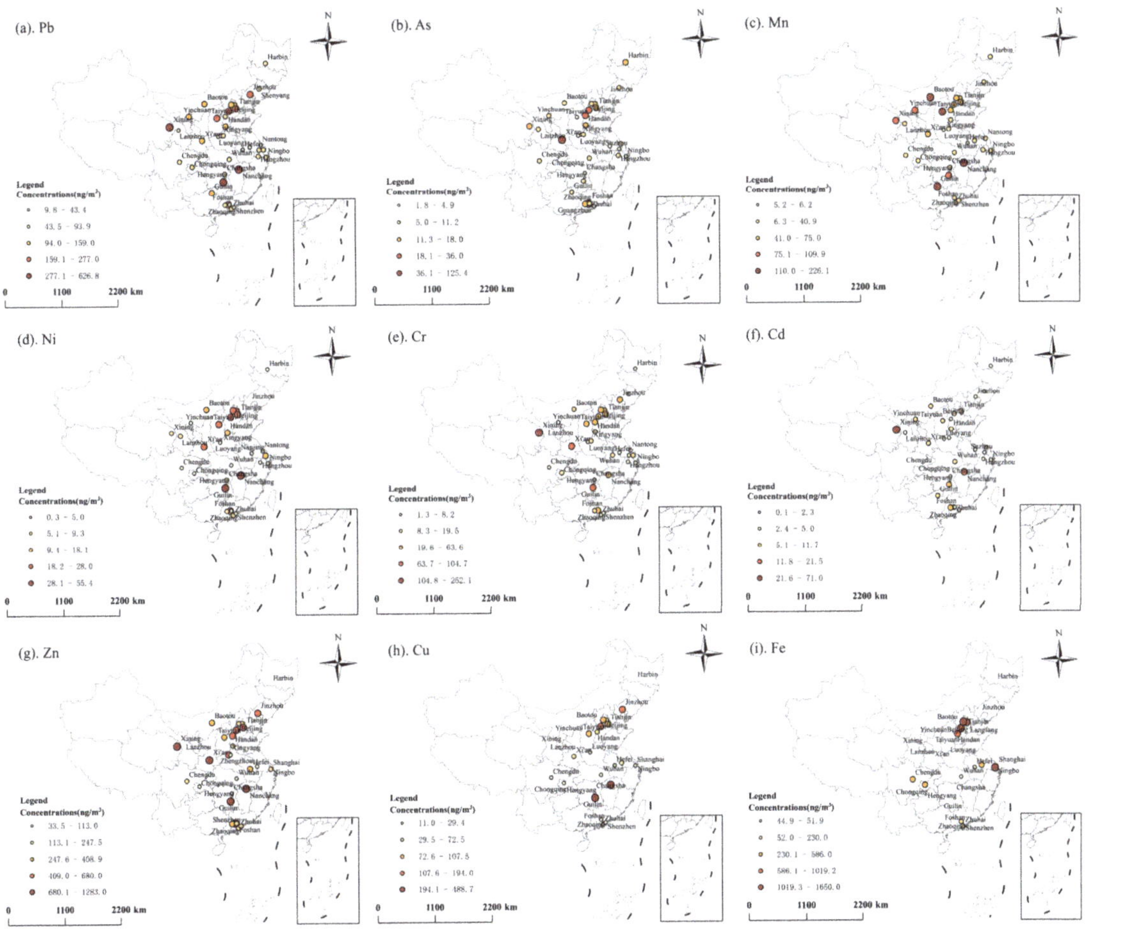

Figure 1. Spatial distribution of the comprehensive concentration of heavy metals in $PM_{2.5}$ before 2018.

In comparison, $PM_{2.5}$ concentrations' GCs declined by 29.4% from 71.1 $\mu g \cdot m^{-3}$ pre-2018 to 50.3 $\mu g \cdot m^{-3}$ post-2018 (Table A1). The average mass concentrations of the heavy metal elements in GCs before 2018 were Fe (594.4 $ng \cdot m^{-3}$), Zn (350.2 $ng \cdot m^{-3}$), Pb (87.6 $ng \cdot m^{-3}$), Cu (60.5 $ng \cdot m^{-3}$), Mn (46.9 $ng \cdot m^{-3}$), Cr (17.0 $ng \cdot m^{-3}$), As (14.2 $ng \cdot m^{-3}$), Ni (9.8 $ng \cdot m^{-3}$), and Cd (2.9 $ng \cdot m^{-3}$) (Figure 3). After 2018, these levels declined for all nine heavy metal elements: Cu by 73.2%, Ni by 71.4%, Cr by 65.9%, Cd by 65.5%, As by 59.9%, Pb by 55.4%, Zn by 55.2%, Fe by 23.2%, and Mn by 14.9%. The proportions of other heavy metals in $PM_{2.5}$, except iron, all decreased after 2018 compared with those before 2018. The proportion of Fe in $PM_{2.5}$ increased from 0.46% before 2018 to 0.91% after 2018. This could be attributed to the accumulation of heavy metals emitted from neighboring industrial areas through long-distance atmospheric transportation [35]. Moreover, the GCs include megacities such as Beijing, Shanghai, Guangzhou, etc., with high population density and vehicular traffic, which, combined with increased coal combustion during the residential heating season, may explain the relatively smaller decline in Mn levels [16,36].

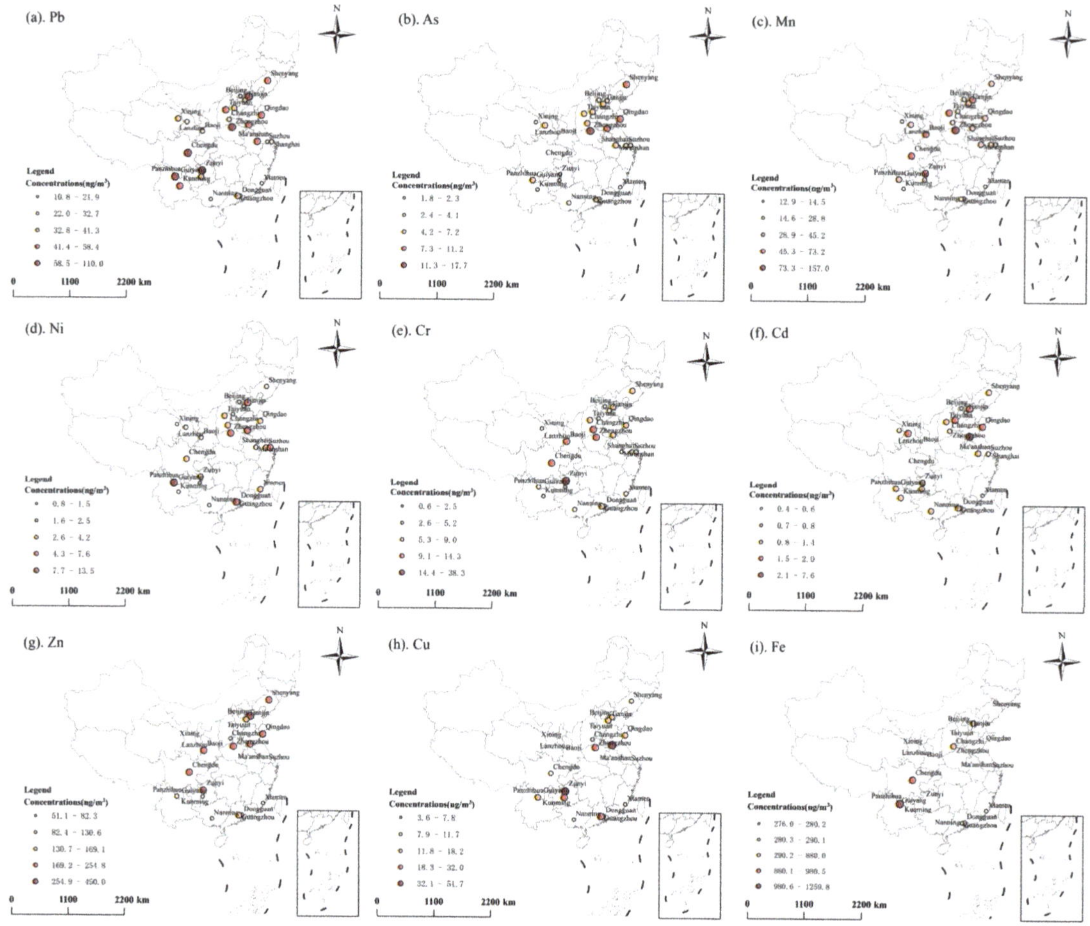

Figure 2. Spatial distribution of the comprehensive concentration of heavy metals in $PM_{2.5}$ after 2018.

Before and after 2018, atmospheric $PM_{2.5}$ concentrations and atmospheric heavy metal levels varied significantly between RICs and GCs. Prior to 2018, the atmospheric $PM_{2.5}$ concentration in RICs exceeded those in GCs, with an average of 86.7 $\mu g \cdot m^{-3}$ and 71.1 $\mu g \cdot m^{-3}$, respectively. Similarly, RICs exhibited a significantly higher concentration ratio of heavy metal in $PM_{2.5}$. Specifically, the average mass concentrations of Cd, Cu, Cr, Pb, Fe, Mn, Ni, Zn, and As in the RICs were 248.3%, 197.2%, 178.8% 135.0%, 115.3%, 103.2%, 103.1%, 69.9%, and 38.7% higher, respectively, compared to GCs (Table A1). Additionally, heavy metals accounted for a higher proportion of $PM_{2.5}$ in RICs, reflecting the region's industrial structure, which is dominated by mining, ore transportation, and slag accumulation, activities known to emit particulate matter enriched with heavy metals [37]. Additionally, energy extraction and transportation produce substantial soil dust, increasing Mn and Cr levels in RICs [38]. Smelting operations further intensified heavy metals like Pb, Cd, As, and Cr in industrial areas [35]. Therefore, the higher concentrations of $PM_{2.5}$ and heavy metal elements in RICs can be attributed to their energy-intensive and high-pollution activities and elevated demand for resources and energy compared to GCs.

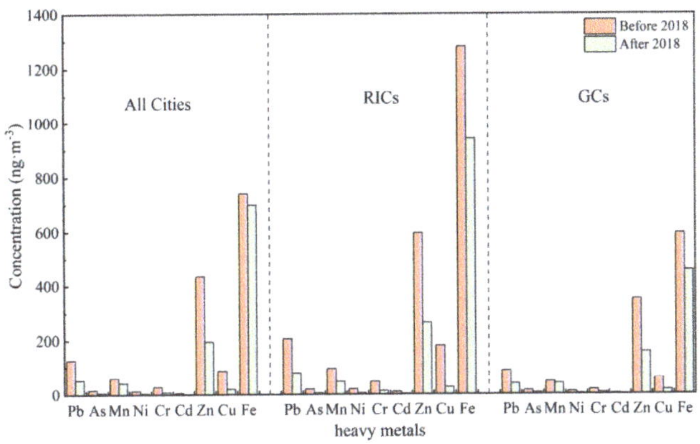

Figure 3. Comparison of the mass concentration of heavy metal elements in atmospheric $PM_{2.5}$ before and after 2018.

After 2018, the $PM_{2.5}$ concentration in RICs (70.5 μg·m^{-3}) remained higher than in GCs (50.3 μg·m^{-3}). The average mass concentrations of Cr, Cd, Fe, Pb, Ni, Zn, Cu, and Mn were persistently higher in RICs than in GCs by 122.5%, 110.0%, 106.2%, 102.8%, 78.6%, 68.5%, 64.2%, and 19.0%, respectively. However, GCs exhibited a 3.5% higher As compared to RICs. This anomaly was driven by cities such as Shenyang, Zhengzhou, and Qingdao, where levels exceed 10 ng·m^{-3} (Table A1). As serves as a marker pollutant for coal combustion emissions [36]. In these cities, the sampling points were located in mixed residential and transportation areas, which lacked proximate industrial sources. Consequently, the elevated As levels may instead originate from residential coal heating, vehicular emissions, and long-distance transportation. To improve urban air quality and protect public health, many large cities have relocated coal-burning enterprises to suburbs or small adjacent cities. However, cities like Shenyang, Zhengzhou, and Qingdao remain affected by nearby industrial cities such as Anshan, Luoyang, and Weifang. Wind and atmospheric turbulence can transport mineral dust and industrial fumes to these sampling sites [39]. For instance, Shenyang, although categorized as a GC based on the geographic location of the sampling site, retains industrial influence from equipment manufacturing, internal industrial sources, and coal use in the heating season. Coastal cities like Qingdao and Tianjin also face contributions from ship emissions at nearby ports, which elevated Pb, As, and Cr levels [40–42]. Therefore, it can be inferred that the high As level in GCs is related to residential coal combustion, transportation-related emissions, and regional atmospheric dynamics.

3.3. *Comparison of Variation in Human Health Risk Assessments for Heavy Metals*

Due to their small particle size and large specific surface area, $PM_{2.5}$ exhibits a strong adsorption of harmful substances, including heavy metals, and poses a significant risk to human health. These particles can penetrate deep into the respiratory system, potentially triggering respiratory, immune, and cardiovascular diseases [33]. Heavy metals attached to $PM_{2.5}$ can enter the human body through inhalation, oral ingestion, and dermal contact, accumulating over time and adversely affecting health [13]. Consequently, assessing the health risk associated with atmospheric heavy metals has become a prominent research area. To quantify these risks, this study employed the U.S. EPA health risk evaluation model [25], evaluating combined non-carcinogenic HI and combined CRT for adults and children across all cities, as well as RICs and GCs, before and after 2018 through three exposure pathways: oral ingestion, inhalation, and dermal contact. The HI and CRT of the three uptake pathways are presented in Figures 4–6 and Table 5.

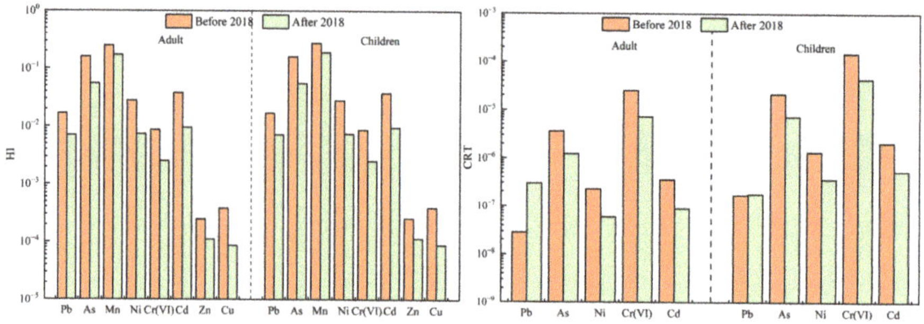

Figure 4. Comparison of HI and CRT of heavy metals in all cities before and after 2018.

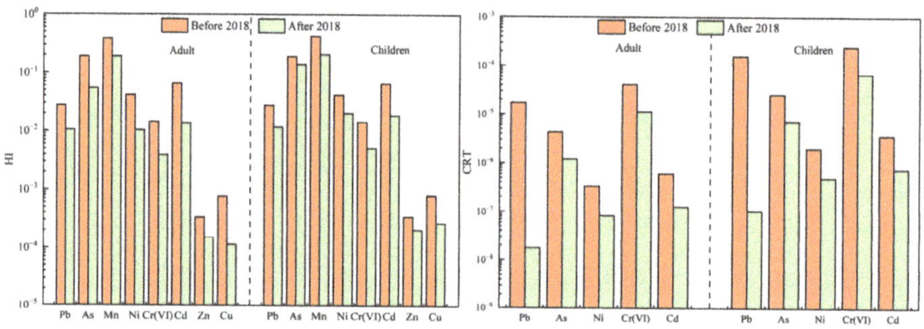

Figure 5. Comparison of HI and CRT of heavy metals in RICs before and after 2018.

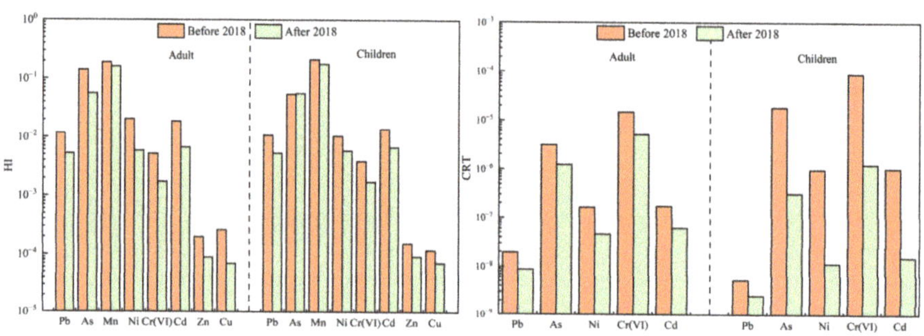

Figure 6. Comparison of HI and CRT of heavy metals in GCs before and after 2018.

For non-carcinogenic risk, the HI values for adults and children in all cities remained below the acceptable threshold (HI = 1) after 2018. However, the HI values for both groups were approximately double before 2018 compared to after 2018, indicating a marked decline in non-carcinogenic risk due to decreased heavy metal concentrations. Carcinogenic risk assessments revealed similar trends; CRT values for adults decreased from 2.9×10^{-5} to 9.0×10^{-6}, while CRT values for children declined from 1.8×10^{-4} to 5.2×10^{-5}. Prior to 2018, the CRT values for children exceeded 1×10^{-4}, indicating significant carcinogenic risk, whereas adults' risk fell within the 10^{-4}–10^{-6} range, suggesting lower but still notable risk. These findings align with previous studies. For example, the CRT values for As, Cd, Co, Cr(VI), and Ni in Beijing's $PM_{2.5}$ decreased from 1.08×10^{-5} in 2016 to 6.50×10^{-6} in 2021–2022 [43]. Similarly, Wang et al., and Zhao et al. revealed that the carcinogenic risks of

Cd, Pb and Ni in $PM_{2.5}$ in Tianjin ranged between 10^{-6} and 10^{-4} before 2018 but dropped below 10^{-6} after 2018 [27,44]. Collectively, these results demonstrate a consistent decline in both HI and CRT for both adults and children after 2018. This reduction correlates with the implementation of the policy, which strengthened emission control and pollution mitigation measurements. The policy intends to reduce industrial emissions, improve energy efficiency, and regulate coal combustion, effectively mitigating heavy metal concentrations in $PM_{2.5}$ and thereby lowering the associated health risks posed by atmospheric heavy metals.

Table 5. Heavy metal CRT and HI values before and after 2018.

Type of City	Time	Age	HI	CRT
All Cities	Pre-2018	Adults	0.5	2.9×10^{-5}
		Children	0.5	1.8×10^{-4}
	Post-2018	Adults	0.3	9.0×10^{-6}
		Children	0.3	5.2×10^{-5}
RICs	Pre-2018	Adults	0.7	6.4×10^{-5}
		Children	0.8	4.4×10^{-4}
	Post-2018	Adults	0.3	1.3×10^{-5}
		Children	0.3	7.6×10^{-5}
GCs	Pre-2018	Adults	0.4	1.8×10^{-5}
		Children	0.4	1.1×10^{-4}
	Post-2018	Adults	0.2	1.6×10^{-6}
		Children	0.2	6.4×10^{-6}

In RICs, the HI values of adults and children remained below acceptable levels both before and after 2018, with CRT values decreasing from 6.4×10^{-5} to 1.3×10^{-5} for adults and from 4.4×10^{-4} to 7.6×10^{-5} for children, respectively. GCs followed a similar pattern: adult CRT values decreased from 1.8×10^{-5} to 1.6×10^{-6}, and children's CRT values dropped from 1.1×10^{-4} to 6.4×10^{-6}. HI was 1.5–2.0 times higher in RICs than in GCs before and after 2018. Across both city types, HI and CRT values declined after 2018, with CRT decreasing by 69.0–71.1%, aligning with reduced heavy metal levels, yet children consistently faced a higher risk than adults, likely due to physiological vulnerability and behavioral factors (e.g., higher inhalation rates per body weight). Residents of RICs also face elevated health risks compared to GCs, aligning with Li et al.'s finding that the cancer risk levels correlate with heavy metal concentrations, particularly in industrial zones [32].

Mn and As posed the highest non-carcinogenic risk levels among all heavy metals throughout the whole study period, while As and Cr(VI) dominated carcinogenic risk. The contributions of As and Cr(VI) to the CRT were 11.7–14.0% and 81.1–86.2%, respectively. The concentrations of As (16.4 $ng \cdot m^{-3}$) and Cr(VI) (4.1 $ng \cdot m^{-3}$) before 2018 exceeded the standard limits (6 $ng \cdot m^{-3}$ for As, 0.025 $ng \cdot m^{-3}$ for Cr(VI)) set by the China Ambient Air Quality Standard (GB 3095-2012) [45], with exceedance rates of 78% for As and 97% for Cr(VI), respectively. Even after 2018, the exceedance rates remained at 17% and 84%, respectively. The exceedances of As and Cr(VI) in the RICs were greater than in the GCs, as were the carcinogenic risk values for As and Cr(VI). These trends mirror findings by Yu et al. [20], underscoring the persistent threat of these metals. The sustained exceedance of As and Cr(VI) thresholds, even after regulatory interventions, highlights the need for stricter emission controls on high-risk carcinogens, particularly in RICs. Prioritizing industrial source regulation, enhancing air quality monitoring, and addressing transboundary pollution are critical to mitigating health risks.

4. Conclusions

The implementation of the Three-Year Action Plan to Win the Blue Sky Defense War has significantly reduced atmospheric $PM_{2.5}$ and associated heavy metals concentrations (Pb, As, Mn, Ni, Cr, Cd, Zn, Cu, and Fe), underscoring the policy's success in improving air quality. However, correlation analyses show that metals linked to industrial and vehicular emission, such as Fe, Mn, Zn, and Pb, exhibited smaller reductions compared to others, likely due to rising steel production and motor vehicle usage. Geographically, heavy metals hotspots before 2018 were concentrated in northwest, northern, and southern China. Post-2018, these high-concentrations areas shifted southwestward while remaining prevalent in northern and southern regions. Although both RICs and GCs exhibited declinations in heavy metal levels after 2018, the reductions were more pronounced in RICs.

The health risk assessment demonstrated that HI remained within acceptable thresholds (HI < 1) for all populations throughout the whole study period. CRT, however, posed greater concern: pre-2018 CRT values for adults fell within the 10^{-6}–10^{-4} range (indicating potential risk), while the CRT for children exceeded 10^{-4} (definitive risk). These risks were consistently higher in RICs than in GCs, aligning with regional disparities in industrial activity. Post-2018 declines in HI and CRT paralleled reductions in heavy metal concentrations, affirming the policy's effectiveness in safeguarding public health.

However the main carcinogens, As and Cr(VI), persist as critical threats, with pre-2018 concentrations exceeding China standards by 78% (As) and 97% (Cr(VI)). Post-2018 exceedance rates remained elevated at 17% and 84%, respectively, particularly in RICs. These findings underscore the urgency of controls on industrial and combustion-related emissions of high-risk carcinogens. This study provides a scientific foundation for refining air quality policies, emphasizing targeted emission regulations, enhancing monitoring in high-risk regions, and employing adaptive strategies to mitigate residual health threats. Sustained efforts to curb toxic pollutants are essential to consolidating gains from the Three-Year Action Plan to Win the Blue Sky Defense War and ensuring long-term environmental and public health resilience.

Author Contributions: Z.C.: Formal analysis, Data curation, Writing—original draft. X.Z.: Writing—review and editing. K.Z.: Conceptualization, Project administration, Writing—review and editing. G.Z.: Writing—review and editing. J.G.: Resources, Funding acquisition. S.S.: Data curation, Visualization. Y.G.: Supervision, Software. H.L.: Resources, Project administration. All authors have read and agreed to the published version of the manuscript.

Funding: This work was funded by the National Key Research and Development Program of China (grant number 2024YFC3713500, 2024YFC3713501), Major Science and Technology Project of the Xinjiang Uygur Autonomous Region (2024A03012, 2024A03012-1), and Yinchuan Air Pollution Control Project (No.2024NCZ(YC)001836).

Institutional Review Board Statement: Not Applicable.

Informed Consent Statement: Not Applicable.

Data Availability Statement: No new data were created or analyzed in this study. Data sharing is not applicable to this article.

Conflicts of Interest: The authors declare that they have no known competing financial interests or personal relationships that could have appeared to influence the work reported in this paper.

Appendix A

Table A1. $PM_{2.5}$ and heavy metal concentrations ($ng{\cdot}m^{-3}$) in All Cities, RICs, and GCs before and after 2018 in China.

Year Range	City	Type of City	$PM_{2.5}$	Pb	As	Mn	Ni	Cr	Cd	Zn	Cu	Fe	Data Source
2014–2018	Hengyang	RICs	111,180	456.9	8.4	109.9	55.4	104.7	11.7	898.8	488.7	-	[46]
	Baotou		53,400	143.8	8.5	211.4	16.6	19.5	2.5	280.9	-	-	[47]
	Handan		91,180	115	13.5	28.3	18.1	12.6	4.9	187.8	-	-	[48]
	Suzhou		66,200	76.2	4.1	-	-	6.9	1.5	-	-	-	[49]
	Ningbo		84,000	61.9	4.9	36.8	4.1	3.4	1.5	-	-	-	[34]
	Foshan		47,000	277	91.6	-	28	36	4.4	1283	-	-	[47]
	Zhaoqing		39,375	93.9	17.6	-	8.8	10.2	10.2	408.9	-	-	[47]
	Baodinng		180,000	235	30	55	27.5	32.5	5	622.5	194	1340	[50]
	Taiyuan		97,300	241.3	1.8	171.9	25.3	63.6	2.8	338.2	83.1	-	[51]
	Xining		49,688	626.8	18	97.7	9.3	252.1	70.9	958.9		-	[47]
	Shijiazhuang		99,475	84.3	35.9	71.4	-	31.4	1.9	634.7	60.5	1019.2	[52]
	Yinchuan		75,000	124.1	9.2	96.1	1.2	2.9	3.3	-	-	-	[53]
	Langfang		133,000	140	12.5	75	25	40	-	337.5	72.5	1480	[54]
	Luoyang	GCs	72,236	9.8	3.1	6.2	0.3	1.3	0.1	-	11	-	[55]
	Xingyang		78,310	60	4.5	-	-	10	-	510	-	-	[56]
	Jinzhou		41,250	225.5	-	-	-	49	0.7	679.8	158.5	-	[57]
	Guangzhou		37,000	40	8.6	25	3.2	-	1.4	190	22	230	[7]
	Zhuhai		45,000	59	-	28	7	6	-	149	20	212	[58]
	Shenzhen		39,100	25.6	4.9	19.5	4.2	2.3	0.8	163.3	29.4	-	[59]
	Hangzhou		92,000	78.8	8.1	34	2.5	3.1	1.9	-	-	-	[34]
	Guilin		66,000	102	11.2	173	-	-	4.9	-	-	-	[60]
	Xi'an		50,600	159	125.4	61.5	21.5	100.9	3.3	1264.8	43.4	-	[61]

Table A1. *Cont.*

Year Range	City	Type of City	$PM_{2.5}$	Pb	As	Mn	Ni	Cr	Cd	Zn	Cu	Fe	Data Source
2014–2018	Nanchang	GCs	29,740	468.7	-	226.1	51.2	32	21.5	1141.1	343.4	-	[62]
	Beijing		126,000	120	10	55	25	32.5	-	247.5	107.5	1650	[34]
	Tianjin		133,000	190	13.3	60	23.3	36.7	3.3	680	100	-	[34]
	Shanghai		94,600	69.7	-	-	14.9	16.9	-	215	24.2	1340	[63]
	Chongqing		60,695	50.4	-	37.7	4.2	11.1	-	113	11.3	586	[1]
	Shenyang		76,225	62.3	10.5	28.3	-	8.2	1.3	-	-	-	[64]
	Harbin		93,250	57.3	14.7	25.4	2.8	3.8	0.9	-	-	-	[65]
	Nanjing		39,250	23.5	2	20.8	4.9	6.7	1	99	50.9	455.2	[66]
	Zhengzhou		165,200	57.7	8.2	40.9	-	-	2.2	72.7	-	-	[67]
	Changsha		46,900	55.8	3.8	18.8	7.6	5.5	0.5	33.5	-	-	[46]
	Chengdu		64,184	55.4	10.8	33.8	2.1	5.6	-	238	18.7	456	[1]
	Nantong		58,400	28.3	6.7	40.6	4.7	2.5	1	-	-	-	[68]
	Hefei		81,000	12.6	-	5.2	-	10	-	273.5	11.3	44.9	[26]
	Lanzhou		50,699	43.4	8.4	40.5	6.6	-	2.3	-	-	-	[69]
	Wuhan		57,027	90.9	8.2	25.9	3.73	-	3	67.7	15.5	51.9	[45]
Avg. ± std		RICs	86,676.8 ± 39,372.6	205.9 ± 167.2	19.7 ± 23.8	95.3 ± 57.8	19.9 ± 15.1	47.4 ± 67.7	10.1 ± 19.5	595.1 ± 354.6	179.8 ± 180.8	1279.7 ± 236.2	
		GCs	71,155.2 ± 33,133.5	87.6 ± 93.3	14.2 ± 27.1	46.9 ± 50.9	9.8 ± 12.2	17.0 ± 23.3	2.9 ± 4.8	350.2 ± 353.5	60.5 ± 83.7	594.4 ± 544.2	
		All Cities	76,329.1 ± 35,591.3	127.0 ± 133.2	16.4 ± 25.0	61.6 ± 56.8	13.4 ± 14.0	28.6 ± 47.1	5.7 ± 13.0	434.7 ± 367.1	87.6 ± 119.1	741.2 ± 567.0	

Table A1. *Cont.*

Year Range	City	Type of City	$PM_{2.5}$	Pb	As	Mn	Ni	Cr	Cd	Zn	Cu	Fe	Data Source
2019–2021	Tangshan	RICs	97,500	110	4.1	60	5.3	8.6	1.9	450	17.7	880	[70]
	Zaozhuang		91,000	46.4	9.4	45.2	6.8	6.5	7.6	215	51.7	-	[71]
	Panzhihua		33,000	82.4	7.2	34.8	13.5	5.1	1.2	130.6	18	1259.8	[72]
	Zunyi		47,600	78.5	2.1	64.4	2.9	38.3	1.2	213.3	49.5	-	[73]
	Taiyuan		87,130	57.6	5.4	51.7	3.1	6.9	1	-	-	-	[74]
	Xining		34,331	36	2.1	12.9	1	1.6	0.7	-	-	-	[75]
	Caofeidian		89,680	230.4	9.3	107.8	7.1	32.7	2.8	495.4	14.7	762.3	[76]
	Changzhi		56,100	30.8	4.9	21.5	4.2	14.3	0.7	82.3	7.8	864.5	[22]
	Shijiazhuang		98,130	41.3	5.4	28.8	1.1	2.5	1.6	-	-	-	[77]
	Maanshan	GCs	51,750	50.1	6.2	35.3	1.8	1.7	1.4	-	-	-	[78]
	Baoji		51,500	27.5	-	73.2	2.5	13.4	-	234.8	-	-	[79]
	Dongguan		35,800	17.9	4.7	19	1.9	4.6	0.7	109.8	11.7	290.1	[80]
	Guangzhou		40,300	37	3.1	20	5.8	9	1.1	161	32	276	[81]
	Suzhou		46,760	21.9	3.5	30.5	3.2	4.6	0.7	-	-	-	[82]
	Beijing		48,000	21.7	2.8	21.1	0.9	0.6	0.5	-	-	-	[83]
	Tianjin		59,000	27.1	6.9	34.3	1.5	3.3	0.6	169.1	18.2		[83]
	Shanghai		40,610	32.7	3.3	22.3	5	5.2	-	-	-	-	[84]
	Shenyang		85,100	58.4	11.2	36.1	2.1	6.6	1	206.4	9.4		[85]
	Zhengzhou		52,000	100	17.7	157	7.6	11.7	-	209.8	29.2	-	[86]
	Chengdu		115,300	73.7	-	60	3.8	11.1	-	254.8	11	980.5	[39]
	Kunming		26,670	48.9	1.8	13.4	0.8	1.9	1.1		-	-	[87]
	Guiyang		26,750	37.8	2.1	19.1	0.8	1.8	1.1	51.1	25.1	-	[88]

Table A1. *Cont.*

Year Range	City	Type of City	$PM_{2.5}$	Pb	As	Mn	Ni	Cr	Cd	Zn	Cu	Fe	Data Source
2019–2021	Qingdao	GCs	85,500	52	10.8	42	3	8	1.8	211.1	14.6	-	[44]
	Nanning		34,000	19.8	3.2	-	1.3	3.7	1	64.2	3.6	-	[89]
	Lanzhou		34,514	27.9	5.5	41.7	2	-	2	-	-	-	[68]
	Xiamen		21,620	10.8	2.3	14.5	3.6	5.1	0.4	53.4	7.3	280.2	[40]
Avg. ± std		RICs	70,496.8 ± 27,393.7	79.3 ± 62.3	5.5 ± 2.7	47.5 ± 28.5	5.0 ± 3.9	12.9 ± 13.4	2.1 ± 2.2	264.4 ± 169.7	26.6 ± 19	941.7 ± 218.4	
decline range			18.7%	61.5%	72.1%	50.2%	74.9%	72.8%	79.2%	55.6%	85.2%	26.4%	
Avg. ± std		GCs	50,304.4 ± 24,532.6	39.1 ± 22.8	5.7 ± 4.4	39.9 ± 35.3	2.8 ± 1.0	5.8 ± 3.9	1.0 ± 0.5	156.9 ± 75.3	16.2 ± 9.6	456.7 ± 349.3	
decline range			29.4%	55.4%	59.9%	14.9%	71.4%	65.9%	65.5%	55.2%	73.2%	23.2%	
Avg. ± std		All Cities	57,294 ± 26,856.8	53 ± 44.2	5.6 ± 3.8	42.7 ± 32.6	3.6 ± 2.9	8.4 ± 9.0	1.5 ± 1.5	194.8 ± 123.9	20.1 ± 14.2	699.2 ± 374.1	
decline range			24.9%	58.3%	65.9%	30.7%	73.1%	70.6%	73.7%	55.2%	77.1%	5.7%	

References

1. Wang, H.; Qiao, B.; Zhang, L.; Yang, F.; Jiang, X. Characteristics and sources of trace elements in $PM_{2.5}$ in two megacities in Sichuan Basin of southwest China. *Environ. Pollut.* **2018**, *242*, 1577–1586. [CrossRef] [PubMed]
2. Wu, L.; Luo, X.; Li, H.; Long, C.; Tang, M. Seasonal Levels, Sources, and Health Risks of Heavy Metals in Atmospheric $PM_{2.5}$ from Four Functional Areas of Nanjing City, Eastern China. *Atmosphere* **2019**, *10*, 419. [CrossRef]
3. Wang, P.; Huang, W.; Ren, F. Pollution evaluation and source identification of heavy metals in soil around steel factories located in Lanshan District, Rizhao City, eastern China. *Environ. Monit. Assess.* **2023**, *195*, 657. [CrossRef] [PubMed]
4. World Health Organization (WHO). *Health Effects of Particulate Matter: Policy Implications for Countries in Eastern Europe, Caucasus and Central Asia*; WHO Regional Office for Europe: Copenhagen, Denmark, 2013; p. 20. Available online: https://iris.who.int/handle/10665/344854 (accessed on 3 June 2024).
5. Zhou, X.; Strezov, V.; Jiang, Y.; Yang, X.; He, Y.; Evans, T. Life Cycle Impact Assessment of Airborne Metal Pollution near Selected Iron and Steelmaking Industrial Areas in China. *Aerosol Air Qual. Res.* **2019**, *20*, 1582–1590. [CrossRef]
6. Lyu, T.; Tang, Y.; Cao, H.; Gao, Y.; Zhou, Y.; Zhang, W.; Zhang, R.; Jiang, Y. Estimating the geographical patterns and health risks associated with $PM_{2.5}$-bound heavy metals to guide $PM_{2.5}$ control targets in China based on machine-learning algorithms. *Environ. Pollut.* **2023**, *337*, 122558. [CrossRef]
7. Guo, L.; Lv, Z.; Ma, W. Contribution of heavy metals in $PM_{2.5}$ to cardiovascular disease mortality risk, a case study in Guangzhou, China. *Chemosphere* **2020**, *297*, 134102. [CrossRef]
8. Yuan, Y.; Wu, Y.; Ge, X.; Nie, D.; Wang, M.; Zhou, H.; Chen, M. In vitro toxicity evaluation of heavy metals in urban air particulate matter on human lung epithelial cells. *Sci. Total Environ.* **2019**, *678*, 301–308. [CrossRef]
9. Hsieh, C.; Jung, C.; Lin, C.; Hwang, B. Combined exposure to heavy metals in $PM_{2.5}$ and pediatric asthma. *J. Allergy Clin. Immun.* **2020**, *147*, 2171–2180. [CrossRef]
10. Cogliano, V.; Baan, R.; Straif, B.; Grosse, B. Preventable exposures associated with human cancers. *J. Natl. Cancer I.* **2011**, *103*, 1827–1839. [CrossRef]
11. Tang, Y.; Han, G. Characteristics of major elements and heavy metals in atmospheric dust in Beijing, China. *J. Geochem. Explora.* **2017**, *176*, 114–119. [CrossRef]
12. Xu, J.; Jia, C.; Yu, H.; Xu, H.; Ji, D.; Wang, C.; Xiao, H.; He, J. Characteristics, sources, and health risks of $PM_{2.5}$-bound trace elements in representative areas of Northern Zhejiang Province. *China Chemosphere* **2021**, *272*, 129632–129643. [CrossRef] [PubMed]
13. Liu, T.; Zhao, C.; Chen, Q.; Li, L.; Si, G.; Li, L.; Guo, B. Characteristics and health risk assessment of heavy metal pollution in atmospheric particulate matter in different regions of the Yellow River Delta in China. *Environ. Geochem. Health* **2022**, *45*, 2013–2030. [CrossRef] [PubMed]
14. Aldabe, J.; Elustondo, D.; Santamaría, C.; Lasheras, E.; Pandolfi, M.; Alastuey, A.; Querol, X.; Santamaría, J. Chemical characterisation and source apportionment of $PM_{2.5}$ and PM_{10} at rural, urban and traffic sites in Navarra (North of Spain). *Atmos. Res.* **2011**, *102*, 191–205. [CrossRef]
15. Patrick, A.; Pierre, S.; Alessandra, M. Long-term exposure to ambient $PM_{2.5}$ and impacts on health in Rome, Italy. *Clin. Epidemiol. Glob.* **2020**, *8*, 531–553.
16. Qu, J.; Zheng, L.; Tang, Q.; Liu, M.; Zhang, S. Source analysis of heavy metals in atmospheric particulate matter in a mining city. *Environ. Geochem. Health* **2022**, *44*, 979–991.
17. Liu, Z.; Zhou, J.; Zhang, J.; Mao, Y.; Huang, X.; Qian, G. Evaluation for the heavy metal risk in fine particulate matter from the perspective of urban energy and industrial structure in China: A meta analysis. *J. Clean. Prod.* **2020**, *244*, 118597. [CrossRef]
18. Duan, J.; Tan, J. Atmospheric heavy metals and Arsenic in China: Situation, sources and control policies. *Atmos. Environ.* **2013**, *74*, 93–101. [CrossRef]
19. Li, F.; Yan, J.; Wei, Y.; Lv, G. $PM_{2.5}$-bound heavy metals from the major cities in China:Spatiotemporal distribution, fuzzy exposure assessment and health risk management. *J. Clean. Prod.* **2021**, *286*, 124967. [CrossRef]
20. Yu, P.; Han, Y.; Wang, M.; Zhu, Z.; Tong, Z.; Shao, X.; Peng, J.; Hamid, Y.; Yang, X.; Deng, Y.; et al. Heavy metal content and health risk assessment of atmospheric particles in China: A meta-analysis. *Sci. Total Environ.* **2023**, *867*, 161556. [CrossRef]
21. Csavina, J.; Field, J.; Taylor, M.; Gao, S.; Landa'zuri, A.; Eric, A.; Eduardo, A. A review on the importance of metals and metalloids in atmospheric dust and aerosol from mining operations. *Sci. Total Environ.* **2012**, *433*, 58–73. [CrossRef]
22. Duan, X.; Yan, Y.; Li, R.; Deng, M.; Peng, L. Seasonal variations, source apportionment, and health risk assessment of trace metals in PM2.5 in the typical industrial city of changzhi, China. *Atmos. Pollut. Res.* **2020**, *12*, 365–374. [CrossRef]
23. Behrooz, R.; Kaskaoutis, D.; Grivas, G.; Mihalopoulos, N. Human health risk assessment for toxic elements in the extreme ambient dust conditions observed in Sistan, Iran. *Chemosphere* **2021**, *262*, 127835. [CrossRef] [PubMed]
24. Zhang, X.; Yuko, E.; Masahide, A. Risk assessment and management of $PM_{2.5}$-bound heavy metals in the urban area of Kitakyushu, Japan. *Sci. Total Environ.* **2021**, *795*, 148748. [CrossRef]

25. US EPA. *Guidelines for the Health Risk Assessment of Chemical Mixtures*; US EPA: Cincinnati, OH, USA, 1986; Volume 51, pp. 34014–34025. Available online: https://www.epa.gov/risk/guidelines-health-risk-assessment-chemical-mixtures (accessed on 3 June 2024).
26. Xue, H.; Liu, G.; Zhang, H.; Hu, R.; Wang, X. Similarities and differences in PM_{10} and $PM_{2.5}$ concentrations, chemical compositions and sources in Hefei City, China. *Chemosphere* **2019**, *220*, 760–765. [CrossRef] [PubMed]
27. Zhao, M.; Ren, L.; Li, G.; Liu, Y.; Zhao, G.; Gao, Y.; Yang, X. Pollution Characteristics and Health Risk Assessment of $PM_{2.5}$ Heavy Metals in Tianjin and Qingdao in Winter of 2018–2019. *Environ Chem.* **2020**, *43*, 12. (In Chinese)
28. Ministry of Ecology and Environment of the People's Republic of China. Bulletin of the State of China's Ecological Environment in 2021. 2022. Available online: https://www.mee.gov.cn/hjzl/sthjzk/zghjzkgb/202205/P020220608338202870777.pdf (accessed on 3 June 2024).
29. National Bureau of Statistics of China. Available online: https://www.stats.gov.cn/sj/ (accessed on 3 June 2024).
30. World Health Organization (WHO). *Health Risks of Heavy Metals from Long-Range Transboundary Air Pollution*; WHO Regional Office for Europ: Copenhagen, Denmark, 2007; p. 144.
31. Kabir, E.; Ray, S.; Kim, K.; Yoon, H.; Jeon, E.; Kim, Y.; Chao, Y.; Yun, S.; Brown, R. Current status of trace metal pollution in soils affected by industrial activities. *Sci. World J.* **2012**, *2012*, 916705. [CrossRef]
32. Li, Y.; Zhou, S.; Jia, Z.; Wang, G. Temporal and spatial distributions and sources of heavy metals in atmospheric deposition in western Taihu Lake, China. *Environ. Pollut.* **2021**, *284*, 117465. [CrossRef]
33. Hao, Y.; Luo, B.; Simayi, M.; Zhang, W.; Xie, S. Spatiotemporal patterns of $PM_{2.5}$ elemental composition over China and associated health risks. *Environ. Pollut.* **2020**, *265*, 114910. [CrossRef]
34. Wang, X.; He, S.; Chen, S.; Zhang, Y.; Wang, A.; Luo, J.; Ye, X.; Mo, Z.; Wu, L.; Xu, P. Spatiotemporal Characteristics and Health Risk Assessment of Heavy Metals in $PM_{2.5}$ in Zhejiang Province. *Environ. Res. Public Health* **2018**, *15*, 583. [CrossRef]
35. Lv, W.; Wang, Y.; Querol, X.; Zhuang, X.; Alastuey, A.; López, A.; Viana, M. Geochemical and statistical analysis of trace metals in atmospheric particulates in Wuhan, central China. *Environ. Geo.* **2006**, *51*, 121–132. [CrossRef]
36. Deng, L.; Bi, C.; Jia, J.; Zeng, Y.; Chen, Z. Effects of heating activities in winter on characteristics of $PM_{2.5}$-bound Pb, Cd and lead isotopes in cities of China. *J. Clean. Prod.* **2020**, *265*, 121826. [CrossRef]
37. Casimiro, P.; Mirante, F.; Oliveira, C.; Matos, M.; Caseiro, A.; Oliveira, C.; Querol, X.; Alves, C.; Martins, N.; Cerqueira, M. Size-segregated chemical composition of aerosol emissions in an urban road tunnel in Portugal. *Atmos. Environ.* **2013**, *71*, 15–25.
38. Chang, S.; Wang, K.; Chang, H.; Ni, W.; Wu, B.; Wong, R.; Lee, H. Comparison of Source Identification of Metals in Road-Dust and Soil. *Soil. Sediment. Contam.* **2009**, *18*, 669–683. [CrossRef]
39. Unal, Y.; Toros, H.; Deniz, A.; Incecik, S. Influence of meteorological factors and emission sources on spatial and temporal variations of PM_{10} concentrations in Istanbul metropolitan area. *Atmos. Environ.* **2011**, *45*, 5504–5513. [CrossRef]
40. Guo, J.; Li, Y.; Liu, H.; Li, M.; Liu, M. Elemental composition and source analysis of atmospheric $PM_{2.5}$ in winter in Chengdu. *Environ. Dvpt.* **2020**, *32*, 143–146+148. (In Chinese)
41. Xiao, S.; Cai, M.; Li, X.; Huang, W.; Wang, J.; Zhu, Q.; Wu, S. Characteristics and health risk assessment of heavy metal pollution in atmospheric $PM_{2.5}$ in Xiamen Port. *Environ. Sci.* **2022**, *43*, 3403–3415. (In Chinese)
42. Brehmer, C.; Norris, C.; Karoline, K.; Bergin, M.; Schauer, J. The impact of household air cleaners on the chemical composition and children's exposure to $PM_{2.5}$ metal sources in suburban Shanghai. *Environ. Pollut.* **2019**, *253*, 190–198. [CrossRef]
43. Zhang, W.; Lv, T.; Liu, J.; Gao, Y.; Zhou, X.; Cao, H. Characteristics, sources and health risks of heavy metal pollution in $PM_{2.5}$ carrier belts in Beijing and assessment of policy effects. *Environ. Sci.* **2024**, *45*, 6229–6237. (In Chinese)
44. Wang, Y.; Li, F.; Liu, Y.; Deng, X.; Yu, H.; Li, J.; Xue, T. Risk Assessment and Source Analysis of Atmospheric Heavy Metals Exposure in Spring of Tianjin, China. *Aerosol Sci. Enginee* **2022**, *7*, 87–95. [CrossRef]
45. Department of Environmental Protection. *GB 3095-2012*; Ambient Air Quality Standard. China National Environmental Science Press: Beijing, China, 2012. Available online: https://www.mee.gov.cn/ywgz/fgbz/bz/bzwb/dqhjbh/dqhjzlbz/201203/t20120302_224165.shtml (accessed on 9 February 2025).
46. Tian, R.; Liu, Y.; Chen, P.; Zhang, H.; Jiang, Y. Heavy metal pollution levels and health risks in $PM_{2.5}$ in Hengyang city. *Environ. Eng.* **2017**, *35*, 127–130. (In Chinese)
47. Wang, L.; Yang, H.; Li, X. Analysis of heavy metal levels and influencing factors in $PM_{2.5}$ in some provincial capitals of China. *Environ. Chem.* **2017**, *36*, 72–83. (In Chinese)
48. Cai, A.; Zhang, H.; Wang, L. Source Apportionment and Health Risk Assessment of Heavy Metals in $PM_{2.5}$ in Handan: A Typical Heavily Polluted City in North China. *Atmosphere* **2021**, *12*, 1232. [CrossRef]
49. Zou, H.; Zhang, B.; Wang, Z.; Li, X.; Dai, M.; Chen, H. Characteristics and health risk assessment of heavy metal pollution in atmospheric $PM_{2.5}$ in Suzhou Industrial Park. *Adm. Techn Environ. Monit.* **2017**, *29*, 37–41. (In Chinese)
50. Gao, J.; Tian, H.; Cheng, K.; Zheng, M.; Wang, S.; Hao, J.; Wang, K.; Hua, S.; Zhu, C. The variation of chemical characteristics of $PM_{2.5}$ and PM_{10} and formation causes during two haze pollution events in urban Beijing, China. *Atmos. Environ.* **2015**, *107*, 1–8. [CrossRef]

51. Liu, K.; Shang, Q.; Wan, C. Sources and Health Risks of Heavy Metals in $PM_{2.5}$ in a Campus in a Typical Suburb Area of Taiyuan, North China. *Atmosphere* **2018**, *9*, 46. [CrossRef]
52. Yan, L.; Zuo, H.; Zhang, J.; Li, Z.; Li, S. Comparative study on distribution characteristics and sources of heavy metal elements in atmospheric PM_1, $PM_{2.5}$ and PM_{10} in Shijiazhuang. *Earth Sci. Front.* **2019**, *26*, 263–270. (In Chinese)
53. Qi, A.; Zhang, Y.; Ding, Y.; Yang, H. Characteristics and sources of metal element pollution in atmospheric $PM_{2.5}$ in Yinchuan City. *J. Environ. Health* **2017**, *34*, 591–594. (In Chinese)
54. Gao, J.; Wang, K.; Wang, Y.; Liu, S.; Hao, J.; Liu, H.; Hua, S.; Tian, H. Temporal-spatial characteristics and source apportionment of $PM_{2.5}$ as well as its associated chemical species in the Beijing-Tianjin-Hebei region of China. *Environ. Pollut.* **2017**, *233*, 714–724. [CrossRef]
55. Jin, Z. Characteristic sources and potential ecological risk assessment of heavy metal elements in atmospheric $PM_{2.5}$ in Luoyang City. *Environ. Prot. Technol.* **2020**, *26*, 11–15. (In Chinese)
56. Liu, Y.; Wang, H.; Guo, E.; Zhang, H.; He, M.; He, X.; Wang, M. Health risk assessment of heavy metals in particulate matter during spring and autumn observation in Xingyang City. *Environ. Chem.* **2019**, *38*, 1012–1918. (In Chinese)
57. Gu, J.; Liu, L.; Liu, Z.; Cong, Q.; Zhao, G. Morphological analysis and bioavailability evaluation of heavy metals in atmospheric particulate matter in Jinzhou City. *Chem. Res. Appl.* **2016**, *28*, 1136–1140. (In Chinese)
58. Tao, J.; Zhang, L.; Cao, J.; Zhong, l.; Chen, D.; Yang, Y.; Chen, D.; Chen, L.; Zhang, Z.; Wu, Y. 2017. Source apportionment of $PM_{2.5}$ at urban and suburban areas of the Pearl River Delta region, south China-With emphasis on ship emissions. *Sci Total Environ.* **2017**, *574*, 1559–1570. [CrossRef] [PubMed]
59. Jiang, J.; Li, R.; Qiu, H.; Wang, C.; Ruan, S.; Liu, G.; Peng, C.; Zhang, H. Characteristics and health risk assessment of heavy metal pollution in atmospheric $PM_{2.5}$ in Shenzhen. *Pract. Prev. Med.* **2019**, *26*, 781–785. (In Chinese)
60. Mo, Z.; Du, J.; Liu, H.; Chen, Z.; Liang, G.; Huang, J.; Li, H.; Lin, H.; Zhu, K. Health risk assessment of heavy metal pollutants in atmospheric $PM_{2.5}$ in winter in Guilin City. *Adm. Tech. Environ. Monit.* **2019**, *31*, 23–27. (In Chinese)
61. Liu, P.; Ren, H.; Xu, H. Assessment of heavy metal characteristics and health risks associated with $PM_{2.5}$ in Xi'an, the largest city in northwestern China. *Air Qual. Atmos. Health* **2018**, *11*, 1037–1047. [CrossRef]
62. Zheng, Q.; Hu, G.; Yu, R.; Zhao, Y.; Zhang, Z. Source analysis and health risk assessment of heavy metal elements in atmospheric $PM_{2.5}$ in winter in Nanchang City. *Earth Environ.* **2018**, *46*, 6–312. (In Chinese)
63. Ming, L.; Li, X.; Jin, L.; Fu, P.; Jun, L. $PM_{2.5}$ in the Yangtze River Delta, China: Chemical compositions, seasonal variations, and regional pollution events. *Environ. Pollut.* **2017**, *223*, 200–212. [CrossRef]
64. Nie, L.; Li, Y.; Hua, Z.; Cui, Z. Characteristics and health risk assessment of heavy metal pollution in atmospheric $PM_{2.5}$ in Shenyang, Liaoning Province. *Chin. J. Public Health* **2018**, *34*, 574–576. (In Chinese)
65. Kang, Z.; Bai, Y.; Yang, G.; Wang, Y.; Yu, T.; Hong, Q.; Liu, X. Characteristics and health risk assessment of heavy metal pollution in atmospheric $PM_{2.5}$ in Harbin City. *J. Environ. Health* **2018**, *35*, 504–507. (In Chinese)
66. Li, H.; Qian, X.; Leng, X.; Dai, Q. Characteristics and health risks of metal element pollution in $PM_{2.5}$ in Nanjing. *Environ. Monit. Forewarning* **2021**, *13*, 7–13. (In Chinese)
67. Yan, G.; Zhang, P.; Wang, C.; Song, X.; Gao, Y.; Zhang, J.; Jiang, J.; Cao, Z.; Zhu, G.; Wang, Y. Sources and potential health risks of heavy metals in $PM_{2.5}$ during heating and non-heating periods in Zhengzhou City. *Acta Sci. Circumstantiae* **2019**, *39*, 2811–2820. (In Chinese)
68. He, Z.; Chen, F. Characteristics and health risk assessment of heavy metal pollution in atmospheric $PM_{2.5}$ in Nantong City. *Mod. Prev. Med.* **2020**, *47*, 233–236. (In Chinese)
69. Du, M.; Yin, X.; Li, Y.; Ke, T.; Zhu, H.; Wu, J.; Zheng, G. Time Trends and Forecasts of Atmospheric Heavy Metals in Lanzhou, China, 2015–2019. *Water Air Soil Pollut.* **2022**, *233*, 233–305. [CrossRef]
70. Yang, T.; Yu, H.; He, Y.; Miao, Y.; Gao, Y.; Li, N.; Wang, W. Element composition and source analysis of $PM_{2.5}$ in Tangshan City in autumn and winter of 2017–2018. *Res. Environ. Sci.* **2020**, *33*, 2030–2039. (In Chinese)
71. Wei, Q.; Chen, W.Y.; Jin, L.X. Health risk assessment and pollution source analysis of atmospheric $PM_{2.5}$ heavy metal elements in Zaozhuang City. *China Powder Sci. Technol.* **2020**, *26*, 69–78. (In Chinese)
72. Xu, X.; Feng, X.; Chen, J.; Yin, H.; Qian, J. Metal element pollution characteristics and health risk assessment of $PM_{2.5}$ in Panzhihua City. *Environ. Chem.* **2021**, *40*, 2780–2788. (In Chinese)
73. Zhang, Y.; Chen, R.; Bao, Y.; Chen, Z. Characteristics and health risk assessment of heavy metal pollution in $PM_{2.5}$ in Zunyi City. *Environ. Prot. Sci.* **2020**, *46*, 179–184. (In Chinese)
74. Li, C.; Lv, Q.; Wang, L. Analysis of heavy metal element pollution in $PM_{2.5}$ in Taiyuan City from 2018 to 2020. *Prev. Med. Trib.* **2021**, *27*, 842–844. (In Chinese)
75. Ji, X.; Yang, J.; Xie, X.; Guo, X. Characteristics and health risk assessment of heavy metal pollution in urban $PM_{2.5}$ in Xining City in 2019. *Mod. Prev. Med.* **2021**, *47*, 4256–4259. (In Chinese)

76. Fang, B.; Zeng, H.; Zhang, L.; Wang, H.; Liu, J.; Hao, K.; Zheng, G.; Wang, M.; Wang, Q.; Yang, W. Toxic metals in outdoor/indoor airborne $PM_{2.5}$ in port city of Northern, China: Characteristics, sources, and personal exposure risk assessment. *Environ. Pollut.* **2021**, *279*, 116937. [CrossRef]
77. Tang, D.; Chang, H.; Zhang, Y.; Sun, C.; Chen, F.; Guan, M.; Zhao, C. Heavy metal pollution and health risk assessment of $PM_{2.5}$ in Shijiazhuang City from 2017 to 2019. *Mod. Prev. Med.* **2021**, *48*, 1177–1180+1197. (In Chinese)
78. Fan, Z.; Wang, Y.; Zhang, R.; Qiu, G. Characteristics and health risk assessment of heavy metal pollution in atmospheric $PM_{2.5}$ in Maanshan City. *J. Environ. Health Sci.* **2019**, *36*, 1064–1068. (In Chinese)
79. Zhou, B.; Wang, J.; Cao, X.; Xu, D.; Feng, Q.; Liu, W.; Yang, Z.; Wang, Y.; Li, J. Pollution characteristics and sources of metal elements in $PM_{2.5}$ in spring in Baoji City. *Environ. Sci. Technol.* **2021**, *44*, 198–206. (In Chinese)
80. Xu, J.W.; Lin, M.; Yue, D.L.; Zhou, Z.; Huang, J. The pollution characteristic and ecological risk of atmospheric fine-particle-bound metals in Dongguan. *Environ. Sci. Technol.* **2021**, *44*, 155–160. (In Chinese)
81. Zhou, X.; Xie, M.; Zhao, M.; Wang, Y.; Luo, J.; Lu, S.; Li, J.; Liu, Q. Pollution characteristics and human health risks of $PM_{2.5}$-bound heavy metals: A 3 year observation in Suzhou, China. *Environ. Geochem. Health* **2023**, *45*, 5145–5162. [CrossRef]
82. Li, W.; Zhang, M.; Wang, B.; Sun, L. Risk assessment of health effects of heavy metals in atmospheric $PM_{2.5}$ on residents in Chaoyang District, Beijing. *Mod. Prev. Med.* **2021**, *48*, 416–419. (In Chinese)
83. Li, L.; Deng, X.; Xiao, Z.; Yuan, J.; Yang, N.; Guo, X.; Bai, Y. The characteristics and health risks of heavy metal pollution in $PM_{2.5}$ from different air masses in Tianjin during heating season. *Environ. Sci.* **2023**, *44*, 30–37. (In Chinese)
84. Luo, R.; Dai, H.; Zhang, Y.; Qiao, L.; Ma, Y.; Zhou, M.; Xia, B.; Zhu, Q.; Zhao, Y.; Huang, C. Exposure levels, sources and health risks of heavy metal components of $PM_{2.5}$ among domestic women in suburban Shanghai. *Environ. Sci.* **2019**, *40*, 5224–5233. (In Chinese)
85. Ren, W.; Li, Y.; Su, C.; Wang, G.; Yu, X.; Kang, N. Characteristics, source analysis and health risk assessment of heavy metal pollution in atmospheric $PM_{2.5}$ in Shenyang. *Environ. Chem.* **2021**, *40*, 1029–1037.
86. He, R.; Zhang, T.; Chen, Y.; Jin, Z.; Han, S.; Zhao, J.; Zhang, R.; Yan, Q. Characteristics and ecological and health risk assessment of atmospheric $PM_{2.5}$ heavy metal pollution in a living area of Zhengzhou City. *Environ. Sci.* **2019**, *40*, 4774–4782. (In Chinese)
87. Su, X.; Li, X.; Han, R.; Dong, L.; Sun, H.; Chen, H. Distribution characteristics and sources of heavy metals in fine particulate matter in two urban areas of Kunming from 2017 to 2019. *Occup. Health* **2023**, *39*, 239–242+246. (In Chinese)
88. Wang, X.; Fei, X.; Yang, Y.; Li, Y.; Gui, J.; Yang, A.; Xu, P. Pollution characteristics, sources and health risk assessment of heavy metal elements in $PM_{2.5}$ in Huaxi district, Guiyang City. *Acta Sci. Circumstantiae* **2023**, *43*, 110–118. (In Chinese)
89. Qin, J.; Zhang, X.; Huang, J.; Mo, Z.; Chen, Z.; Zhang, D.; Liu, H.; Li, H. Characteristics and health risk assessment of heavy metal pollution in atmospheric $PM_{2.5}$ in Nanning City. *Environ. Sci. Technol.* **2020**, *43*, 35–44.

toxics

MDPI

Review

A Review of Biogenic Volatile Organic Compounds from Plants: Research Progress and Future Prospects

Rongrong Luo [1], Xiaoxiu Lun [2], Rui Gao [3], Le Wang [1], Yuan Yang [4], Xingqian Su [5], Md Habibullah-Al-Mamun [6], Xiaohang Xu [7], Hong Li [3,*] and Jinjuan Li [1,7,*]

1 College of Resources and Environmental Engineering, Guizhou University, Guiyang 550025, China; 18285485626@163.com (R.L.); wangle0602@foxmail.com (L.W.)
2 College of Environmental Science and Engineering, Beijing Forestry University, Beijing 100083, China; lunxiaoxiu@bjfu.edu.cn
3 State Key Laboratory of Environmental Criteria and Risk Assessment, Chinese Research Academy of Environmental Sciences, Beijing 100012, China; gaorui@craes.org.cn
4 Guizhou Research and Designing Institute of Environmental Sciences, Guizhou Academy of Environmental Science and Design, Guiyang 550081, China; gzhky_yy@163.com
5 North Alabama International College of Engineering and Technology, Guizhou University, Guiyang 550025, China; 19078812302@163.com
6 Department of Fisheries, Faculty of Biological Sciences, University of Dhaka, Dhaka 1000, Bangladesh; almamunhabib@gmail.com
7 Key Laboratory of Karst Georesources and Environment, Ministry of Education, College of Resources and Environmental Engineering, Guizhou University, Guiyang 550025, China; xuxh@gzu.edu.cn
* Correspondence: summy_lee@163.com (J.L.); lihong@craes.org.cn (H.L.)

Abstract: Biogenic volatile organic compounds (BVOCs) emitted by plants contribute to secondary air pollution through photochemical reactions in sunlight. Due to the influence of multiple factors, accurately characterizing and quantifying the emission of BVOCs from plant sources is challenging, which poses significant obstacles to the effective management and control of BVOCs. Therefore, this paper summarizes the emission mechanisms of BVOCs from plants, explores the primary factors influencing variations in the emission rates of these compounds, and evaluates the advantages and limitations of contemporary "measurement-modeling" methods for characterizing BVOC emissions. It is concluded that current measurement techniques still need to be further developed to meet the criteria of simplicity, affordability, and high precision simultaneously, and in terms of modeling and prediction studies, there is a lack of in-depth research on the atmospheric chemistry of BVOCs and the synergistic effects of multiple factors. Finally, it is suggested to leverage interdisciplinary strengths to develop advanced measurement technologies and high-resolution models for monitoring volatile compounds. Additionally, strategically selecting low-BVOC tree species in pollution-vulnerable urban areas—contingent on rigorous ecological assessments—combined with stringent controls on anthropogenic precursors (e.g., anthropogenic volatile organic compounds (AVOCs)) could serve as a complementary measure to mitigate secondary pollution.

Keywords: biogenic volatile organic compounds (BVOCs); air pollution; ozone (O_3); secondary organic aerosol (SOA); sampling analysis; reaction mechanisms

1. Introduction

As public awareness of environmental protection increases, attention to environmental issues has intensified, particularly concerning air pollution, which directly affects human health. Among various air pollutants, volatile organic compounds (VOCs) stand out not

 https://doi.org/10.3390/toxics13050364

only for their potential toxicity but also for their propensity to react with other pollutants, leading to secondary pollution. In 1990, the U.S. enacted the Clean Air Act Amendments, which listed 189 hazardous air pollutants (HAPs), including anthropogenic volatile organic compounds (AVOCs) such as benzene and toluene. The subsequent European Union Industrial Emissions Directive (2010/75/EU) also stipulates that industries emitting VOCs must implement Best Available Technology (BAT) to achieve a 60–90% reduction rate. And China also began to regulate VOCs at the beginning of the 21st century with the promulgation of the Volatile Organic Compounds (VOCs) Emission Control Standard. However, significant challenges remain for VOCs in non-regulated areas and emerging emission sources. Notably, biogenic volatile organic compounds (BVOCs) account for 90% of global VOC emissions [1–3]. The primary source of global BVOCs is plants, which naturally emit these compounds. However, their photochemical derivatives, like ozone (O_3) and secondary organic aerosols (SOAs), may pose environmental and health risks in regions with elevated anthropogenic influences [4]. Although emissions from individual plants are relatively minor, forests, grasslands, wetlands, and agricultural lands—covering 31% of the Earth's land surface—substantially amplify BVOC emissions [5–7]. Therefore, understanding the emission mechanisms of BVOCs and their synergistic effects with anthropogenic pollutants is critical for developing integrated pollution prevention and control strategies.

BVOCs are the main precursors of ozone (O_3), accounting for 20% of total O_3 production [8]. In addition, due to the photosynthetic characteristics of plants, the photochemical reaction activity of plant-generated BVOCs is significantly higher than that of anthropogenic volatile organic compounds (AVOCs) [9,10]. Under sunlight, BVOCs readily undergo photochemical reactions, contributing to the formation of biogenic secondary organic aerosols (BSOAs), which play a significant role in the global secondary organic aerosol (SOA) budget [11–14]. Although BVOCs are naturally emitted and not directly anthropogenic, their photochemical derivatives—tropospheric ozone (O_3) and secondary organic aerosol (SOA)—are recognized as hazardous pollutants under elevated anthropogenic influences. Furthermore, a comprehensive understanding of BVOCs is vital for advancing insights into Earth's ecological and atmospheric systems.

The historical evolution of BVOCs reveals significant progress driven by advancements in mechanistic studies and measurement techniques. In the 1950s, Sanadze [15] first observed the emission of gaseous organic substances from plants, initially identifying isoprene as a key compound [16]. With the application of gas chromatography, it became evident that BVOCs include a diverse array of around 30,000 hydrocarbons and oxides, including isoprene (C_5H_8), monoterpenes ($C_{10}H_{16}$), sesquiterpenes ($C_{15}H_{24}$), and other BVOCs [17–19]. These compounds exhibit diverse chemical structures and properties, necessitating specialized detection methods for their identification and quantification. By gas chromatography–mass spectrometry (GC-MS) technology, Kesselmeier and Staudt [20] provided insights into BVOCs biosynthesis, emission inventories, and the relationship between emissions, plant physiology, temperature, radiation, and ecophysiological functions. Helmig et al. [21] further advanced BVOCs detection by employing thermal desorption–gas chromatography–mass spectrometry (TDS-GC-MS), which enabled the analysis of a broader range of polar compounds in BVOCs and facilitated deeper mechanistic studies of BVOCs.

Despite these advances, challenges still persist in detecting and identifying the types of BVOCs from plants. Atmospheric concentrations of BVOCs from plants are typically very low, and their high reactivity leads to rapid chemical transformation or adsorption within sampling equipment, complicating accurate measurement. Additionally, temporal and spatial variations in BVOC emissions—driven by factors such as industrial activity, regional environmental differences, and seasonal climate changes—introduce further complexities [22]. On larger scales, studies by Papiez et al. [23] and Kramshøj et al. [24] revealed that cumulative emissions of isoprene, monoterpenes, and other BVOCs, amplified by anthropogenic climate perturbations, contribute to atmospheric processes (e.g., O_3 formation) and carbon–climate feedback, thereby indirectly influencing climate warming trends.

Recent advancements in computational methodologies have revolutionized the study of BVOC emissions from vegetation. Research paradigms have transitioned from foundational investigations into biochemical mechanisms and empirical sampling analyses to the incorporation of advanced machine learning architectures. These computational innovations facilitate the predictive modeling of BVOCs' multifactorial impacts, including climate adaptation dynamics, O_3 generation pathways, and SOA formation kinetics. Guenther et al. [3,25,26] pioneered BVOC emissions modeling, developing global models that laid the foundation for quantifying BVOC emissions. On this basis, Pierce [27] studied the aerosol impacts on the climate systems by the improved atmospheric chemistry models. More recently, Ng et al. [28] explored interactions between nitrate radicals ($\bullet NO_3$) and BVOCs during SOA formation and highlighted the importance of such interactions in improving atmospheric model predictions. Tang et al. [29] combined dynamic vegetation models, such as LPJ-GUESS v4.1, with global chemical transport models like TM5-MP, demonstrating that the spatial pattern of the BVOCs-driven SOA impact was strongly influenced by climate-induced vegetation changes.

Now, the research on BVOCs is no longer limited to a single discipline. Studies by Lun et al. [30] and Yang et al. [31] have contributed significantly to initiating BVOCs research by reviewing sampling and analysis methods, emission inventories, and emission factors. As disciplines such as chemistry, biology, mathematics, and computer science intersect with environmental science, new perspectives in BVOCs from plant research are emerging. This paper reviews recent studies on BVOCs emission sources, chemical composition, and reaction mechanisms; discusses their environmental impact factors; and evaluates effective mitigation strategies. Additionally, it highlights future interdisciplinary research trends, providing valuable insights for addressing BVOCs-related atmospheric pollution.

2. Methodology

To comprehensively review and summarize the research progress of BVOCs, this study adopted an interdisciplinary narrative review framework (Figure 1). A systematic literature search was conducted across multiple databases, including Web of Science, Elsevier ScienceDirect, and the China National Knowledge Infrastructure (CNKI). Keywords such as "BVOCs", "BVOCs emission", and "atmospheric oxidation" were employed for thematic searches, yielding an initial collection of 2483 articles. Following the screening and analysis of search results, duplicate records were removed, resulting in 2152 articles selected for detailed evaluation. Subsequently, 654 articles unrelated to the research scope were excluded based on titles and abstracts. Finally, through full-text precision assessments, 991 articles were further excluded due to data redundancy, incomplete methodological descriptions, or lack of environmental mechanism relevance, ultimately retaining 507 research articles for in-depth analysis.

As shown in Figure 2, the reviewed literature comprises 132 domestic Chinese articles and 375 international articles, spanning from 1999 to 2024. The earliest relevant articles originated in the United States, followed by contributions from South Korea. Between 1999 and 2009, only 28 articles were published. Since 2010, however, the number of publications has steadily increased, reaching an explosion in 2020 and reaching its peak in 2024. This trend highlights the growing global interest in BVOCs research, particularly in understanding their atmospheric interactions and environmental impacts.

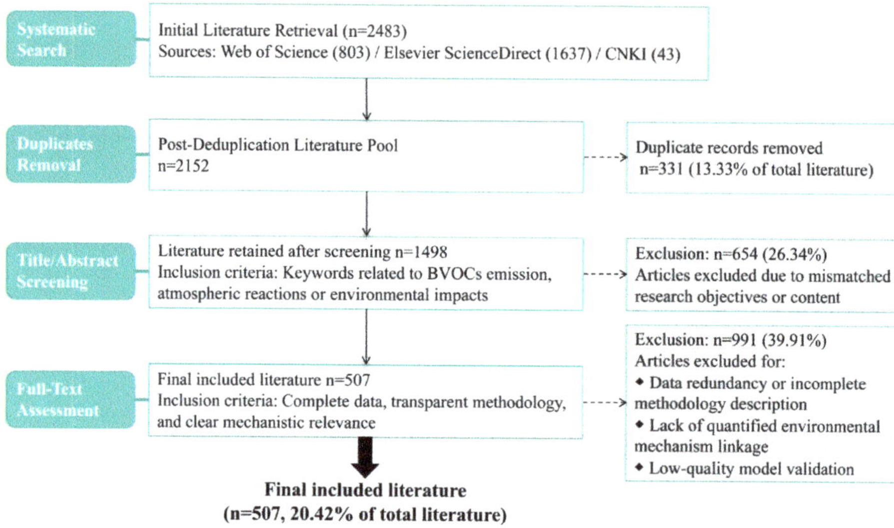

Figure 1. Narrative review flow diagram.

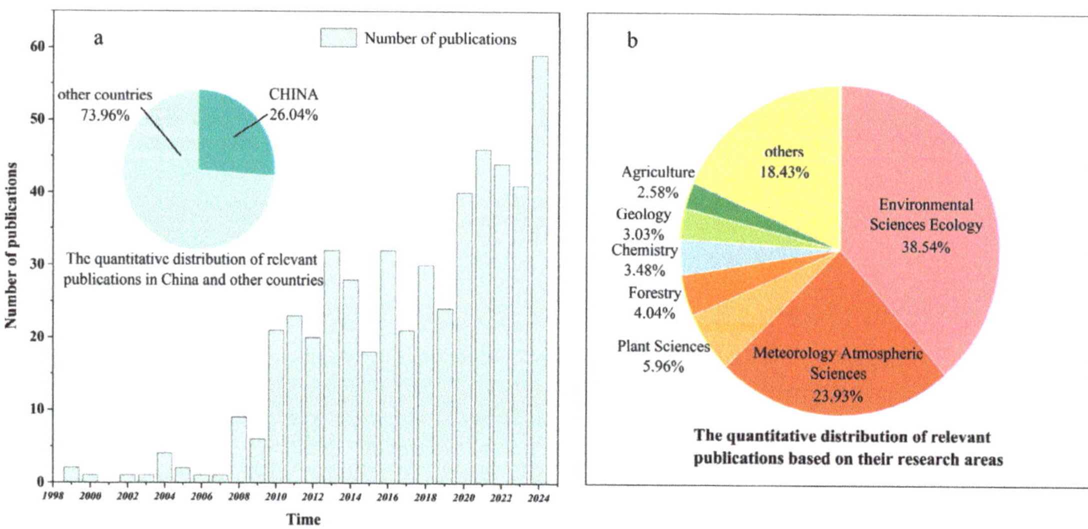

Figure 2. Distribution of 507 studies on biogenic volatile organic compounds (BVOCs) since 1999. (**a**) The change in the number of publications related to BVOCs in the world and the proportion of the number of articles published in China and other countries. (**b**) The quantitative distribution of relevant publications based on research areas.

3. BVOC Emissions Characterization Methods

3.1. Methods of Emission Measurement

3.1.1. Single-Plant Scale Sampling Methods

To comprehensively understand the impact of BVOC emissions on atmospheric chemistry and the environment, it is essential to accurately monitor their emission flux. Currently, emission measurements of BVOCs from a single plant focus on the leaf scale, and the main methods are divided into on-line and off-line techniques (Table 1), with each category encompassing various measurement techniques and technologies, which involve analytical techniques as shown in Table 2.

Table 1. Comprehensive comparisons of biogenic volatile organic compounds (BVOCs) emission measurement methods.

Measurement Methods	On-Line	Off-Line	
Main technologies	Direct analysis: PTR-MS [32] FTIR [33] GC-FID [34] Portable GC/MS [35] E-noses [36]	Sampling: sampling bags, adsorption tubes, sampling tanks	Analysis: GC-MS [37] TDS-GC/MS [38] TCT-GC/MS [39] TD-GC/TOF-MS [40] SPME-GC/MS [41]
Advantages	Real-time monitoring; high temporal resolution; high sensitivity; high degree of automation	Lower equipment costs; high flexibility; diverse and detailed analysis	
Disadvantages	Expensive equipment; complexity; high on-site requirements	Time delay; lower temporal resolution; human error	
Comprehensive comparison	• Precision and sensitivity: on-line > off-line • Operational complexity: on-line > off-line		

Table 2. Biogenic volatile organic compounds (BVOCs) analytical techniques comparison (notes: Ref., references).

Analysis Technique	Detector Type	Sensitivity	Time Resolution	Advantages and Disadvantages	Applicable BVOC Types	Ref.
PTR-MS	Mass spectrometry	ppt- ppm	Real-time analysis	Sensitivity and cost	Most BVOCs	[42,43]
FTIR	Pyroelectric detector or photoconductive detector	ppb- ppm	Minutes to hours	Versatility and sensitivity	Polar BVOCs	[44,45]
GC-FID	Flame ionization detection	ppt	Minutes	High speed and limitations	Nonpolar BVOCs	[34,46]
E-noses	Gas sensors	ppb- ppm	Minutes	Speed and resolution	Trace BVOCs	[36,47]
Portable GC/MS	Mass spectrometer	ppb	Minutes	Portability and limitations	Most BVOCs	[48,49]
GC-MS	Mass spectrometer	ppt	Minutes to hours	Precision and complexity	Most BVOCs	[37,50]
TD-GC/MS	Mass spectrometer	ppb	Minutes	Sensitivity and time-consuming	Trace BVOCs	[38,51]
TCT-GC/MS	Mass spectrometer	ppb	Minutes	Precision and requirements	Complex BVOCs	[39,52]
TD-GC/TOF-MS	Time-of-flight mass spectrometer	ppb	Minutes	Resolution and cost	Trace BVOCs	[40,53]
SPME-GC/MS	Mass spectrometer	ppb	Minutes	Simplicity and sensitivity	Most BVOCs	[41,54]

(1) On-line measurement

On-line measurement methods are frequently employed for monitoring environmental conditions in the field. On-line detection can analyze the emission concentration of BVOCs in the atmosphere and their temporal variations in real time. This method typically involves direct air sampling and immediate analysis using advanced instruments [55–57].

Representative on-line measurement methods include proton transfer reaction–mass spectrometry (PTR-MS) [58,59], Fourier transform infrared spectroscopy (FTIR) [60,61], portable gas chromatography–flame ionization detector (GC-FID), and portable GC-MS [62,63]. PTR-MS is particularly useful for trace gas analysis and is widely used in outdoor atmospheric measurements. To study the response of temperate deciduous tree species to flooding stress, Copolovici et al. [64] used proton transfer reaction–mass spectrometer (PTR-MS, high sensitivity version) to measure VOC concentration. Gómez et al. [34] routinely determined VOCs and qualitatively identified monoterpene compounds in ambient air using a gas chromatography system with GC-FID. You et al. [65] upgraded the traditional portable gas chromatograph and used it for continuous and on-site determination of trace volatile organic compounds, especially aromatic volatile organic compounds, in air samples.

The FTIR device can identify and quantify multiple BVOCs in air by measuring the infrared absorption spectrum of the gas without sample pretreatment and can simultaneously detect multiple compounds. Kalalian et al. [35] used FTIR and solid-phase microextraction sampling combined with a gas chromatography–mass spectrometry (SPME-GC/MS) in situ analysis of the reaction products of O_3 and three different BVOCs to predict the formation of photooxidants such as SOA. In order to check the health of plants, Sharma et al. [66] developed and applied a fully automated portable gas phase color spectrum for a rapid and in situ analysis of BVOCs from plants. Cicciol et al. [67] measured the vertical fluxes of volatile organic compounds from vegetation in a Mediterranean forested area using PTR-MS.

In addition to these methods, electronic noses (E-noses) represent an emerging technology for real-time environmental monitoring. E-noses offer high sensitivity and rapid response and eliminate the need for the complex pretreatment processes associated with gas chromatography sensors [36,68]. Feng et al. [69] used E-noses, headspace solid-phase microextraction–gas chromatography–mass spectrometry, and headspace–gas chromatography–ion mobility spectrometry to identify and characterize volatile organic compounds in eight kinds of *Zanthoxylum bungeanum* with geographical indications in China, providing a useful basis for the study of the differences in floral aroma. However, while E-noses are advantageous for qualitative and semi-quantitative analysis, the high complexity of environmental backgrounds can complicate the practical application of the E-noses, and their analytical accuracy and repeatability remain limited.

(2) Off-line measurement

Off-line measurement methods involve air sample collection and laboratory analysis. These methods are advantageous for their simplicity, high accuracy, and high sensitivity, but they lack real-time data.

(1) Sample collection

Off-line measurement methods separate sampling from analysis, where samples are collected on-site and then transported to a laboratory for analysis. Sample preservation methods for the different biogenic volatile organic compound (BVOC) collection containers are shown in Table 3. These methods are simple to operate, highly accurate, and sensitive. However, because samples need to be transported, they cannot reflect real-time data.

Typical collection containers for BVOCs include sampling bags (such as Tedlar® Bags and Teflon Bags), adsorption tubes (such as Tenax® Tubes and Activated Charcoal Tubes), and sampling canisters (such as SUMMA Canisters, Stainless Steel Canisters, and Silonite Canisters). Of these, the most commonly used containers for ambient air sampling are SUMMA Canisters, which have special treatments such as silanization or fused silica coatings that minimize the adsorption or reaction of VOCs on the walls of the canisters, thus maintaining the chemical stability of the sample. Sampling methods mainly include static closed sampling and dynamic headspace sampling. Static closed sampling, a classic method for calculating gas emission rates, involves measuring the concentration of BVOCs in a chamber or bag containing the plant species over time to determine the emission rate. Lin et al. [70] demonstrated that static enclosure techniques and in situ sampling can accurately capture BVOC emissions from entire leaves, which can then be estimated using remote sensing data. To address the limitations of static enclosure techniques, a semi-static method has been developed. Li et al. [71] conducted repeated experiments to evaluate the performance of the semi-static method for measuring BVOC emissions, confirming the feasibility of semi-static closed techniques as a screening tool for characterizing monoterpene emissions.

While static methods allow for rapid measurements and lower detection limits, they are susceptible to environmental conditions; factors such as temperature, pressure, and humidity can easily alter the equilibrium of BVOCs between the sample and the gas phase [72]. To address these limitations and meet the measurement requirements of various different scenarios, researchers have developed a dynamic headspace sampling method. This dynamic system features air circulation capabilities, effectively reducing discrepancies in environmental parameters inside and outside the system, enabling comprehensive environmental control for accurate BVOCs emission measurements. Vedel-Petersen et al. [73] utilized dynamic enclosure techniques to collect BVOC emissions from Arctic plants; this approach reduced environmental influence and allowed for the study of emission characteristics and seasonal variations in BVOCs. Jing et al. [74] also used dynamic headspace sampling to investigate BVOC emissions in northern China, concluding that different tree species emit different components at different rates, further demonstrating the advantages of dynamic headspace sampling. Dynamic headspace sampling is the most commonly used method for measuring BVOC emissions from living plants.

(2) Sample analysis

After field samples are collected, they are preserved and transported to the laboratory using specific methods to maintain sample integrity, as shown in Table 3. Once in the laboratory, these samples must typically be analyzed using various detection instruments within a specific timeframe to ensure accurate results.

At present, the most widely used separation and identification instrument is GC-MS; TDS-GC-MS is often used in the qualitative analysis study of BVOC emissions. Yuan et al. [75] quantified isoprene emissions from young hybrid poplar trees using TDS-GC-MS, and for the first time, investigated the combined effects of O_3 and drought on isoprene emissions. In order to detect the main BVOC species emitted by Mediterranean plant species, Fattobene et al. [38] employed thermal desorption gas chromatography–mass spectroscopy (TD-GC-MS) to analyze the air on-site sampled in the field.

Based on GC-MS analysis methods, researchers have also improved and developed automatic thermal desorption–gas chromatography–mass spectrometry (TCT-GC-MS), thermal desorption gas chromatography–time-of-flight mass spectrometry (TD-GC/TOF-MS) [40], etc.

Typical collection containers for BVOCs include sampling bags (such as Tedlar® Bags and Teflon Bags), adsorption tubes (such as Tenax® Tubes and Activated Charcoal Tubes), and sampling canisters (such as SUMMA Canisters, Stainless Steel Canisters, and Silonite Canisters).

Table 3. Sample preservation methods for different biogenic volatile organic compound (BVOC) collection containers.

Collection Containers	Pretreatment Requirements	Storage Conditions	Maximum Preservation Time	Applicable Compound Range	Ref.
Tedlar Bags	Nitrogen flushed 3 times, light-proof aluminum foil wrapped	4 °C, <30% humidity	48 h	Nonpolar terpenes (C_5-C_{10})	[76,77]
Tenax Tubes	Aging (320 °C, 2 h), vacuum sealed in aluminum foil pouch	−20 °C with desiccant	7–30 days	Semi-volatile VOCs (C_{10}-C_{15})	[78,79]
Activated Charcoal Tubes	Aged 1 h at 400 °C, sealed in moisture-proof container	4 °C, avoid condensation	14 days	Highly volatile VOCs (C_2-C_8)	[80,81]
SUMMA Canisters	Vacuum baked at 160 °C for 12 h, nitrogen filled to 30 psi	Protected from light at room temperature, pressure monitoring	30 days	Broad spectrum VOCs (C_2-C_{15})	[77,82]

3.1.2. Canopy Scale Flux Measurement Methods

The emissions of BVOCs exhibit high spatiotemporal heterogeneity. Traditional leaf-scale studies are constrained by the limited spatial representativeness of sampling methods. However, canopy-scale flux measurement methods such as Eddy Covariance (EC) and Relaxed Eddy Accumulation (REA), which directly capture integrated canopy-level fluxes, significantly enhance spatial representativeness, temporal continuity, and data comprehensiveness [83,84]. While early EC technology provides direct turbulent flux data, its reliability for BVOCs—due to their ultra-low concentrations and adsorption losses—requires advanced sensors like PTR-MS [85]. For example, Spirig et al. [86] combined EC and PTR-MS to measure BVOCs canopy fluxes in a mixed deciduous forest in northwestern Germany, validating EC's reliability under complex heterogeneous canopy conditions, with isoprene and monoterpenes identified as dominant emissions. Misztal et al. [87] employed EC-PTR-MS to directly quantify BVOC fluxes in Southeast Asian oil palm ecosystems, revealing isoprene's predominant role (84% of midday fluxes) and significant contributions from floral emissions. To enhance signal-to-noise ratios, REA integrates airflow separation and offline analysis (e.g., GC-MS) for trace BVOC detection. Bai et al. [88] applied REA and gradient techniques to conduct long-term continuous canopy flux measurements of BVOCs in a subtropical pine plantation, uncovering pronounced diurnal, seasonal, and interannual variations in isoprene and monoterpene emissions.

Subsequent advancements include Disjunct Eddy Covariance (DEC), which combines EC's turbulence-phase data with REA's intermittent sampling, broadening EC's applicability [89–91]. Emerging methods like Disjunct Eddy Accumulation (DEA) leverage ultra-low-frequency sampling and automated accumulation to reduce power consumption and costs, enabling long-term field observations [92]. Current trends focus on integrating multiple methods and utilizing Unmanned Aerial Vehicle (UAV) platforms to expand

3D flux monitoring capabilities, thereby comprehensively elucidating the spatiotemporal patterns of BVOC emissions and their impacts on atmospheric chemistry [93–95].

3.2. Model-Based Estimation of BVOCs Emission

Although canopy-scale flux measurement techniques provide high-precision real-time flux data through direct observation, effectively reflecting plant responses to environmental changes, this method has limitations in spatial coverage and long-term monitoring, making it unable to comprehensively capture the dynamic variations in BVOCs.

Emission models bridge critical knowledge gaps by synthesizing empirical observations with environmental drivers (e.g., meteorology and land use patterns), enabling dynamic predictions of BVOC fluxes across heterogeneous ecosystems and under varying climatic scenarios. These tools are particularly vital for scaling plot-level measurements to regional or global assessments, where direct monitoring becomes logistically impractical. The development of BVOC emission models serves a dual purpose: (1) to extrapolate mechanistic insights from laboratory and field studies into predictive algorithms, and (2) to inform policy-relevant scenarios such as air quality management and climate mitigation strategies.

Thus, in the study of BVOC emissions in a certain area and at a large scale, the model-based estimation method is mainly used [96,97]. The model simulation method is based on remote sensing data to obtain the regional leaf area index and leaf biomass, combined with emission factors, meteorological data, and other information to use model simulation to obtain the total emission of plant-derived volatile organic pollutants, including the Guenther model (Guenther), model of emissions of gases and aerosol from nature (MEGAN), biological source emission inventory system model (BEIS), and so on. Their main characteristics and applicable scenarios are summarized in Figure 3. In recent years, the various models have been continuously upgraded and developed and are widely used in regional and global BVOC emission research.

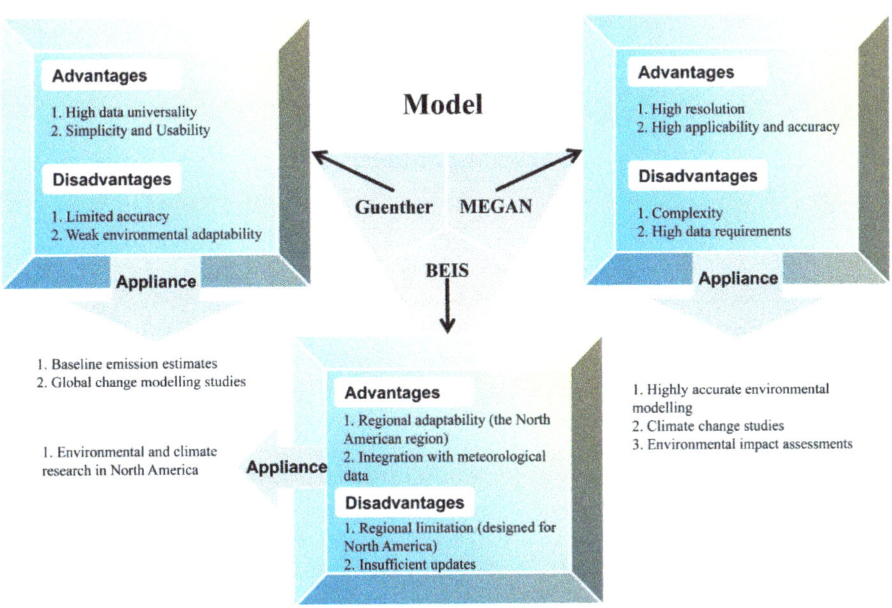

Figure 3. The characteristics of different model simulation methods of biogenic volatile organic compounds (BVOCs) and their applicable scenarios.

3.2.1. Guenther

The isoprene emission model-G91, constructed by Guenther et al. [26] based on the photosynthetic electron transport model developed by Farquhar and the physiological characteristics of plants, provides an important framework for estimating inventories of biogenic non-methane hydrocarbon emission rates. In order to simulate the effects of light and leaf temperature on the emission of plant isoprene and monoterpenes, Guenther et al. [25] developed G93 on the basis of G91. Subsequently, based on the G91 and G93 algorithms, Guenther et al. [3] developed a global natural VOCs emission model, G95, based on relevant data, which has a high-resolution spatial grid to estimate hourly averaged emissions of organic compounds from natural sources. Chi et al. [98] used the Guenther model to convert vegetation volume and yield data, such as the forest and grassland resource inventory and vegetation maps, into leaf biomass data to estimate the space–time and source distribution of the emissions of BVOCs in China, and established a BVOC emissions inventory in mainland China. In order to explore the emission of BVOCs in Qinghai Province and its impact on air quality, Wang et al. [99] used the G95 method to estimate the total emissions of isoprene, monoterpenes, and other VOCs emitted by vegetation in Qinghai Province in 2021.

3.2.2. BEIS

Pierce et al. [27] established a biological emission inventory system model-BEIS1 based on the data of leaf biomass, land use, emission factors, and meteorological parameters obtained by observation or calculation to estimate VOCs emissions in vegetation and NO emissions in soil. Subsequently, Pierce et al. [100] developed the second version of the bio-emission inventory system BEIS2 on the basis of BEIS1. BEIS2 updated the land use inventory, emission factors, and environmental correction formulas following the G95 algorithm. Considering the factors such as plant growth environment, the National Center for Atmospheric Research (NCAR) and the United States Environmental Protection Agency (EPA) developed the global biosphere emission and interaction system—GloBEIS [30]. With the development of geographic information system technology, researchers have successively updated and developed BEIS3, BEIS4, etc. Feldman and Zheng et al. [101,102] used vegetation data and the GloBEIS model to estimate the total emission of plant-derived volatile organic compounds in southeastern Texas of the United States and in the Pearl River Delta of China.

3.2.3. MEGAN

Based on the G95 model and multi-source information such as satellite and ground observation data, Guenther et al. [2] constructed the natural gas and aerosol emission model, MEGAN, to estimate global terrestrial isoprene emissions. This model is suitable for regional- and global-scale chemical and transport modeling. With the deepening of research, researchers have successively developed MEGANv2.02 [103], MEGANv2.1 [104], MEGANv3.0 [105], and so on.

Sindelarova et al. [106] used the MEGANv2.1 model and MERRA meteorological field to establish a monthly global BVOC emissions dataset from 1980 to 2010. The model estimated the average annual BVOC emissions to be 760 Tg (C) yr^{-1}. Using the improved MEGAN-BEIS361 model combination based on the MEGAN model, Wang et al. [107] simulated and predicted isoprene emissions and environmental concentrations in the United States. Jiang et al. [108] simulated biogenic isoprene emissions using the MEGAN3 response to drought and its impact on global atmospheric chemistry. Compared to the Guenther model and BEIS models, the MEGAN model is more widely used and is one of the commonly used models.

The operation of these three models requires various types of remote sensing data, meteorological data, and emission factors based on laboratory or field observations. They are widely used to obtain BVOC emissions across different temporal and spatial scales, providing effective methods for large-scale BVOC research. In recent years, building upon these three major models, researchers have successively developed new models such as the empirical model of BVOC emissions (EMBE) [109] and the Plant Specific Emission Model (PSEM) [67], offering new approaches for BVOC emission studies.

4. Emission Mechanisms of BVOCs from Plants

4.1. Emission Sources

The sources of BVOCs are highly diverse, both in urban and rural environments. In urban areas, both natural vegetation remnants (e.g., preserved woodlands) and cultivated green infrastructure (e.g., parks and street trees) serve as dominant sources of BVOCs. Notably, in certain cities, the spatial coverage of landscaped vegetation surpasses that of natural ecosystems, which may generate a distinct urban BVOCs signature compared to natural landscapes. Furthermore, the synergistic effects of environmental stressors—particularly the urban heat island effect and industrial pollutants—alter the compositional characteristics (e.g., shifts in isoprene-to-monoterpene ratios) and elevate the emission fluxes of plant-derived BVOCs, thereby intensifying their role in urban atmospheric chemistry [110,111]. Jing et al. [74] investigated BVOC emissions from dominant forest vegetation species in Beijing, finding that urban forest trees primarily emit isoprene and monoterpenes. Bao et al. [112] estimated BVOC emissions from 23 urban plants in Shenzhen and discovered that BVOC emissions from urban green spaces account for approximately 20.07% of the total BVOCs in the study region, with their estimated emissions in built-up areas (1.65 kt) nearly doubling conventional estimates (0.86 kt). This result challenges the historical marginalization of urban vegetation in BVOC emission inventories.

In rural areas, natural forest vegetation, agricultural crops, and wetlands are primary BVOC sources. Forests, in particular, are the largest contributors, as their emissions are highly influenced by natural environmental conditions [104,113,114]. Smiatek and Steinbrecher [115] studied the temporal and spatial variation in VOC emissions from German forests and found that the average annual BVOC emissions from forests during 1994–2003 were 366 Gg. Liu et al. [116] identified isoprene and monoterpenes as the primary emissions from pine and birch forests, with these compounds presenting the highest BVOC mixing ratios in forest ecosystems. Bamboo forests are also a major BVOC source [117]. Okumura et al. [118] found that 86% of bamboo species emit isoprene at rates ranging from 0.7 to 99.1 nmol m^{-2} s^{-1}, with *Phyllostachys glauca*, *Semiarundinaria fastuosa*, and *Bambusa pervariabilis* being prominent sources.

Wetland plants, such as *Phragmites australis*, *Typha angustifolia*, *Iris pseudacorus*, and *Cyperus rotundus*, also produce BVOCs [119,120]. In agricultural landscapes, crops such as wheat, corn, and canola can release BVOCs, with corn and canola emitting methanol and acetaldehyde [121,122]. Additionally, microbial activity in soil produces small amounts of BVOCs, which, while minor in magnitude, are significant on a global scale [123–126]. Wester-Larsen et al. [127] reported variation in soil BVOC (SBVOC) concentrations based on different vegetation types, with areas of high vegetation cover and biodiversity, such as the Amazon rainforest and tropical forests in Southeast Asia, exhibiting higher BVOC emissions [128–130].

In both urban and rural areas, human activities have solidified the plant species in those regions (landscape plants in urban areas and economic crops in rural areas). If the emission sources of BVOCs in human-affected areas are not taken into account and higher precision emission estimates are not employed, there will be significant deviations in the understanding of BVOC emission characteristics. Therefore, studies on BVOC emissions require regional representation, with global research necessitating multi-regional composite studies that account for vegetation cover and topography. This paper primarily focuses on plant-derived BVOC emissions.

4.2. Compositions

BVOCs are broadly categorized into four main groups: isoprene, monoterpenes, sesquiterpenes, and oxygenated biogenic volatile organic compounds (OBVOCs), which include alcohols, aldehydes, ketones, and esters (Table 4). Among these, isoprenes and monoterpenes dominate, accounting for over 70% of the total annual BVOC emissions, particularly during the summer in temperate forests [131]. Liu et al. [132] revealed that isoprene emissions from *Ficus virens* could account for up to 85% of the total plant emissions. Similarly, monoterpene from *Camphora officinarum* contributes over 75% to total emissions and is highly reactive in forming O_3 and SOA. To further illustrate the variability in BVOC emissions and concentrations, we present Table 5, which provides a comparative analysis of the mean concentrations of key BVOC components, such as isoprene and monoterpenes, across different study areas and climatic conditions [133].

Although sesquiterpenes and oxygenated biogenic volatile organic compounds (OBVOCs) are emitted in smaller quantities, they still significantly influence atmospheric chemistry. Liu et al. [134] confirmed that 2-methyl-3-buten-2-ol and methanol from pine forests undergo rapid oxidation, leading to SOA formation. While sesquiterpenes are emitted in lower amounts compared to isoprene and monoterpenes, they dominate in specific plant species such as *Empetrum hermaphroditum* and *Cassiope tetragona* [135]. Sesquiterpenes exhibit high oxidation reactivity with hydroxyl radicals (OH) and ozone (O_3), resulting in rapid degradation and short atmospheric residence times (minutes to hours). Nevertheless, their oxidation products are important precursors of secondary organic aerosols (SOAs) [136–138]. Similarly, OBVOCs, such as esters and ketones, have short atmospheric lifetimes. For instance, certain ketones (e.g., methyl vinyl ketone) exhibit lifetimes of less than 1 h, where their low persistence is governed more by photolysis than by "lower chemical reactivity". This limits their direct contribution to ozone formation but may enhance nighttime nitrate radical (NO_3) chemistry [139–142]. The rapid dissociation of various BVOCs, including isoprene [143], α-pinene [144], β-pinene [145], d-limonene [146], and linalool [147], plays a critical role in O_3 and SOA formation.

Table 4. Major categories of biogenic volatile organic compounds (BVOCs) and their global average emission values (notes: Ref., references).

BVOCs Categories	Synthesis Method	Main Emission Plants Category	Contribution Rate	Global c (Tg C yr^{-1})	Ref.
1. Isoprenoids	MEP	Broadleaf trees	40–60%	400–600	[148–150]
2. Monoterpenes	MVA	Coniferous trees	25–40%	40–180	[151,152]
3. Sesquiterpenes	MEP and MVA	Tropical rainforest plants Herbaceous plants	5–15%	20–40	[153–155]
4. OBVOCs	LOX/MEP/MVA	Herbaceous plants; trees; grasses crops; flowers	10–25%	–280	[156,157]

Table 5. Mean concentrations of key biogenic volatile organic compounds (BVOCs) in different study areas and climatic conditions (Ref: references; -: no data).

Study Area	Seasonal Mean Concentrations (μg/m^3) Spring	Summer	Autumn	Winter	Annual Mean Concentrations (μg/m^3)	Ref.
			Isoprene			
Urban area of Beijing	1.83	2.82	1.48	0.43	1.66	[133]
Suburban areas of Tianjin	-	-	1.61	-	-	[158]
Urban area of Shenzhen	2.07	4.26	1.09-	1.06	-	[159]
Village of Guangzhou	6.69	-	4.26	0.40	3.41	[160]
Urban area of Shanghai	-	-	-	-	0.10	[161]
Urban area of Houston	-	-	-	-	1.92	[162]
Urban area of Lille	-	0.88	-	0.33	0.58	[162]
Zurich	-	0.49	-	0.24	-	[162]
			α-Pinene			
Typical urban area of Beijing	0.05	0.11	0.18	0.04	0.10	[133]
Mountainous areas of Sichuan	-	-	-	-	0.30	[163]
Guangzhou Forest Park	-	1.12	-	-		[164]
			β-Pinene			
Typical urban area of Beijing	0.05	0.003	0.006	0.002	0.02	[133]
Mountainous areas of Sichuan	-	-	-	-	0.30	[163]

4.3. Synthetic Pathways

Biogenic volatile organic compounds (BVOCs) in plants originate from three principal biosynthetic routes: the methylerythritol phosphate (MEP) pathway localized in plastids, the mevalonate (MVA) pathway operating in cytoplasmic compartments, and the lipoxygenase (LOX) pathway associated with peroxisomal membranes [165–168]. The MEP pathway predominates in higher plant terpenoid biosynthesis, generating monoterpenoids and diterpenoids via sequential condensation of isopentenyl diphosphate (IPP) and dimethylallyl diphosphate (DMAPP) precursors synthesized from glyceraldehyde-3-phosphate and pyruvate [169–173]. Counterintuitive to previous assumptions, the MVA pathway—traditionally associated with fungal sesquiterpenogenesis—exhibits functional conservation in plant cytoplasmic metabolism [174,175]. This pathway converts acetyl-CoA through enzymatic cascades to yield IPP/DMAPP, primarily fueling sesquiterpene and phytosterol biosynthesis, with emerging evidence confirming its collaborative role with the MEP pathway in generating specific monoterpenes under stress conditions [176,177]. The LOX pathway initiates the oxidative cleavage of polyunsaturated fatty acids (e.g., linolenic acid), producing jasmonate phytohormones and green leaf volatiles that orchestrate systemic defense responses against biotic stressors [178–181]. Therefore, studies on the components and formation pathways of BVOCs should be analyzed on a case-by-case basis, and the synergistic effects of multiple factors may result in emission scenarios that are quite different from a single analysis.

5. Influencing Factors of BVOC Emissions

The release of BVOCs is a complex, multifactor process influenced by both intrinsic plant factors (plant types, physiological differences, and biotic stresses) and environmental factors (temperature, light, humidity, CO_2 concentration, and O_3 stress), which may affect the release rate and release characteristics of BVOCs in plants [182]. These factors and their influence on BVOC emissions are shown in Figure 4.

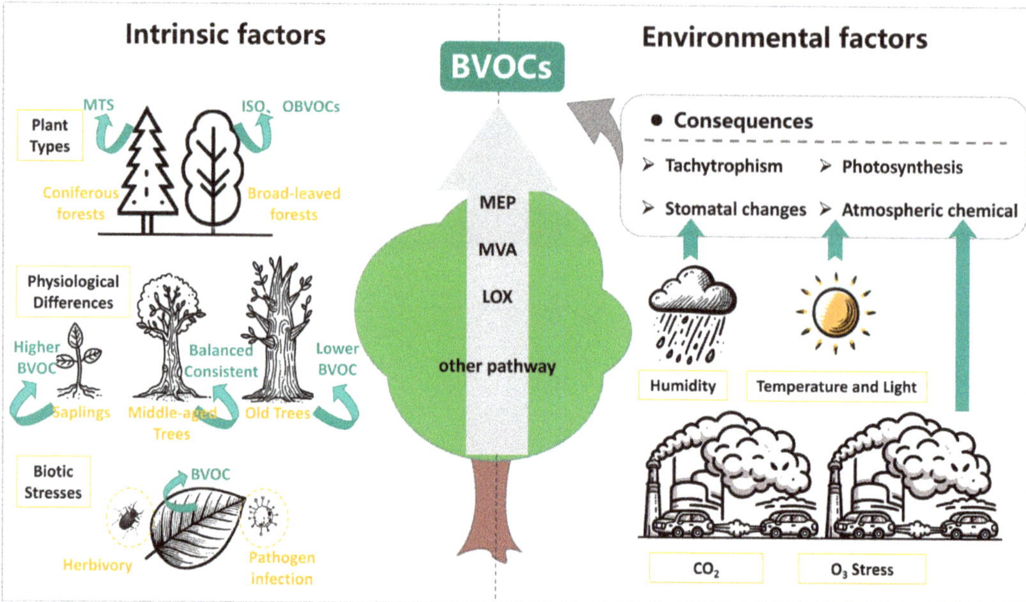

Figure 4. The effects of plant internal and environmental factors on biogenic volatile organic compounds (BVOCs)(notes: MTS, monoterpenes; ISO, isoprene; OBVOCs, oxygenated biogenic volatile organic compounds; MEP, Methylerythritol Phosphate Pathway; MVA, Mevalonate Pathway; LOX, Lipoxygenase Pathway; AVOCs, anthropogenic volatile organic compounds; CO_2, carbon dioxide; O_3, ozone).

5.1. Intrinsic Plant Factors Influencing BVOC Emissions

5.1.1. Plant Types

Plant species are a key determinant of BVOC synthesis and release due to the differences in genetic control mechanisms and metabolic pathways, and the composition and release rate of BVOCs vary from different vegetation types [99]. Coniferous forests and broadleaf forests are two typical plant types, and their emission characteristics of BVOCs are different [183,184]. Coniferous forests mainly release monoterpenes and sesquiterpenes that may have more significant impacts on SOAs and the global climate system [185,186], while broadleaf forests mainly release short-chain BVOCs such as isoprene that are highly volatile and mainly affect photochemical smog and local O_3 concentrations [187–189].

5.1.2. Physiological Differences

The physiological state and development stage of plants also play a critical role in BVOC emissions. Younger trees generally emit BVOCs at higher rates, while mature trees emit higher absolute quantities of these compounds. Schade et al. [190] and Street et al. [191] found that young pine branches and leaves release BVOCs at a rate 2–3 times higher than mature branches and leaves. Kuzma and Fall [192] and Rasulov et al. [193] observed lower isoprene emissions from young leaves to mature leaves both in *Vicia faba* and *Populus alba*, respectively.

Stomatal conductance is another critical physiological trait that influences BVOCs' synthesis and emission. As the primary pathway for BVOC emission, stomatal conductance affects the partitioning of BVOCs between the gas and liquid phases within leaves' release [194–197]. Leaf structure also plays a role in modulating the effect of stomata on VOC emission [198,199].

5.1.3. Biotic Stresses

Biotic stress, particularly pathogen infection and herbivore activity, can elevate BVOC emissions by 2–10-fold compared to unstressed plants, with emission profiles diverging based on stressor type (e.g., pathogen-induced oxygenated compounds vs. herbivore-induced terpenoids) [200,201]. Insect herbivory often increases BVOC emissions [202]; Joutsensaari et al. [203] noted that insect outbreaks could elevate BVOCs and potentially even BSOA emissions. However, severe pest or disease damage may impair physiological functions, such as reduced photosynthesis, resulting in a decrease in the overall production of BVOCs [204]. However, Litvak and Monson [205] found that while monoterpene synthesis in ponderosa pine increased following herbivore induction, the volatile losses from damage issues offset this increase, resulting in no significant net increase in the total monoterpene pool within the damaged needles.

5.2. *Environmental Factors*

5.2.1. Temperature and Light

Temperature and light are critical environmental factors influencing the emission of BVOCs, as they influence plant enzymatic activity and physiological processes [206]. Generally, increases in temperature and photosynthetically active radiation (PAR) activate the enzymes responsible for BVOCs synthesis [207]. For instance, Li et al. [208] found that high temperatures could triple the monoterpenes emission flux. However, extreme temperatures may disrupt metabolic activities, leading to a decrease in emission rates. Loreto et al. [209] found that the isoprene emission from reed leaves increases with temperatures until 45 °C, after which emissions decline.

Different BVOC components exhibit varying dependencies on temperature and light. The synthesis of isoprene, which primarily occurs in chloroplasts, is directly dependent on the production of glyceraldehyde-3-phosphate (G3P) and energy from photosynthesis. Consequently, isoprene emissions are most influenced by light exposure [210–212]. Seasonal and latitudinal variations also affect emissions, with higher light intensities in summer and lower latitudes leading to stronger BVOC emissions.

5.2.2. Humidity

Humidity indirectly influences BVOC emissions by influencing stomata conductance [213,214]. Higher humidity levels promote stomatal opening, facilitating BVOC emissions [215,216]. However, prolonged high humidity can reduce transpiration rates and induce stomatal closure, thereby reducing BVOC emissions [217]. BVOCs' responses to humidity also vary by components. For instance, Ning et al. [218] found a positive correlation between α-pinene emissions and humidity, while isoprene emissions exhibited a negative correlation.

5.2.3. Concentration of CO_2

Changes in carbon dioxide (CO_2) concentration influence photosynthesis and metabolic processes, subsequently affecting BVOC emissions. Elevated CO_2 concentrations can inhibit BVOC emissions by suppressing photosynthesis, as observed in the reductions in monoterpenes and sesquiterpenes found by Constable et al. [219]. However, certain plants may exhibit increased BVOC synthesis under high CO_2 concentrations, as found by Jasoni et al. [220] in onions. The duration of CO_2 exposure is also critical. Rapparini et al. [221] found that short-term exposure can significantly suppress the emission of isoprenoids and monoterpenes, whereas long-term exposure had minimal effects.

5.2.4. O_3 Stress

The concentration of BVOCs plays a significant role in the formation of O_3. Similarly, O_3, as a potent oxidant, can influence the release of BVOCs through a series of complex physiological, biochemical, and genetic regulatory mechanisms [222–225]. Under increased O_3 stress, plants may mitigate oxidative damage by increasing the release of BVOCs with strong antioxidant capacities. For example, isoprene emissions increase by 20–50% under acute O_3 exposure (100 ppb for 4 h), as demonstrated in Populus tremula fumigation experiments [187]. Llusià et al. [226] also found that plants subjected to O_3 fumigation exhibited increased net photosynthetic and total VOC emission rates. Li et al. [227] also discovered that exposure to O_3 pollution increased BVOC emissions in the Yellow River Delta and the Yangtze River Delta by 13.9% and 4.9%, respectively. However, plant responses to O_3 stress are also influenced by factors such as the exposure duration and the composition of BVOCs. Vo and Faiola [228] found that terpenoid emission rates decreased or remained unchanged after acute O_3 exposure, and that sesquiterpenes exhibited a significantly higher response concentration to O_3 than monoterpenes. Yuan et al. [75] found that under short-term high-concentration ozone stimulation, the emission of isoprene from leaves that were not severely damaged was promoted. However, prolonged exposure to high concentrations of O_3 significantly inhibited isoprene emissions, resulting in a reduction of 40.4%. Additionally, the decrease in isoprene emissions from middle-layer leaves (61.1%) was significantly greater than that from upper-layer leaves (20.9%). Furthermore, this study also found that isoprene emissions were unrelated to stomatal conductance. The interaction between BVOC emissions and O_3 is a dynamic process influenced by multiple factors. Understanding the impact of O_3 on BVOC emissions requires consideration of multiple intrinsic and environmental variables.

6. Atmospheric Oxidation Mechanisms of BVOCs

The release of BVOCs has a significant impact on atmospheric chemistry. Once released into the atmosphere, these BVOCs undergo a complex series of photochemical reactions that produce and affect the corresponding free radicals, which influence the formation of O_3 and SOA in the atmosphere. Understanding the role of BVOCs in atmospheric chemistry and their transformation processes (Figure 5) is critical for comprehending atmospheric chemistry, air quality, and climate change implications [229,230].

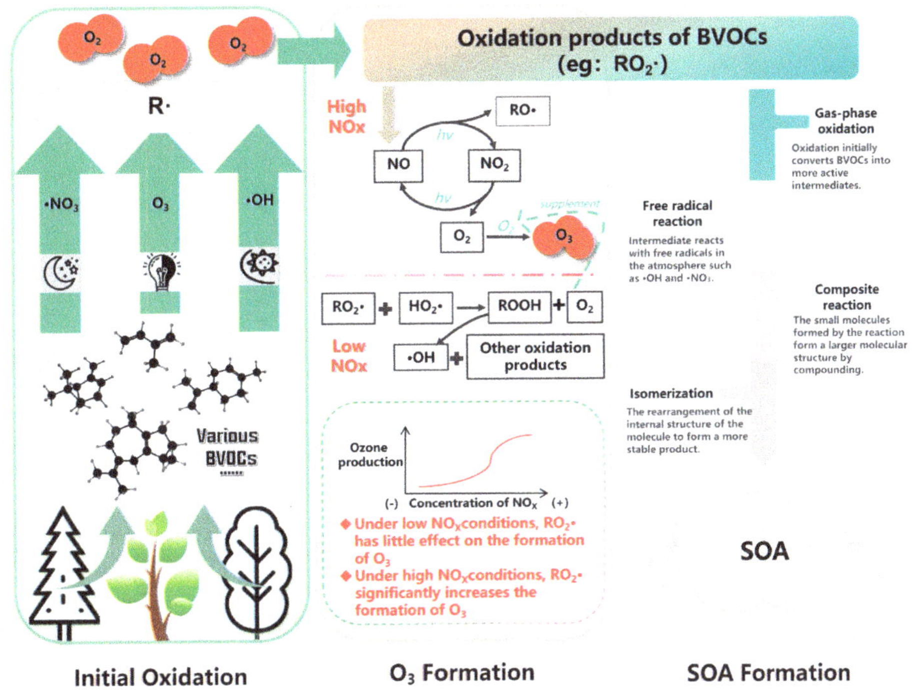

Figure 5. The initial oxidation reaction of biogenic volatile organic compounds (BVOCs) and the pathway diagram of their formation of ozone (O_3) and secondary organic aerosols (SOAs) (notes: for the corresponding abbreviations in the chart, please refer to nomenclature (Table S1)).

6.1. Initial Oxidation Reactions

The initial oxidation reaction pathways of BVOCs in the atmosphere include reactions with hydroxyl radicals (•OH), $•NO_3$, and O_3 [231–234].

The primary source of •OH in the troposphere is from the decomposition of O_3 under ultraviolet radiation, which generates excited oxygen atoms $O(^1D)$. This process results in high concentrations of •OH during the daytime [235–237]. Consequently, during the day when ultraviolet intensity is high, highly reactive BVOCs frequently undergo addition and oxidation reactions with •OH. The hydrogen atoms in carbon-based BVOCs are abstracted by •OH to produce alkyl radicals (R•) and H_2O [238,239]. At night, in the absence of light, the reactions between •OH and BVOCs are greatly inhibited, while $•NO_3$, having high stability, becomes the primary radical reacting with BVOCs during nighttime [240–242]. Mogensen [243], Xu [244], and Waring and Wells [245] corroborate this mechanism. Since the reaction between O_3 and BVOCs is not dependent on light and is highly reactive, at night, $•NO_3$ radicals (derived from O_3 + NO_2) are typically the primary oxidants for BVOCs in NO_x-rich environments. However, in regions with extremely low NO_x levels (e.g., pristine forests), the scarcity of $•NO_3$ shifts the dominant oxidation pathway to reactions between O_3 and unsaturated BVOCs (e.g., isoprene and monoterpenes), generating carbonyl compounds and sustaining nighttime radical chemistry [246–248]. Atkinson [249], Docherty [250], and Kroll [251] found that the oxidation of organic compounds by O_3 differs from radical oxidation because BVOCs can be cleaved by O_3 to form carbonyls and energetically excited carbonyl oxides. These intermediates then decompose via the hydroperoxide channel to produce •OH and R•. The organic peroxyl radicals (RO_2•) from the primary oxidation of BVOCs are part of the radical cycling that controls atmospheric oxidative capacity, leading to the subsequent formation of O_3 and SOA [252].

6.2. O_3 Formation

Numerous studies have demonstrated that BVOCs play a critical role in the formation of O_3 through photochemical reactions, and this role largely depends on the concentration of nitrogen oxides (NO_x) [231,253–255]. According to the studies by Xie [256] and Kim et al. [257], biogenic and anthropogenic VOCs, especially the photochemical reactions of isoprene, significantly increase the formation of local O_3.

Under high-NO_x environments, $RO_2\bullet$ reacts mainly with NO to produce new oxidation products ($RO\bullet$, NO_2). The newly formed NO_2 undergoes photolysis ($NO_2 + h\nu \rightarrow NO + O(^3P)$), followed by $O(^3P) + O_2 \rightarrow O_3$, while $RO\bullet$ decomposes to yield carbonyl compounds (e.g., HCHO) that further sustain radical cycling [258,259]. Simon et al. [260] found that to a certain extent, the more NO_x there is, the higher the rate of O_3 production, which is largely dependent on the NO_2/NO ratio and the amount of BVOCs available. Zhang et al. [261] found that when in high-NO_x conditions, the increase in BVOCs will effectively promote radical cycling and enhance the level of radicals, which to some extent may lead to an increase in O_3.

In contrast, the reaction of $RO_2\bullet$ with hydroperoxy radicals ($HO_2\bullet$) is more important in low NO_x concentration environments, where NO_x is not a major limiting factor. $RO_2\bullet$ reacts with $HO_2\bullet$ to produce peroxides (e.g., organic hydroperoxides (ROOH)), and these reactions do not usually result in the direct production of NO_2, and therefore contribute less to O_3 production. In this environment, the rate of O_3 production is limited by the concentration of VOCs and the rate of reaction [262,263]. Orlando and Tyndall [252] and Iyer et al. [264] found that under low NO_x conditions, $RO_2\bullet$ reacts mainly with $HO_2\bullet$, a reaction that is very important in the atmosphere and is the end of O_3 production, manifesting itself in the fact that BVOCs produce less O_3 or even consume O_3.

6.3. SOA Formation

BVOCs are crucial precursors for the formation of SOA [265–267]. SOA consists of fine particulate matter formed through a series of complex chemical transformations of BVOCs, significantly impacting atmospheric climate and human health [268,269]. Among these, isoprene, terpenes, and sesquiterpenes are the most common BVOC precursors [270,271].

The formation process of SOA involves multiple reaction steps. BVOCs first undergo gas-phase oxidation, generating $RO_2\bullet$. These radicals can react through different pathways to produce secondary products, such as reacting with NO to form NO_2 and $R\bullet$, reacting with $HO_2\bullet$ to produce ROOH, or reacting with other $RO_2\bullet$ to form peroxides (ROOR) and other organics [272,273]. These organics include low-volatility organic compounds (LVOCs) or extremely low-volatility organic compounds (ELVOCs) [274,275]. Concurrently, these LVOCs and ELVOCs can further grow into larger sizes through processes like complexation, isomerization, and condensation nucleation, ultimately forming SOAs [276,277]. This process is also influenced by environmental factors such as temperature, humidity, and NO_x concentrations [278,279]. BVOCs are transformed into SOA through complex chemical processes in the atmosphere, influenced by multiple environmental factors. BVOCs such as isoprene, terpenes, and sesquiterpenes undergo oxidation, radical reactions, condensation and adsorption complex reactions, and isomerization, forming SOAs through multiple steps. The formation and evolution of SOA not only affect the balance of atmospheric chemistry but also have significant implications for air quality and human health.

6.4. Synergistic Effects of BVOCs with AVOCs

Both AVOCs (such as NO_x and toluene) and BVOCs (such as isoprene and α-pinene) participate in atmospheric photochemical reactions that exacerbate O_3 production [280] and promote the condensation of ELVOCs into SOA [281]. In high-NO_x environments, the reactivity of BVOCs is enhanced; for instance, BVOCs accounted for 56% of emissions during summer nights in Taipei, thereby extending the photochemical chain when combined with AVOCs. In semi-urban areas of Turkey, the cumulative impact of traffic emissions (AVOCs at 16%) and vegetation releases (BVOCs at 38.6%) has resulted in peak summer O_3 levels, with both contributing synergistically to over 60% of ozone formation potential [282,283]. Li et al. [284] found, through analyzing results from different simulation scenarios, that AVOCs promote O_3 production by interacting with BVOCs, with the effect being most pronounced from noon to evening. Choi et al. [285] found, through combining Proton Transfer Reaction Time of Flight Mass Spectrometer (PTR-ToF-MS) and the Aerodyne High Resolution Time of Flight Aerosol Mass Spectrometer (HR-ToF-AMS) data, that BVOCs (e.g., monoterpenes) and AVOCs (e.g., toluene) cooperatively drive SOA production through daytime OH oxidation and nighttime NO_3 chemistry, with aqueous-phase reactions dominating under high humidity and gas-phase partitioning prevailing in polluted air masses. Ghirardo et al. [286] found that BVOCs (e.g., sesquiterpenes) synergistically enhance SOA formation with AVOCs through NO_x-mediated oxidation pathways, where anthropogenic NO_x promotes the conversion of both BVOCs and AVOCs into low-volatility products and aerosol aging processes involving multi-phase reactions (e.g., acid-catalyzed polymerization), with their combined contributions amplified under urban pollution and plant stress conditions.

In high-NO_x environments, the O_3 and SOA formation efficiency of BVOCs is significantly enhanced. Moreover, emission control strategies targeting anthropogenic sources, such as reducing VOCs or NO_x, can either amplify or diminish the contribution of biogenic sources. Notably, under carbon neutrality pathways, biogenic contributions may surpass those from anthropogenic sources, highlighting the dual challenges of regional collaborative emission reduction and climate change.

7. Atmospheric Impacts of BVOCs

As critical mediators of biosphere–atmosphere interactions, biogenic volatile organic compounds (BVOCs) profoundly influence air quality, climate systems, and public health through complex atmospheric oxidation pathways [287,288].

7.1. Ozone Formation Potential (OFP) and Climate Feedbacks

The OFP quantifies the mass-specific contribution of BVOCs to ozone production. The O_3 formation potentials of isoprene and monoterpenes were 0.05–57.21 and 0.003–72.35 ppb [289]. In addition to this, aromatic hydrocarbons such as toluene and o-xylene and olefins such as ethylene and propylene also contribute significantly to OFP [290]. This potential is amplified in the urban–rural boundary layer, where BVOC-NO_x synergies elevate surface O_3 by 13–16 ppb [291]. For instance, emissions of reactive olefins (e.g., isoprene) in Zhengzhou during the summer combine with traffic NO_x to result in O_3 concentrations that exceed China's Grade II standard (93.3 ppbv) by more than 50% [292].

7.2. Secondary Organic Aerosol Formation Potential (SOAP) and Environmental Impacts

The SOAP characterizes the nucleation efficiency of BVOC oxidation products. The SOA values for isoprene and monoterpenes are 0.0002–0.21 and 0.0006–10.46 $\mu g/m^3$, respectively [289]. Sesquiterpene oxidation products play an important role in the formation and oxidative chemistry of atmospheric SOA, especially during the spring recovery period [229]. BSOAs exhibit different diurnal formation mechanisms. NO_x-induced SOA production dominates at night, and its impact on the environment is influenced by solar radiation, temperature, and humidity [293]. It has been shown that nocturnal terpene ozonolysis with high NO_x levels can significantly contribute to secondary organic aerosol formation and organic nitrate formation through anthropogenic–biotic interactions [244].

From 2000 to 2019, the contribution of secondary species to 2.5-micrometer particulate matter (PM2.5) in the Pearl River Delta, China, stabilized at about 80%, with SOAs and secondary inorganic aerosols (SIAs) contributing equally to fine particulate pollution [294]. The presence of SOAs in the atmosphere has implications for climate and health. Respiratory diseases, reduced visibility, haze, deterioration of air quality leading to environmental pollution, etc., are the result of the presence of organic species in atmospheric aerosols [295].

8. Future Research Challenges

While significant progress has been made in understanding BVOC emission mechanisms, characterization methods, and their atmospheric impacts, several challenges remain:

(1) Modeling methods such as Guenther, MEGAN, and BEIS are research methods that overcome spatial and temporal limitations. However, accurate and credible emission prediction models of BVOCs require a substantial amount of high-precision and readily accessible measurement data.

(2) Free radical reactions play a key role in the atmospheric transformation process of BVOCs, directly influencing the levels of O_3 and SOA. Nevertheless, the specific reaction pathways in the oxidation stage of BVOCs, as well as the details and mechanisms of the dynamic changes in free radicals, remain poorly understood.

(3) The emission of BVOCs is affected by a multitude of factors, many of which do not individually influence the emission of BVOCs. There is a dearth of systematic studies on the multiple synergistic effects of the internal and external factors on the emission mechanisms of BVOCs and their long-term impacts.

Given these challenges, the following three research areas are particularly important for future BVOC studies: (1) the development of high-precision, low-cost, and easy-to-operate methods for rapid measurements of BVOCs; (2) an in-depth study of the atmospheric chemical reaction mechanism between BVOCs and free radicals; and (3) an investigation into the synergistic effects of multiple factors on the emission mechanism of BVOCs.

9. Recommendations and Prospects

9.1. Recommendations of BVOC Prevention and Control

(1) Formulate BVOC management strategies or guidelines:

Among the world's major economies, the EU's Clean Air Quality Directive, the U.S. Clean Air Act, and China's Comprehensive Control Program for Volatile Organic Compounds in Key Industries all have stringent controls on the emission of anthropogenic VOCs. However, all of these programs merge and unify the management of anthropogenic sources of VOCs, such as vehicles, industry, and construction materials, and do not provide effective guidance on BVOCs. Meanwhile, the guidance of each country is basically formulated based on its own situation and lacks universality and popularization. It is recommended that developed countries with high forest cover, such as the U.S. and Canada, and developing countries with high forest resources, such as China and Brazil, cooperate to incorporate the management of BVOCs into an integrated air quality management framework, focusing on their synergistic effects with anthropogenic pollutants, and to further improve the global VOCs pollution situation.

(2) Establish BVOC emission control indicators based on multi-dimensional parameters:

In contrast to the relatively straightforward and predictable emissions observed in laboratory settings, the release of BVOCs in natural environments is a complex and variable process. It is therefore evident that a single indicator cannot be relied upon for the quantitative control of BVOC emissions. By cross-analyzing multi-dimensional parameters through big data, accurately identifying the emission characteristics of different BVOC sources, and highlighting the major items, a more scientific and targeted BVOC emission control indicator can be formulated. This approach helps to avoid the inaccuracy and significant errors of single indicators, optimize the use of resources, and reduce unnecessary economic losses and control costs.

(3) Dynamically adjust BVOC management based on temporal and spatial characteristics of BVOC emissions:

BVOC emissions exhibit significant seasonality and diurnal variations over time and are influenced by geographical and climatic conditions spatially. Additionally, the synergistic pollution effects with other pollutants like anthropogenic NO_x vary. Therefore, it is essential to implement differentiated control strategies based on the temporal and spatial characteristics of specific regions and adjust BVOC management strategies dynamically according to local conditions. This makes policies more flexible and scientific, enhancing the effectiveness and specificity of regional management.

9.2. Prospects of BVOC Monitor Strategies

(1) Develop more efficient, convenient, and low-cost BVOC measurement technologies:

As previously stated, a rapid sampling method for BVOCs that is highly accurate, straightforward to operate, and cost-effective is the foundation for obtaining high-quality detection data and conducting related research. The Guangzhou Institute of Geochemistry, Chinese Academy of Sciences (GIGC) devised a semi-open dynamic chamber system for measuring BVOCs. This system can reduce the time required to reach a steady state at higher flow rates, thereby minimizing adsorption loss. Nevertheless, the adsorption loss for macromolecular compounds remains above 30%, indicating a need for further optimization [296]. Electrochemical sensors are employed for the measurement of atmospheric NO_2 due to their low cost, portability, and flexibility, which can also be applied to the measurement of BVOCs [297]. Laser-induced fluorescence tracking has also been employed for the sampling and measurement of BVOCs, although the technology is not yet fully developed [298]. It is therefore feasible to pursue the optimization of existing methods or the adoption of techniques from other fields as avenues for the development of more efficient, convenient, and cost-effective measurement techniques for BVOCs.

(2) Develop BVOC monitoring and prediction technologies based on high temporal and spatial resolution big data models:

The advent of new information science and technology has opened up new avenues for research in the field of environmental management. By extending the aforementioned laser-induced fluorescence tracking technology, the regional dissemination of environmentally friendly fluorescent probes, and the laser-induced generation of remotely sensed data through the use of UAVs, in conjunction with ground-based observation data, a comprehensive database of high temporal and spatial resolution BVOC emissions can be formed. The application of machine learning and deep learning algorithms can enhance the synergistic effects of multiple factors, optimize existing BVOC emission models, and facilitate the development of new prediction models. A map of BVOC emission characteristics in different regions can be constructed, allowing region-specific control strategies to be formulated and precise management to be achieved.

(3) Promote the selection of low-BVOC-emitting tree species in urban greening and ecological restoration:

In urban greening and ecological restoration, besides considering the ornamental value and ecological function of trees, selecting low-BVOC-emitting tree species can help optimize air quality while maintaining ecological function and aesthetic values.

(4) Artificially control precursor substance emissions:

The strengthening of emission controls of precursors such as O_3 and SOA can indirectly reduce the emission level of BVOCs, reducing secondary chemical reactions that lead to O_3 and SOA formation.

Supplementary Materials: The following Supporting Information can be downloaded at: https://www.mdpi.com/article/10.3390/toxics13050364/s1, Table S1: Nomenclature.

Author Contributions: Conceptualization, J.L.; methodology, R.L., R.G. and Y.Y.; software, Y.Y.; validation, L.W. and M.H.-A.-M.; formal analysis, X.X.; investigation, R.L., L.W. and X.S.; resources, R.G. and H.L.; data curation, R.L., X.S. and X.X.; writing—original draft preparation, R.L.; writing—review and editing, X.L. and R.G.; visualization, M.H.-A.-M.; supervision, X.L. and H.L.; project administration, J.L.; funding acquisition, J.L. All authors have read and agreed to the published version of the manuscript.

Funding: This work was supported by the National Natural Science Foundation of China (Grant No. 42307149) "Studies on gradient observation of O_3 and its precursors based on plateau background mountain stations", awarded to Dr. Yuan Yang (Principal Investigator).

Institutional Review Board Statement: Not applicable.

Informed Consent Statement: Not applicable.

Data Availability Statement: No new data were created or analyzed in this study. Data sharing is not applicable to this article.

Acknowledgments: This work is supported by the National Natural Science Foundation of China (Grant No. 42307149).

Conflicts of Interest: The authors declare that they have no known competing financial interests or personal relationships that could have appeared to influence the work reported in this paper.

References

1. Cai, M.; An, C.; Guy, C. A scientometric analysis and review of biogenic volatile organic compound emissions: Research hotspots, new frontiers, and environmental implications. *Renew. Sustain. Energy Rev.* **2021**, *149*, 111317. [CrossRef]
2. Guenther, A.; Karl, T.; Harley, P.; Wiedinmyer, C.; Palmer, P.I.; Geron, C. Estimates of global terrestrial isoprene emissions using MEGAN (Model of Emissions of Gases and Aerosols from Nature). *Atmos. Chem. Phys.* **2006**, *6*, 3181–3210. [CrossRef]
3. Guenther, A.; Hewitt, C.N.; Erickson, D.; Fall, R.; Geron, C.; Graedel, T.; Harley, P.; Klinger, L.; Lerdau, M.; Mckay, W.A.; et al. A global model of natural volatile organic compound emissions. *J. Geophys. Res.* **1995**, *100*, 8873. [CrossRef]
4. World Health Organization. *WHO Global Air Quality Guidelines: Particulate Matter (PM2.5 and PM10), Ozone, Nitrogen Dioxide, Sulfur Dioxide and Carbon Monoxide*; World Health Organization: Geneva, Switzerland, 2021; Available online: https://www.who.int/publications/i/item/9789240034228 (accessed on 27 April 2025).
5. Eastman, J.R.; Sangermano, F.; Machado, E.A.; Rogan, J.; Anyamba, A. Global Trends in Seasonality of Normalized Difference Vegetation Index (NDVI), 1982–2011. *Remote Sens.* **2013**, *5*, 4799–4818. [CrossRef]
6. Huang, L.; Zhao, X.; Chen, C.; Tan, J.; Li, Y.; Chen, H.; Wang, Y.; Li, L.; Guenther, A.; Huang, H. Uncertainties of biogenic VOC emissions caused by land cover data and implications on ozone mitigation strategies for the Yangtze river Delta region. *Atmos. Environ.* **2024**, *337*, 120765. [CrossRef]
7. Steinbrecher, R.; Smiatek, G.; Köble, R.; Seufert, G.; Theloke, J.; Hauff, K.; Ciccioli, P.; Vautard, R.; Curci, G. Intra- and inter-annual variability of VOC emissions from natural and semi-natural vegetation in Europe and neighbouring countries. *Atmos. Environ.* **2009**, *43*, 1380–1391. [CrossRef]
8. Wang, P.; Chen, Y.; Hu, J.; Zhang, H.; Ying, Q. Source apportionment of summertime ozone in China using a source-oriented chemical transport model. *Atmos. Environ.* **2019**, *211*, 79–90. [CrossRef]
9. Gu, Y.; Liu, B.; Meng, H.; Song, S.; Dai, Q.; Shi, L.; Feng, Y.; Hopke, P.K. Source apportionment of consumed volatile organic compounds in the atmosphere. *J. Hazard. Mater.* **2023**, *459*, 132138. [CrossRef]

10. Hallquist, M.; Wenger, J.C.; Baltensperger, U.; Rudich, Y.; Simpson, D.; Claeys, M.; Dommen, J.; Donahue, N.M.; George, C.; Goldstein, A.H.; et al. The formation, properties and impact of secondary organic aerosol: Current and emerging issues. *Atmos. Chem. Phys.* **2009**, *9*, 5155–5236. [CrossRef]
11. Cao, J.; Situ, S.; Hao, Y.; Xie, S.; Li, L. Enhanced summertime ozone and SOA from biogenic volatile organic compound (BVOC) emissions due to vegetation biomass variability during 1981–2018 in China. *Atmos. Chem. Phys.* **2022**, *22*, 2351–2364. [CrossRef]
12. Qin, M.; Hu, Y.; Wang, X.; Vasilakos, P.; Boyd, C.M.; Xu, L.; Song, Y.; Ng, N.L.; Nenes, A.; Russell, A.G. Modeling biogenic secondary organic aerosol (BSOA) formation from monoterpene reactions with NO3: A case study of the SOAS campaign using CMAQ. *Atmos. Environ.* **2018**, *184*, 146–155. [CrossRef]
13. Rap, A.; Scott, C.E.; Reddington, C.L.; Mercado, L.; Ellis, R.J.; Garraway, S.; Evans, M.J.; Beerling, D.J.; MacKenzie, A.R.; Hewitt, C.N.; et al. Enhanced global primary production by biogenic aerosol via diffuse radiation fertilization. *Nature Geosci.* **2018**, *11*, 640–644. [CrossRef]
14. Yang, K.; Llusià, J.; Preece, C.; Ogaya, R.; Márquez Tur, L.; Mu, Z.; You, C.; Xu, Z.; Tan, Y.; Peñuelas, J. Impacts of seasonality, drought, nitrogen fertilization, and litter on soil fluxes of biogenic volatile organic compounds in a Mediterranean forest. *Sci. Total Environ.* **2024**, *906*, 167354. [CrossRef] [PubMed]
15. Sanadze, G.A. Emission of Gaseous Organic Sub-stances from Plants. *Rep. Akad. Nauk. GruzSSR* **1956**, *17*, 429–433.
16. Sanadze, G.A. Biogenic Isoprene (A Review). *Russ. J. Plant Physiol.* **2004**, *51*, 729–741. [CrossRef]
17. Laothawornkitkul, J.; Taylor, J.E.; Paul, N.D.; Hewitt, C.N. Biogenic volatile organic compounds in the Earth system. *New Phytol.* **2009**, *183*, 27–51. [CrossRef]
18. Peng, Q.; Yang, Y.; Ou, W.; Wei, L.; Li, Z.; Deng, X.; Gao, Q. The characteristics and environmental significance of BVOCs released by aquatic macrophytes. *Chemosphere* **2024**, *361*, 142574. [CrossRef]
19. Peñuelas, J.; Llusià, J. BVOCs: Plant defense against climate warming? *Trends Plant Sci.* **2003**, *8*, 105–109. [CrossRef]
20. Kesselmeier, J.; Staudt, M. Biogenic Volatile Organic Compounds (VOC): An Overview on Emission, Physiology and Ecology. *J. Atmos. Chem.* **1999**, *33*, 23–88. [CrossRef]
21. Helmig, D.; Klinger, L.F.; Guenther, A.; Vierling, L.; Geron, C.; Zimmerman, P. Biogenic volatile organic compound emissions (BVOCs) II. Landscape flux potentials from three continental sites in the U.S. *Chemosphere* **1999**, *38*, 2189–2204. [CrossRef]
22. Otter, L.B.; Guenther, A.; Greenberg, J. Seasonal and spatial variations in biogenic hydrocarbon emissions from southern African savannas and woodlands. *Atmos. Environ.* **2002**, *36*, 4265–4275. [CrossRef]
23. Papiez, M.R.; Potosnak, M.J.; Goliff, W.S.; Guenther, A.B.; Matsunaga, S.N.; Stockwell, W.R. The impacts of reactive terpene emissions from plants on air quality in Las Vegas, Nevada. *Atmos. Environ.* **2009**, *43*, 4109–4123. [CrossRef]
24. Kramshøj, M.; Vedel-Petersen, I.; Schollert, M.; Rinnan, Å.; Nymand, J.; Ro-Poulsen, H.; Rinnan, R. Large increases in Arctic biogenic volatile emissions are a direct effect of warming. *Nature Geosci.* **2016**, *9*, 349–352. [CrossRef]
25. Guenther, A.B.; Zimmerman, P.R.; Harley, P.C.; Monson, R.K.; Fall, R. Isoprene and monoterpene emission rate variability: Model evaluations and sensitivity analyses. *J. Geophys. Res. Atmos.* **1993**, *98*, 12609–12617. [CrossRef]
26. Guenther, A.B.; Monson, R.K.; Fall, R. Isoprene and monoterpene emission rate variability: Observations with eucalyptus and emission rate algorithm development. *J. Geophys. Res. Atmos.* **1991**, *96*, 10799–10808. [CrossRef]
27. Pierce, T.E. PC-BEIS: A personal computer version of the biogenic emissions inventory system. *J. Air Waste Manag. Assoc.* **1991**, *41*, 937–941. [CrossRef]
28. Ng, N.L.; Brown, S.S.; Archibald, A.T.; Atlas, E.; Cohen, R.C.; Crowley, J.N.; Day, D.A.; Donahue, N.M.; Fry, J.L.; Fuchs, H.; et al. Nitrate radicals and biogenic volatile organic compounds: Oxidation, mechanisms, and organic aerosol. *Atmos. Chem. Phys.* **2017**, *17*, 2103–2162. [CrossRef]
29. Tang, J.; Zhou, P.; Miller, P.A.; Schurgers, G.; Gustafson, A.; Makkonen, R.; Fu, Y.H.; Rinnan, R. High-latitude vegetation changes will determine future plant volatile impacts on atmospheric organic aerosols. *npj Clim. Atmos. Sci.* **2023**, *6*, 1–13. [CrossRef]
30. Lun, X.; Lin, Y.; Chai, F.; Fan, C.; Li, H.; Liu, J. Reviews of emission of biogenic volatile organic compounds (BVOCs) in Asia. *J. Environ. Sci.* **2020**, *95*, 266–277. [CrossRef]
31. Yang, W.; Cao, J.; Wu, Y.; Kong, F.; Li, L. Review on plant terpenoid emissions worldwide and in China. *Sci. Total Environ.* **2021**, *787*, 147454. [CrossRef]
32. Smith, D.; Španěl, P. Direct, rapid quantitative analyses of BVOCs using SIFT-MS and PTR-MS obviating sample collection. *TrAC Trends Anal. Chem.* **2011**, *30*, 945–959. [CrossRef]
33. Bacsik, Z.; Mink, J.; Keresztury, G. FTIR Spectroscopy of the Atmosphere. I. Principles and Methods. *Appl. Spectrosc. Rev.* **2004**, *39*, 295–363.
34. Gómez, M.C.; Durana, N.; García, J.A.; de Blas, M.; Sáez de Cámara, E.; García-Ruiz, E.; Gangoiti, G.; Torre-Pascual, E.; Iza, J. Long-term measurement of biogenic volatile organic compounds in a rural background area: Contribution to ozone formation. *Atmos. Environ.* **2020**, *224*, 117315.
35. Kalalian, C.; Roth, E.; El Dib, G.; Singh, H.J.; Rao, P.K.; Chakir, A. Product investigation of the gas phase ozonolysis of 1-penten-3-ol, *cis*-2-penten-1-ol and *trans*-3-hexen-1-ol. *Atmos. Environ.* **2020**, *238*, 117732. [CrossRef]

36. Shi, Y.; Wang, B.; Yin, C.; Li, Z.; Yu, Y. Performance improvement: A lightweight gas information classification method combined with an electronic nose system. *Sens. Actuators B Chem.* **2023**, *396*, 134551. [CrossRef]
37. Das, P.P. Glass microfiber based method for extraction and detection of BVOCs of *Camellia sinensis* (L.) O. Kuntze. *J. Appl. Res. Med. Aromat. Plants* **2021**, *24*, 100319.
38. Fattobene, M.; Papa, F.; Russo, R.E.; Zamponi, S.; Conti, P.; Taffetani, F.; Sorci, A.; Liu, F.; Berrettoni, M. ON-SITE monitoring OF BVOCS emission in Tremiti island, Italy. *Heliyon* **2024**, *10*, e23822. [CrossRef]
39. Baroja, O.; Unceta, N.; Sampedro, M.C.; Goicolea, M.A.; Barrio, R.J. Optimization and validation of a method of analysis for fenitrothion and its main metabolites in forestry air samples using sorbent tubes with thermal desorption cold trap injection and gas chromatography–mass spectrometry. *J. Chromatogr. A* **2004**, *1059*, 165–170. [CrossRef]
40. Wu, K.; Yang, X.; Chen, D.; Gu, S.; Lu, Y.; Jiang, Q.; Wang, K.; Ou, Y.; Qian, Y.; Shao, P.; et al. Estimation of biogenic VOC emissions and their corresponding impact on ozone and secondary organic aerosol formation in China. *Atmos. Res.* **2020**, *231*, 104656. [CrossRef]
41. Gonçalves, O.C.; Cerqueira, J.S.R.F.; Mestre, A.S.; Neng, N.R.; Nogueira, J.M.F. HS-BAμE: A New Alternative Approach for VOCs Analysis—Application for Monitoring Biogenic Emissions from Tree Species. *Molecules* **2023**, *28*, 1179. [CrossRef]
42. Kim, S.; Karl, T.; Guenther, A.; Tyndall, G.; Orlando, J.; Harley, P.; Rasmussen, R.; Apel, E. Emissions and ambient distributions of Biogenic Volatile Organic Compounds (BVOC) in a ponderosa pine ecosystem: Interpretation of PTR-MS mass spectra. *Atmos. Chem. Phys.* **2010**, *10*, 1759–1771. [CrossRef]
43. Lindinger, W.; Jordan, A. Proton-transfer-reaction mass spectrometry (PTR–MS): On-line monitoring of volatile organic compounds at pptv levels. *Chem. Soc. Rev.* **1998**, *27*, 347–375. [CrossRef]
44. D'Arco, A.; Mancini, T.; Paolozzi, M.C.; Macis, S.; Mosesso, L.; Marcelli, A.; Petrarca, M.; Radica, F.; Tranfo, G.; Lupi, S.; et al. High Sensitivity Monitoring of VOCs in Air through FTIR Spectroscopy Using a Multipass Gas Cell Setup. *Sensors* **2022**, *22*, 5624. [CrossRef] [PubMed]
45. Fathy, A.; Sabry, Y.M.; Amr, M.; Gnambodoe-Capo-chichi, M.; Anwar, M.; Ghoname, A.O.; Amr, A.; Saeed, A.; Gad, M.; Haron, M.A.; et al. MEMS FTIR optical spectrometer enables detection of volatile organic compounds (VOCs) in part-per-billion (ppb) range for air quality monitoring. In *MOEMS and Miniaturized Systems XVIII, Proceedings of the SPIE OPTO, San Francisco, CA, USA, 2–7 February 2019*; SPIE: Bellingham, WA, USA; Volume 10931, pp. 69–75.
46. Mermet, K.; Sauvage, S.; Dusanter, S.; Salameh, T.; Léonardis, T.; Flaud, P.-M.; Perraudin, É.; Villenave, E.; Locoge, N. Optimization of a gas chromatographic unit for measuring biogenic volatile organic compounds in ambient air. *Atmos. Meas. Tech.* **2019**, *12*, 6153–6171. [CrossRef]
47. Chang, Y.; Tang, N.; Qu, H.; Liu, J.; Zhang, D.; Zhang, H.; Pang, W.; Duan, X. Detection of Volatile Organic Compounds by Self-assembled Monolayer Coated Sensor Array with Concentration-independent Fingerprints. *Sci. Rep.* **2016**, *6*, 23970. [CrossRef]
48. Fair, J.D.; Bailey, W.F.; Felty, R.A.; Gifford, A.E.; Shultes, B.; Volles, L.H. Method for rapid on-site identification of VOCs. *J. Environ. Sci.* **2009**, *21*, 1005–1008. [CrossRef]
49. Stierlin, É.; Michel, T.; Fernandez, X. Field analyses of lavender volatile organic compounds: Performance evaluation of a portable gas chromatography–mass spectrometry device. *Phytochem. Anal.* **2020**, *31*, 778–785. [CrossRef]
50. Santos, F.J.; Galceran, M.T. The application of gas chromatography to environmental analysis. *TrAC Trends Anal. Chem.* **2002**, *21*, 672–685. [CrossRef]
51. Santhanakrishnan, V.P.; Shoba, N.; Senthamizh Selvi, B.; Varun, E.; Mohankumar, S.; Raveendran, M. Aromatic profiling of Murraya koenigii leaves by Thermal Desorption Gas chromatography-Mass Spectroscopy (TD-GC-MS). *Heliyon* **2023**, *9*, e17832.
52. Ren, Q.; Jin, Y.; Hu, Y.; Chen, H.; Li, Z. Rapid changes of induced volatile organic compounds in *Pinus massoniana*. *Front. Forest. China* **2007**, *2*, 459–465. [CrossRef]
53. Kim, Y.-H.; Kim, K.-H. Ultimate Detectability of Volatile Organic Compounds: How Much Further Can We Reduce Their Ambient Air Sample Volumes for Analysis? *Anal. Chem.* **2012**, *84*, 8284–8293. [CrossRef] [PubMed]
54. Lee, J.-H.; Hwang, S.-M.; Lee, D.-W.; Heo, G.-S. Determination of Volatile Organic Compounds (VOCs) Using Tedlar Bag/Solid-phase Microextraction/Gas Chromatography/Mass Spectrometry (SPME/GC/MS) in Ambient and Workplace Air. *Bull. Korean Chem. Soc.* **2002**, *23*, 488–496.
55. Feng, J.; Zhang, B.; Zhang, H.; Wu, Z.; Li, M.; Wang, D.; Wang, C. Combining with E-nose, GC-MS, GC-IMS and chemometrics to explore volatile characteristics during the different stages of *Zanthoxylum bungeanum* maxim fruits. *Food Res. Int.* **2024**, *195*, 114964. [CrossRef]
56. Wang, W.; Liu, C.; Clarisse, L.; Van Damme, M.; Coheur, P.-F.; Xie, Y.; Shan, C.; Hu, Q.; Sun, Y.; Jones, N. Ground-based measurements of atmospheric NH3 by Fourier transform infrared spectrometry at Hefei and comparisons with IASI data. *Atmos. Environ.* **2022**, *287*, 119256. [CrossRef]
57. Yuan, B.; Koss, A.R.; Warneke, C.; Coggon, M.; Sekimoto, K.; de Gouw, J.A. Proton-Transfer-Reaction Mass Spectrometry: Applications in Atmospheric Sciences. *Chem. Rev.* **2017**, *117*, 13187–13229. [CrossRef]

58. Hansel, A.; Jordan, A.; Holzinger, R.; Prazeller, P.; Vogel, W.; Lindinger, W. Proton transfer reaction mass spectrometry: On-line trace gas analysis at the ppb level. *Int. J. Mass Spectrom. Ion Process.* **1995**, *149–150*, 609–619. [CrossRef]
59. Zhu, B.; Han, Y.; Wang, C.; Huang, X.; Xia, S.; Niu, Y.; Yin, Z.; He, L. Understanding primary and secondary sources of ambient oxygenated volatile organic compounds in Shenzhen utilizing photochemical age-based parameterization method. *J. Environ. Sci.* **2019**, *75*, 105–114. [CrossRef] [PubMed]
60. Bessala, L.F.B.; Gao, J.; He, Z.; Wang, Z.; Yi, S. Effects of different heat treatment media on odorous constituents, chemical decomposition and mechanical properties of two hardwoods. *RSC Adv.* **2024**, *14*, 7414–7429. [CrossRef]
61. Fathy, A.; Pivert, M.L.; Kim, Y.J.; Ba, M.O.; Erfan, M.; Sabry, Y.M.; Khalil, D.; Leprince-Wang, Y.; Bourouina, T.; Gnambodoe-Capochichi, M. Continuous Monitoring of Air Purification: A Study on Volatile Organic Compounds in a Gas Cell. *Sensors* **2020**, *20*, 934. [CrossRef]
62. Ma, F.; Zhang, G.; Guo, H.; Liao, L.; Huang, X.; Yi, Z. Transient interaction effects of temperature and light intensity on isoprene and monoterpene emissions from Schima superba and Phoebe bournei. *Sci. Total Environ.* **2023**, *894*, 165082. [CrossRef]
63. Rhew, R.C.; Deventer, M.J.; Turnipseed, A.A.; Warneke, C.; Ortega, J.; Shen, S.; Martinez, L.; Koss, A.; Lerner, B.M.; Gilman, J.B.; et al. Ethene, propene, butene and isoprene emissions from a ponderosa pine forest measured by relaxed eddy accumulation. *Atmos. Chem. Phys.* **2017**, *17*, 13417–13438. [CrossRef]
64. Copolovici, L.; Niinemets, Ü. Flooding induced emissions of volatile signalling compounds in three tree species with differing waterlogging tolerance. *Plant Cell Environ.* **2010**, *33*, 1582–1594. [CrossRef]
65. You, D.-W.; Seon, Y.-S.; Jang, Y.; Bang, J.; Oh, J.-S.; Jung, K.-W. A portable gas chromatograph for real-time monitoring of aromatic volatile organic compounds in air samples. *J. Chromatogr. A* **2020**, *1625*, 461267. [CrossRef] [PubMed]
66. Sharma, R.; Zhou, M.; Hunter, M.D.; Fan, X. Rapid In Situ Analysis of Plant Emission for Disease Diagnosis Using a Portable Gas Chromatography Device. *J. Agric. Food Chem.* **2019**, *67*, 7530–7537. [CrossRef] [PubMed]
67. Ciccioli, P.; Silibello, C.; Finardi, S.; Pepe, N.; Ciccioli, P.; Rapparini, F.; Neri, L.; Fares, S.; Brilli, F.; Mircea, M.; et al. The potential impact of biogenic volatile organic compounds (BVOCs) from terrestrial vegetation on a Mediterranean area using two different emission models. *Agric. For. Meteorol.* **2023**, *328*, 109255. [CrossRef]
68. Anisimov, D.S.; Chekusova, V.P.; Trul, A.A.; Abramov, A.A.; Borshchev, O.V.; Agina, E.V.; Ponomarenko, S.A. Fully integrated ultra-sensitive electronic nose based on organic field-effect transistors. *Sci. Rep.* **2021**, *11*, 10683. [CrossRef]
69. Feng, X.; Wang, H.; Wang, Z.; Huang, P.; Kan, J. Discrimination and characterization of the volatile organic compounds in eight kinds of huajiao with geographical indication of China using electronic nose, HS-GC-IMS and HS-SPME-GC–MS. *Food Chem.* **2022**, *375*, 131671. [CrossRef]
70. Lin, C.-Y.; Chang, T.-C.; Chen, Y.-H.; Chen, Y.-J.; Cheng, S.-S.; Chang, S.-T. Monitoring the dynamic emission of biogenic volatile organic compounds from Cryptomeria japonica by enclosure measurement. *Atmos. Environ.* **2015**, *122*, 163–170. [CrossRef]
71. Li, L.; Guenther, A.B.; Xie, S.; Gu, D.; Seco, R.; Nagalingam, S.; Yan, D. Evaluation of semi-static enclosure technique for rapid surveys of biogenic volatile organic compounds (BVOCs) emission measurements. *Atmos. Environ.* **2019**, *212*, 1–5. [CrossRef]
72. Aydin, Y.M.; Yaman, B.; Koca, H.; Dasdemir, O.; Kara, M.; Altiok, H.; Dumanoglu, Y.; Bayram, A.; Tolunay, D.; Odabasi, M.; et al. Biogenic volatile organic compound (BVOC) emissions from forested areas in Turkey: Determination of specific emission rates for thirty-one tree species. *Sci. Total Environ.* **2014**, *490*, 239–253. [CrossRef]
73. Vedel-Petersen, I.; Schollert, M.; Nymand, J.; Rinnan, R. Volatile organic compound emission profiles of four common arctic plants. *Atmos. Environ.* **2015**, *120*, 117–126. [CrossRef]
74. Jing, X.; Lun, X.; Fan, C.; Ma, W. Emission patterns of biogenic volatile organic compounds from dominant forest species in Beijing, China. *J. Environ. Sci.* **2020**, *95*, 73–81. [CrossRef]
75. Yuan, X.; Calatayud, V.; Gao, F.; Fares, S.; Paoletti, E.; Tian, Y.; Feng, Z. Interaction of drought and ozone exposure on isoprene emission from extensively cultivated poplar. *Plant Cell Environ.* **2016**, *39*, 2276–2287. [CrossRef]
76. Beauchamp, J.; Herbig, J.; Gutmann, R.; Hansel, A. On the use of Tedlar® bags for breath-gas sampling and analysis. *J. Breath Res.* **2008**, *2*, 046001. [CrossRef] [PubMed]
77. Hsieh, C.-C.; Horng, S.-H.; Liao, P.-N. Stability of Trace-level Volatile Organic Compounds Stored in Canisters and Tedlar Bags. *Aerosol Air Qual. Res.* **2003**, *3*, 17–28. [CrossRef]
78. Arnts, R.R. Evaluation of adsorbent sampling tube materials and Tenax-TA for analysis of volatile biogenic organic compounds. *Atmos. Environ.* **2010**, *44*, 1579–1584. [CrossRef]
79. Hellén, H.; Tykkä, T.; Schallhart, S.; Stratigou, E.; Salameh, T.; Iturrate-Garcia, M. Measurements of atmospheric C_{10}–C_{15} biogenic volatile organic compounds (BVOCs) with sorbent tubes. *Atmos. Meas. Tech.* **2024**, *17*, 315–333. [CrossRef]
80. Nassiri, P.; Golbabaie, F.; Mehrain, K.; Hematian, A. Evaluation of activated charcoal sampler tubes. *Iran. J. Public Health* **1994**, *23*, 11–20.
81. Prado, C.; Alcaraz, M.J.; Fuentes, A.; Garrido, J.; Periago, J.F. Storage stability of ketones on carbon adsorbents. *J. Chromatogr. A* **2006**, *1129*, 82–87. [CrossRef]

82. Wang, D.K.W.; Austin, C.C. Determination of complex mixtures of volatile organic compounds in ambient air: Canister methodology. *Anal. Bioanal. Chem.* **2006**, *386*, 1099–1120. [CrossRef]
83. Acton, W.J.F.; Schallhart, S.; Langford, B.; Valach, A.; Rantala, P.; Fares, S.; Carriero, G.; Tillmann, R.; Tomlinson, S.J.; Dragosits, U.; et al. Canopy-scale flux measurements and bottom-up emission estimates of volatile organic compounds from a mixed oak and hornbeam forest in northern Italy. *Atmos. Chem. Phys.* **2016**, *16*, 7149–7170. [CrossRef]
84. Karl, T.; Misztal, P.K.; Jonsson, H.H.; Shertz, S.; Goldstein, A.H.; Guenther, A.B. Airborne Flux Measurements of BVOCs above Californian Oak Forests: Experimental Investigation of Surface and Entrainment Fluxes, OH Densities, and Damkohler Numbers. *J. Atmos. Sci.* **2013**, *70*, 3277–3287. [CrossRef]
85. Baldocchi, D.D.; Hincks, B.B.; Meyers, T.P. Measuring Biosphere-Atmosphere Exchanges of Biologically Related Gases with Micrometeorological Methods. *Ecology* **1988**, *69*, 1331–1340. [CrossRef]
86. Spirig, C.; Neftel, A.; Ammann, C.; Dommen, J.; Grabmer, W.; Thielmann, A.; Schaub, A.; Beauchamp, J.; Wisthaler, A.; Hansel, A. Eddy covariance flux measurements of biogenic VOCs during ECHO 2003 using proton transfer reaction mass spectrometry. *Atmos. Chem. Phys.* **2005**, *5*, 465–481. [CrossRef]
87. Misztal, P.K.; Nemitz, E.; Langford, B.; Di Marco, C.F.; Phillips, G.J.; Hewitt, C.N.; MacKenzie, A.R.; Owen, S.M.; Fowler, D.; Heal, M.R.; et al. Direct ecosystem fluxes of volatile organic compounds from oil palms in South-East Asia. *Atmos. Chem. Phys.* **2011**, *11*, 8995–9017. [CrossRef]
88. Bai, J.; Guenther, A.; Turnipseed, A.; Duhl, T.; Greenberg, J. Seasonal and interannual variations in whole-ecosystem BVOC emissions from a subtropical plantation in China. *Atmos. Environ.* **2017**, *161*, 176–190. [CrossRef]
89. Park, J.-H.; Fares, S.; Weber, R.; Goldstein, A.H. Biogenic volatile organic compound emissions during BEARPEX 2009 measured by eddy covariance and flux-gradient similarity methods. *Atmos. Chem. Phys.* **2014**, *14*, 231–244. [CrossRef]
90. Holst, T.; Arneth, A.; Hayward, S.; Ekberg, A.; Mastepanov, M.; Jackowicz-Korczynski, M.; Friborg, T.; Crill, P.M.; Backstrand, K. BVOC ecosystem flux measurements at a high latitude wetland site. *Atmos. Chem. Phys.* **2010**, *10*, 1617–1634. [CrossRef]
91. Bachy, A.; Aubinet, M.; Schoon, N.; Amelynck, C.; Bodson, B.; Moureaux, C.; Heinesch, B. Are BVOC exchanges in agricultural ecosystems overestimated? Insights from fluxes measured in a maize field over a whole growing season. *Atmos. Chem. Phys.* **2016**, *16*, 5343–5356. [CrossRef]
92. Edwards, G.D.; Martins, D.K.; Starn, T.; Pratt, K.; Shepson, P.B. A disjunct eddy accumulation system for the measurement of BVOC fluxes: Instrument characterizations and field deployment. *Atmos. Meas. Tech.* **2012**, *5*, 2115–2132. [CrossRef]
93. Chen, Y.; Zhang, X.; Wu, X.; Li, J.; Qiu, Y.; Wang, H.; Cheng, Z.; Zheng, C.; Yang, F. Volatile Organic Compound Sampling through Rotor Unmanned Aerial Vehicle Technique for Environmental Monitoring. *Atmosphere* **2022**, *13*, 1442. [CrossRef]
94. Li, Y.; Liu, B.; Ye, J.; Jia, T.; Khuzestani, R.B.; Sun, J.Y.; Cheng, X.; Zheng, Y.; Li, X.; Wu, C.; et al. Unmanned Aerial Vehicle Measurements of Volatile Organic Compounds over a Subtropical Forest in China and Implications for Emission Heterogeneity. *ACS Earth Space Chem.* **2021**, *5*, 247–256. [CrossRef]
95. McKinney, K.A.; Wang, D.; Ye, J.; de Fouchier, J.-B.; Guimarães, P.C.; Batista, C.E.; Souza, R.A.F.; Alves, E.G.; Gu, D.; Guenther, A.B.; et al. A sampler for atmospheric volatile organic compounds by copter unmanned aerial vehicles. *Atmos. Meas. Tech.* **2019**, *12*, 3123–3135. [CrossRef]
96. Ashworth, K.; Boissard, C.; Folberth, G.; Lathière, J.; Schurgers, G. Global Modelling of Volatile Organic Compound Emissions. In *Biology, Controls and Models of Tree Volatile Organic Compound Emissions*; Niinemets, Ü., Monson, R.K., Eds.; Tree Physiology; Springer: Dordrecht, The Netherlands, 2013; pp. 451–487. ISBN 978-94-007-6606-8.
97. Müller, J.-F.; Stavrakou, T.; Wallens, S.; De Smedt, I.; Van Roozendael, M.; Potosnak, J.; Rinne, J.; Munger, B.; Goldstein, A.; Guenther, B. Global isoprene emissions estimated using MEGAN, ECMWF analyses and a detailed canopy environment model. *Atmos. Chem. Phys.* **2008**, *8*, 1329–1341. [CrossRef]
98. Chi, Y.Q.; Xie, S.D. Spatiotemporal Inventory of Biogenic Volatile Organic Compound Emissions in China Based on Vegetation Volume and Production. *Adv. Mater. Res.* **2012**, *356–360*, 2579–2582. [CrossRef]
99. Wang, L.; Lun, X.; Wu, J.; Wang, Q.; Tao, J.; Dou, X.; Zhang, Z. Investigation of biogenic volatile organic compounds emissions in the Qinghai-Tibetan Plateau. *Sci. Total Environ.* **2023**, *902*, 165877. [CrossRef]
100. Pierce, T.; Geron, C.; Bender, L.; Dennis, R.; Tonnesen, G.; Guenther, A. Influence of increased isoprene emissions on regional ozone modeling. *J. Geophys. Res. Atmos.* **1998**, *103*, 25611–25629.
101. Feldman, M.S.; Howard, T.; McDonald-Buller, E.; Mullins, G.; Allen, D.T.; Hansel, A.; Wisthaler, A. Applications of satellite remote sensing data for estimating biogenic emissions in southeastern Texas. *Atmos. Environ.* **2010**, *44*, 917–929. [CrossRef]
102. Zheng, J.; Zheng, Z.; Yu, Y.; Zhong, L. Temporal, spatial characteristics and uncertainty of biogenic VOC emissions in the Pearl River Delta region, China. *Atmos. Environ.* **2010**, *44*, 1960–1969. [CrossRef]
103. Sakulyanontvittaya, T.; Duhl, T.; Wiedinmyer, C.; Helmig, D.; Matsunaga, S.; Potosnak, M.; Milford, J.; Guenther, A. Monoterpene and Sesquiterpene Emission Estimates for the United States. *Environ. Sci. Technol.* **2008**, *42*, 1623–1629. [CrossRef]

104. Guenther, A.B.; Jiang, X.; Heald, C.L.; Sakulyanontvittaya, T.; Duhl, T.; Emmons, L.K.; Wang, X. The Model of Emissions of Gases and Aerosols from Nature version 2.1 (MEGAN2.1): An extended and updated framework for modeling biogenic emissions. *Geosci. Model Dev.* **2012**, *5*, 1471–1492. [CrossRef]
105. Guenther, A.; Jiang, X.; Shah, T.; Huang, L.; Kemball-Cook, S.; Yarwood, G. Model of Emissions of Gases and Aerosol from Nature Version 3 (MEGAN3) for Estimating Biogenic Emissions. In *Air Pollution Modeling and its Application XXVI, Proceedings of the ITM 2018, Ottawa, ON, Canada, 14–18 May 2018*; Mensink, C., Gong, W., Hakami, A., Eds.; Springer International Publishing: Cham, Switzerland, 2019; pp. 187–192.
106. Sindelarova, K.; Granier, C.; Bouarar, I.; Guenther, A.; Tilmes, S.; Stavrakou, T.; Müller, J.-F.; Kuhn, U.; Stefani, P.; Knorr, W. Global data set of biogenic VOC emissions calculated by the MEGAN model over the last 30 years. *Atmos. Chem. Phys.* **2014**, *14*, 9317–9341. [CrossRef]
107. Wang, P.; Schade, G.; Estes, M.; Ying, Q. Improved MEGAN predictions of biogenic isoprene in the contiguous United States. *Atmos. Environ.* **2017**, *148*, 337–351. [CrossRef]
108. Jiang, X.; Guenther, A.; Potosnak, M.; Geron, C.; Seco, R.; Karl, T.; Kim, S.; Gu, L.; Pallardy, S. Isoprene emission response to drought and the impact on global atmospheric chemistry. *Atmos. Environ.* **2018**, *183*, 69–83. [CrossRef] [PubMed]
109. Bai, J.; Duhl, T. A primary generalized empirical model of BVOC emissions for some typical forests in China. *Atmos. Pollut. Res.* **2021**, *12*, 101126. [CrossRef]
110. Benjamin, M.T.; Winer, A.M. Estimating the ozone-forming potential of urban trees and shrubs. *Atmos. Environ.* **1998**, *32*, 53–68. [CrossRef]
111. Churkina, G.; Kuik, F.; Bonn, B.; Lauer, A.; Grote, R.; Tomiak, K.; Butler, T.M. Effect of VOC Emissions from Vegetation on Air Quality in Berlin during a Heatwave. *Environ. Sci. Technol.* **2017**, *51*, 6120–6130. [CrossRef]
112. Bao, X.; Zhou, W.; Wang, W.; Yao, Y.; Xu, L. Tree species classification improves the estimation of BVOCs from urban greenspace. *Sci. Total Environ.* **2024**, *914*, 169762. [CrossRef]
113. Antonelli, M.; Donelli, D.; Barbieri, G.; Valussi, M.; Maggini, V.; Firenzuoli, F. Forest Volatile Organic Compounds and Their Effects on Human Health: A State-of-the-Art Review. *Int. J. Environ. Res. Public. Health* **2020**, *17*, 6506. [CrossRef]
114. Meneguzzo, F.; Albanese, L.; Bartolini, G.; Zabini, F. Temporal and Spatial Variability of Volatile Organic Compounds in the Forest Atmosphere. *Int. J. Environ. Res. Public. Health* **2019**, *16*, 4915. [CrossRef]
115. Smiatek, G.; Steinbrecher, R. Temporal and spatial variation of forest VOC emissions in Germany in the decade 1994–2003. *Atmos. Environ.* **2006**, *40*, 166–177. [CrossRef]
116. Liu, Y.; Schallhart, S.; Tykkä, T.; Räsänen, M.; Merbold, L.; Hellén, H.; Pellikka, P. Biogenic volatile organic compounds in different ecosystems in Southern Kenya. *Atmos. Environ.* **2021**, *246*, 118064. [CrossRef]
117. Chang, J.; Ren, Y.; Shi, Y.; Zhu, Y.; Ge, Y.; Hong, S.; Jiao, L.; Lin, F.; Peng, C.; Mochizuki, T.; et al. An inventory of biogenic volatile organic compounds for a subtropical urban-rural complex. *Atmos. Environ.* **2012**, *56*, 115–123. [CrossRef]
118. Okumura, M.; Kosugi, Y.; Tani, A. Biogenic volatile organic compound emissions from bamboo species in Japan. *J. Agric. Meteorol.* **2018**, *74*, 40–44. [CrossRef]
119. Chen, W.; Wang, L.; Wu, J.; Lun, X.; Wang, X.; Li, X. Emission Pattern of Biogenic Volatile Organic Compounds from Wetland Vegetation. *Atmosphere* **2024**, *15*, 651. [CrossRef]
120. Mannisto, E.; Ylanne, H.; Losoi, M.; Keinanen, M.; Yli-Pirila, P.; Korrensalo, A.; Back, J.; Hellen, H.; Virtanen, A.; Tuittila, E.-S. Emissions of biogenic volatile organic compounds from adjacent boreal fen and bog as impacted by vegetation composition. *Sci. Total Environ.* **2023**, *858*, 159809. [CrossRef]
121. Gomez, L.G.; Loubet, B.; Lafouge, F.; Ciuraru, R.; Buysse, P.; Durand, B.; Gueudet, J.-C.; Fanucci, O.; Fortineau, A.; Zurfluh, O.; et al. Comparative study of biogenic volatile organic compounds fluxes by wheat, maize and rapeseed with dynamic chambers over a short period in northern France. *Atmos. Environ.* **2019**, *214*, 116855. [CrossRef]
122. Mozaffar, A.; Schoon, N.; Digrado, A.; Bachy, A.; Delaplace, P.; du Jardin, P.; Fauconnier, M.-L.; Aubinet, M.; Heinesch, B.; Amelynck, C. Methanol emissions from maize: Ontogenetic dependence to varying light conditions and guttation as an additional factor constraining the flux. *Atmos. Environ.* **2017**, *152*, 405–417. [CrossRef]
123. Bäck, J.; Aaltonen, H.; Hellén, H.; Kajos, M.K.; Patokoski, J.; Taipale, R.; Pumpanen, J.; Heinonsalo, J. Variable emissions of microbial volatile organic compounds (MVOCs) from root-associated fungi isolated from Scots pine. *Atmos. Environ.* **2010**, *44*, 3651–3659. [CrossRef]
124. Bourtsoukidis, E.; Behrendt, T.; Yañez-Serrano, A.M.; Hellén, H.; Diamantopoulos, E.; Catão, E.; Ashworth, K.; Pozzer, A.; Quesada, C.A.; Martins, D.L.; et al. Strong sesquiterpene emissions from Amazonian soils. *Nat. Commun.* **2018**, *9*, 2226. [CrossRef]
125. Mäki, M. Volatile Organic Compound Fluxes from Northern Forest Soils. Ph.D. Thesis, Institute for Atmospheric and Earth System Research, Forest Sciences Faculty of Agriculture and Forestry, University of Helsinki, Helsinki, Finland, 2019.
126. Peñuelas, J.; Asensio, D.; Tholl, D.; Wenke, K.; Rosenkranz, M.; Piechulla, B.; Schnitzler, J.P. Biogenic volatile emissions from the soil. *Plant Cell Environ.* **2014**, *37*, 1866–1891. [CrossRef] [PubMed]

127. Wester-Larsen, L.; Kramshøj, M.; Albers, C.N.; Rinnan, R. Biogenic Volatile Organic Compounds in Arctic Soil: A Field Study of Concentrations and Variability with Vegetation Cover. *J. Geophys. Res. Biogeosci.* **2020**, *125*, e2019JG005551. [CrossRef]
128. Sindelarova, K.; Markova, J.; Simpson, D.; Huszar, P.; Karlicky, J.; Darras, S.; Granier, C. High-resolution biogenic global emission inventory for the time period 2000–2019 for air quality modelling. *Earth Syst. Sci. Data* **2022**, *14*, 251–270. [CrossRef]
129. Yáñez-Serrano, A.M.; Bourtsoukidis, E.; Alves, E.G.; Bauwens, M.; Stavrakou, T.; Llusià, J.; Filella, I.; Guenther, A.; Williams, J.; Artaxo, P.; et al. Amazonian biogenic volatile organic compounds under global change. *Global Change Biol.* **2020**, *26*, 4722–4751. [CrossRef] [PubMed]
130. Yee, L.D.; Isaacman-VanWertz, G.; Wernis, R.A.; Meng, M.; Rivera, V.; Kreisberg, N.M.; Hering, S.V.; Bering, M.S.; Glasius, M.; Upshur, M.A.; et al. Observations of sesquiterpenes and their oxidation products in central Amazonia during the wet and dry seasons. *Atmos. Chem. Phys.* **2018**, *18*, 10433–10457. [CrossRef]
131. Wang, X.; Zhang, Y.; Tan, Y.; Tan, Y.; Bai, J.; Gu, D.; Ma, Z.; Du, J.; Han, Z. Effects of light on the emissions of biogenic isoprene and monoterpenes: A review. *Atmos. Pollut. Res.* **2022**, *13*, 101397. [CrossRef]
132. Liu, L.; Seyler, B.C.; Liu, H.; Zhou, L.; Chen, D.; Liu, S.; Yan, C.; Yang, F.; Song, D.; Tan, Q.; et al. Biogenic volatile organic compound emission patterns and secondary pollutant formation potentials of dominant greening trees in Chengdu, southwest China. *J. Environ. Sci.* **2022**, *114*, 179–193. [CrossRef]
133. Cheng, X.; Li, H.; Zhang, Y.; Li, Y.; Zhang, W.; Wang, X.; Bi, F.; Zhang, H.; Gao, J.; Chai, F.; et al. Atmospheric isoprene and monoterpenes in a typical urban area of Beijing: Pollution characterization, chemical reactivity and source identification. *J. Environ. Sci.* **2018**, *71*, 150–167. [CrossRef]
134. Liu, J.; Chu, B.; Chen, T.; Liu, C.; Wang, L.; Bao, X.; He, H. Secondary Organic Aerosol Formation from Ambient Air at an Urban Site in Beijing: Effects of OH Exposure and Precursor Concentrations. *Environ. Sci. Technol.* **2018**, *52*, 6834–6841. [CrossRef]
135. Ndah, F.; Valolahti, H.; Schollert, M.; Michelsen, A.; Rinnan, R.; Kivimäenpää, M. Influence of increased nutrient availability on biogenic volatile organic compound (BVOC) emissions and leaf anatomy of subarctic dwarf shrubs under climate warming and increased cloudiness. *Ann. Bot.* **2022**, *129*, 443–455. [CrossRef]
136. Lugo, P.L.; Straccia, V.G.; Rivela, C.B.; Patroescu-Klotz, I.; Illmann, N.; Teruel, M.A.; Wiesen, P.; Blanco, M.B. Diurnal photodegradation of fluorinated diketones (FDKs) by OH radicals using different atmospheric simulation chambers: Role of keto-enol tautomerization on reactivity. *Chemosphere* **2022**, *286*, 131562. [CrossRef] [PubMed]
137. Frazier, G.; McGlynn, D.F.; Barry, L.E.; Lerdau, M.; Pusede, S.E.; Isaacman-VanWertz, G. Composition, concentration, and oxidant reactivity of sesquiterpenes in the southeastern U.S. *Environ. Sci. Atmos.* **2022**, *2*, 1208–1220. [CrossRef]
138. McGlynn, D.F.; Barry, L.E.R.; Lerdau, M.T.; Pusede, S.E.; Isaacman-VanWertz, G. Measurement report: Variability in the composition of biogenic volatile organic compounds in a Southeastern US forest and their role in atmospheric reactivity. *Atmos. Chem. Phys.* **2021**, *21*, 15755–15770. [CrossRef]
139. Atkinson, R.; Arey, J. Gas-phase tropospheric chemistry of biogenic volatile organic compounds: A review. *Atmos. Environ.* **2003**, *37*, 197–219. [CrossRef]
140. Bouvier-Brown, N.C.; Goldstein, A.H.; Gilman, J.B.; Kuster, W.C.; de Gouw, J.A. In-situ ambient quantification of monoterpenes, sesquiterpenes, and related oxygenated compounds during BEARPEX 2007: Implications for gas- and particle-phase chemistry. *Atmos. Chem. Phys.* **2009**, *9*, 5505–5518. [CrossRef]
141. Calvert, J.; Mellouki, A.; Orlando, J.; Pilling, M.; Wallington, T. *Mechanisms of Atmospheric Oxidation of the Oxygenates*; Oxford University Press: Oxford, UK, 2011.
142. Isidorov, V.A.; Masłowiecka, J. Chemical Composition of Volatile and Extractive Organic Compounds in the Inflorescence Litter of Five Species of Woody Plants. *Plants* **2024**, *13*, 1829. [CrossRef]
143. Sharkey, T.D.; Wiberley, A.E.; Donohue, A.R. Isoprene Emission from Plants: Why and How. *Ann. Bot.* **2008**, *101*, 5–18. [CrossRef]
144. Ma, S.X.; Rindelaub, J.D.; McAvey, K.M.; Gagare, P.D.; Nault, B.A.; Ramachandran, P.V.; Shepson, P.B. α-Pinene nitrates: Synthesis, yields and atmospheric chemistry. *Atmos. Chem. Phys.* **2011**, *11*, 6337–6347. [CrossRef]
145. Dash, M.R.; Rajakumar, B. Experimental and theoretical rate coefficients for the gas phase reaction of β-Pinene with OH radical. *Atmos. Environ.* **2013**, *79*, 161–171. [CrossRef]
146. Chen, J.; Griffin, R.J. Modeling secondary organic aerosol formation from oxidation of α-pinene, β-pinene, and *d*-limonene. *Atmos. Environ.* **2005**, *39*, 7731–7744. [CrossRef]
147. Chen, X.; Hopke, P.K. A chamber study of secondary organic aerosol formation by linalool ozonolysis. *Atmos. Environ.* **2009**, *43*, 3935–3940. [CrossRef]
148. Arneth, A.; Schurgers, G.; Lathiere, J.; Duhl, T.; Beerling, D.J.; Hewitt, C.N.; Martin, M.; Guenther, A. Global terrestrial isoprene emission models: Sensitivity to variability in climate and vegetation. *Atmos. Chem. Phys.* **2011**, *11*, 8037–8052. [CrossRef]
149. Cao, Y.; Yue, X.; Liao, H.; Yang, Y.; Zhu, J.; Chen, L.; Tian, C.; Lei, Y.; Zhou, H.; Ma, Y. Ensemble projection of global isoprene emissions by the end of 21st century using CMIP6 models. *Atmos. Environ.* **2021**, *267*, 118766. [CrossRef]
150. Weng, H.; Lin, J.; Martin, R.; Millet, D.B.; Jaeglé, L.; Ridley, D.; Keller, C.; Li, C.; Du, M.; Meng, J. Global high-resolution emissions of soil NOx, sea salt aerosols, and biogenic volatile organic compounds. *Sci. Data* **2020**, *7*, 148. [CrossRef]

151. Rabin, S.S.; Alexander, P.; Henry, R.; Anthoni, P.; Pugh, T.A.M.; Rounsevell, M.; Arneth, A. Impacts of future agricultural change on ecosystem service indicators. *Earth Syst. Dyn.* **2020**, *11*, 357–376. [CrossRef]
152. Sancho-Knapik, D.; Gil-Pelegrin, E.; Pedro Ferrio, J.; Alonso-Forn, D.; Martin-Sanchez, R.; dos Santos Silva, J.V.; Imanishi, J.; Javier Peguero-Pina, J.; Angeles Sanz, M. Changes in the Abundance of Monoterpenes from Breathable Air of a Mediterranean Conifer Forest: When Is the Best Time for a Human Healthy Leisure Activity? *Forests* **2022**, *13*, 965. [CrossRef]
153. Acosta Navarro, J.C.; Smolander, S.; Struthers, H.; Zorita, E.; Ekman, A.M.L.; Kaplan, J.O.; Guenther, A.; Arneth, A.; Riipinen, I. Global emissions of terpenoid VOCs from terrestrial vegetation in the last millennium. *J. Geophys. Res. Atmos.* **2014**, *119*, 6867–6885. [CrossRef]
154. Llusia, J.; Penuelas, J.; Guenther, A.; Rapparini, F. Seasonal variations in terpene emission factors of dominant species in four ecosystems in NE Spain. *Atmos. Environ.* **2013**, *70*, 149–158. [CrossRef]
155. Wang, H.; Liu, X.; Wu, C.; Lin, G. Regional to global distributions, trends, and drivers of biogenic volatile organic compound emission from 2001 to 2020. *Atmos. Chem. Phys.* **2024**, *24*, 3309–3328. [CrossRef]
156. Dudareva, N.; Negre, F.; Nagegowda, D.A.; Orlova, I. Plant Volatiles: Recent Advances and Future Perspectives. *Crit. Rev. Plant Sci.* **2006**, *25*, 417–440. [CrossRef]
157. Lathière, J.; Hauglustaine, D.A.; Friend, A.D.; De Noblet-Ducoudré, N.; Viovy, N.; Folberth, G.A. Impact of climate variability and land use changes on global biogenic volatile organic compound emissions. *Atmos. Chem. Phys.* **2006**, *6*, 2129–2146. [CrossRef]
158. Xin-min, Z.; Fa-he, C.; Ting-ting, Y.; Kai, Z.; Yi-zhen, C.; Jing-chun, D.; Zhi-gang, X. Photochemical Characteristics and Sources of Volatile Organic Compounds in Wuqing, Tianjin. *Res. Environ. Sci.* **2012**, *25*, 1085–1091.
159. Zhu, S.; Huang, X.; He, L.; Lu, S.; Feng, N. Variation characteristics and chemical reactivity of ambient VOCs in Shenzhen. *China Environ. Sci.* **2012**, *32*, 2140–2148.
160. Zou, Y.; Deng, X.; Li, F.; Wang, B.; Tan, H.; Deng, T.; Mai, B.; Liu, X. Variation characteristics, chemical reactivity and sources of isoprene in the atmosphere of Guangzhou. *Huanjing Kexue Xuebao* **2015**, *35*, 647–655.
161. Ran, L.; Zhao, C.; Geng, F.; Tie, X.; Tang, X.; Peng, L.; Zhou, G.; Yu, Q.; Xu, J.; Guenther, A. Ozone photochemical production in urban Shanghai, China: Analysis based on ground level observations. *J. Geophys. Res.* **2009**, *114*. [CrossRef]
162. Park, C.; Schade, G.W.; Boedeker, I. Characteristics of the flux of isoprene and its oxidation products in an urban area. *J. Geophys. Res. Atmos.* **2011**, 116. [CrossRef]
163. Zhang, J.-K.; Wang, Y.-S.; Wu, F.-K.; Sun, J. Study on atmospheric VOCs in Gongga Mountain base station. *Huanjing Kexue* **2012**, *33*, 4159–4166.
164. Ying-xin, Y.U.; Sheng, W.E.N.; Hui-xiong, L.V.; Xin-ming, W.; Guo-ying, S.; Jia-mo, F.U. Diurnal Variations of VOCs and Relative Contribution to Ozone in Forest of Guangzhou. *Huanjing Kexue Yu Jishu* **2009**, *32*, 94–98.
165. Dudareva, N.; Klempien, A.; Muhlemann, J.K.; Kaplan, I. Biosynthesis, function and metabolic engineering of plant volatile organic compounds. *New Phytol.* **2013**, *198*, 16–32. [CrossRef]
166. Li, Z.; Sharkey, T.D. Molecular and Pathway Controls on Biogenic Volatile Organic Compound Emissions. In *Biology, Controls and Models of Tree Volatile Organic Compound Emissions*; Niinemets, Ü., Monson, R.K., Eds.; Springer: Dordrecht, The Netherlands, 2013; pp. 119–151. ISBN 978-94-007-6606-8.
167. Ma, M.; Li, M.; Wu, Z.; Liang, X.; Zheng, Q.; Li, D.; Wang, G.; An, T. The microbial biosynthesis of noncanonical terpenoids. *Appl. Microbiol. Biotechnol.* **2024**, *108*, 226. [CrossRef]
168. Tholl, D. Biosynthesis and Biological Functions of Terpenoids in Plants. In *Biotechnology of Isoprenoids*; Schrader, J., Bohlmann, J., Eds.; Springer International Publishing: Cham, Switzerland, 2015; pp. 63–106. ISBN 978-3-319-20107-8.
169. Arigoni, D.; Sagner, S.; Latzel, C.; Eisenreich, W.; Bacher, A.; Zenk, M.H. Terpenoid biosynthesis from 1-deoxy-D-xylulose in higher plants by intramolecular skeletal rearrangement. *Proc. Natl. Acad. Sci. USA* **1997**, *94*, 10600–10605. [CrossRef] [PubMed]
170. Bongers, M.; Perez-Gil, J.; Hodson, M.P.; Schrübbers, L.; Wulff, T.; Sommer, M.O.; Nielsen, L.K.; Vickers, C.E. Adaptation of hydroxymethylbutenyl diphosphate reductase enables volatile isoprenoid production. *eLife* **2020**, *9*, e48685. [CrossRef] [PubMed]
171. Niinemets, Ü.; Sun, Z.; Talts, E. Controls of the quantum yield and saturation light of isoprene emission in different-aged aspen leaves. *Plant Cell Environ.* **2015**, *38*, 2707–2720. [CrossRef]
172. Sun, Z.; Shen, Y.; Niinemets, Ü. Responses of isoprene emission and photochemical efficiency to severe drought combined with prolonged hot weather in hybrid Populus. *J. Exp. Bot.* **2020**, *71*, 7364–7381. [CrossRef] [PubMed]
173. Tritsch, D.; Hemmerlin, A.; Bach, T.J.; Rohmer, M. Plant isoprenoid biosynthesis via the MEP pathway: In vivo IPP/DMAPP ratio produced by (*E*)-4-hydroxy-3-methylbut-2-enyl diphosphate reductase in tobacco BY-2 cell cultures. *FEBS Lett.* **2010**, *584*, 129–134. [CrossRef]
174. Rodríguez-Concepción, M.; Boronat, A. Breaking new ground in the regulation of the early steps of plant isoprenoid biosynthesis. *Curr. Opin. Plant Biol.* **2015**, *25*, 17–22. [CrossRef]
175. Liao, P.; Hemmerlin, A.; Bach, T.J.; Chye, M.-L. The potential of the mevalonate pathway for enhanced isoprenoid production. *Biotechnol. Adv.* **2016**, *34*, 697–713. [CrossRef]

176. Bergman, M.E.; Kortbeek, R.W.J.; Gutensohn, M.; Dudareva, N. Plant terpenoid biosynthetic network and its multiple layers of regulation. *Prog. Lipid Res.* **2024**, *95*, 101287. [CrossRef]
177. Lange, B.M.; Ahkami, A. Metabolic engineering of plant monoterpenes, sesquiterpenes and diterpenes—Current status and future opportunities. *Plant Biotechnol. J.* **2013**, *11*, 169–196. [CrossRef]
178. Liu, B.; Zhang, L.; Rusalepp, L.; Kaurilind, E.; Sulaiman, H.Y.; Püssa, T.; Niinemets, Ü. Heat priming improved heat tolerance of photosynthesis, enhanced terpenoid and benzenoid emission and phenolics accumulation in. *Plant Cell Environ.* **2021**, *44*, 2365–2385. [CrossRef]
179. Vellosillo, T.; Martínez, M.; López, M.A.; Vicente, J.; Cascón, T.; Dolan, L.; Hamberg, M.; Castresana, C. Oxylipins Produced by the 9-Lipoxygenase Pathway in Arabidopsis Regulate Lateral Root Development and Defense Responses through a Specific Signaling Cascade. *The Plant Cell* **2007**, *19*, 831–846. [PubMed]
180. Wan, S.; Xin, X.-F. Regulation and integration of plant jasmonate signaling: A comparative view of monocot and dicot. *J. Genet. Genom.* **2022**, *49*, 704–714. [CrossRef]
181. Wang, M.; Fan, X.; Ding, F. Jasmonate: A Hormone of Primary Importance for Temperature Stress Response in Plants. *Plants* **2023**, *12*, 4080. [CrossRef] [PubMed]
182. Guenther, A.; Geron, C.; Pierce, T.; Lamb, B.; Harley, P.; Fall, R. Natural emissions of non-methane volatile organic compounds, carbon monoxide, and oxides of nitrogen from North America. *Atmos. Environ.* **2000**, *34*, 2205–2230. [CrossRef]
183. Potosnak, M.J.; LeStourgeon, L.; Pallardy, S.G.; Hosman, K.P.; Gu, L.; Karl, T.; Geron, C.; Guenther, A.B. Observed and modeled ecosystem isoprene fluxes from an oak-dominated temperate forest and the influence of drought stress. *Atmos. Environ.* **2014**, *84*, 314–322. [CrossRef]
184. Yang, W.; Zhang, B.; Wu, Y.; Liu, S.; Kong, F.; Li, L. Effects of soil drought and nitrogen deposition on BVOC emissions and their O3 and SOA formation for Pinus thunbergii. *Environ. Pollut.* **2023**, *316*, 120693. [CrossRef] [PubMed]
185. Peng, Y.; Mouat, A.P.; Hu, Y.; Li, M.; McDonald, B.C.; Kaiser, J. Source appointment of volatile organic compounds and evaluation of anthropogenic monoterpene emission estimates in Atlanta, Georgia. *Atmos. Environ.* **2022**, *288*, 119324.
186. Xu, L.; Pye, H.O.T.; He, J.; Chen, Y.; Murphy, B.N.; Ng, N.L. Experimental and model estimates of the contributions from biogenic monoterpenes and sesquiterpenes to secondary organic aerosol in the southeastern United States. *Atmos. Chem. Phys.* **2018**, *18*, 12613–12637. [CrossRef]
187. Loreto, F.; Velikova, V. Isoprene Produced by Leaves Protects the Photosynthetic Apparatus against Ozone Damage, Quenches Ozone Products, and Reduces Lipid Peroxidation of Cellular Membranes. *Plant Physiol.* **2001**, *127*, 1781–1787. [CrossRef]
188. Zenone, T.; Hendriks, C.; Brilli, F.; Fransen, E.; Gioli, B.; Portillo-Estrada, M.; Schaap, M.; Ceulemans, R. Interaction between isoprene and ozone fluxes in a poplar plantation and its impact on air quality at the European level. *Sci. Rep.* **2016**, *6*, 32676. [CrossRef]
189. Zou, Y.; Deng, X.J.; Deng, T.; Yin, C.Q.; Li, F. One-Year Characterization and Reactivity of Isoprene and Its Impact on Surface Ozone Formation at A Suburban Site in Guangzhou, China. *Atmosphere* **2019**, *10*, 201. [CrossRef]
190. Schade, G.W.; Goldstein, A.H.; Lamanna, M.S. Are monoterpene emissions influenced by humidity? *Geophys. Res. Lett.* **1999**, *26*, 2187–2190. [CrossRef]
191. Street, R.A.; Owen, S.; Duckham, S.C.; Boissard, C.; Hewitt, C.N. Effect of habitat and age on variations in volatile organic compound (VOC) emissions from Quercus ilex and Pinus pinea. *Atmos. Environ.* **1997**, *31*, 89–100. [CrossRef]
192. Kuzma, J.; Fall, R. Leaf Isoprene Emission Rate Is Dependent on Leaf Development and the Level of Isoprene Synthase. *Plant Physiol.* **1993**, *101*, 435–440. [CrossRef] [PubMed]
193. Rasulov, B.; Bichele, I.; Laisk, A.; Niinemets, Ü. Competition between isoprene emission and pigment synthesis during leaf development in aspen. *Plant Cell Environ.* **2014**, *37*, 724–741. [CrossRef]
194. Lin, P.-A.; Chen, Y.; Ponce, G.; Acevedo, F.E.; Lynch, J.P.; Anderson, C.T.; Ali, J.G.; Felton, G.W. Stomata-mediated interactions between plants, herbivores, and the environment. *Trends Plant Sci.* **2022**, *27*, 287–300. [CrossRef]
195. Loreto, F.; Schnitzler, J.-P. Abiotic stresses and induced BVOCs. *Trends Plant Sci.* **2010**, *15*, 154–166. [CrossRef] [PubMed]
196. Niinemets, Ü.; Reichstein, M. Controls on the emission of plant volatiles through stomata: A sensitivity analysis. *J. Geophys. Res. Atmos.* **2003**, *108*, 4211. [CrossRef]
197. Yin, Q.; Tian, T.; Kou, M.; Liu, P.; Wang, L.; Hao, Z.; Yue, M. The relationships between photosynthesis and stomatal traits on the Loess Plateau. *Glob. Ecol. Conserv.* **2020**, *23*, e01146. [CrossRef]
198. Chuong, B.; Zhang, J.; Donahue, N.M. Cycloalkene Ozonolysis: Collisionally Mediated Mechanistic Branching. *J. Am. Chem. Soc.* **2004**, *126*, 12363–12373. [CrossRef]
199. Nemecek-Marshall, M.; MacDonald, R.C.; Franzen, J.J.; Wojciechowski, C.L.; Fall, R. Methanol Emission from Leaves (Enzymatic Detection of Gas-Phase Methanol and Relation of Methanol Fluxes to Stomatal Conductance and Leaf Development). *Plant Physiol.* **1995**, *108*, 1359–1368. [CrossRef] [PubMed]
200. Niinemets, Ü. Mild versus severe stress and BVOCs: Thresholds, priming and consequences. *Trends Plant Sci.* **2010**, *15*, 145–153. [CrossRef]

201. Riedlmeier, M.; Ghirardo, A.; Wenig, M.; Knappe, C.; Koch, K.; Georgii, E.; Dey, S.; Parker, J.E.; Schnitzler, J.-P.; Vlot, A.C. Monoterpenes Support Systemic Acquired Resistance within and between Plants. *Plant Cell* **2017**, *29*, 1440–1459. [CrossRef] [PubMed]
202. Imbiscuso, G.; Trotta, A.; Maffei, M.; Bossi, S. Herbivory induces a ROS burst and the release of volatile organic compounds in the fern *Pteris vittata* L. *J. Plant Interact.* **2009**, *4*, 15–22. [CrossRef]
203. Joutsensaari, J.; Yli-Pirilä, P.; Korhonen, H.; Arola, A.; Blande, J.D.; Heijari, J.; Kivimäenpää, M.; Mikkonen, S.; Hao, L.; Miettinen, P.; et al. Biotic stress accelerates formation of climate-relevant aerosols in boreal forests. *Atmos. Chem. Phys.* **2015**, *15*, 12139–12157. [CrossRef]
204. Penuelas, J.; Llusia, J. The Complexity of Factors Driving Volatile Organic Compound Emissions by Plants. *Biol. Plant.* **2001**, *44*, 481–487. [CrossRef]
205. Litvak, M.E.; Monson, R.K. Patterns of induced and constitutive monoterpene production in conifer needles in relation to insect herbivory. *Oecologia* **1998**, *114*, 531–540. [CrossRef]
206. Abis, L.; Kalalian, C.; Lunardelli, B.; Wang, T.; Zhang, L.; Chen, J.; Perrier, S.; Loubet, B.; Ciuraru, R.; George, C. Measurement report: Biogenic volatile organic compound emission profiles of rapeseed leaf litter and its secondary organic aerosol formation potential. *Atmos. Chem. Phys.* **2021**, *21*, 12613–12629. [CrossRef]
207. Monson, R.K.; Jaeger, C.H.; Adams, W.W.; Driggers, E.M.; Silver, G.M.; Fall, R. Relationships among Isoprene Emission Rate, Photosynthesis, and Isoprene Synthase Activity as Influenced by Temperature. *Plant Physiol.* **1992**, *98*, 1175–1180. [CrossRef]
208. Li, T.; Holst, T.; Michelsen, A.; Rinnan, R. Amplification of plant volatile defence against insect herbivory in a warming Arctic tundra. *Nat. Plants* **2019**, *5*, 568–574. [CrossRef]
209. Loreto, F.; Barta, C.; Brilli, F.; Nogues, I. On the induction of volatile organic compound emissions by plants as consequence of wounding or fluctuations of light and temperature. *Plant Cell Environ.* **2006**, *29*, 1820–1828. [CrossRef] [PubMed]
210. Lerdau, M.; Gray, D. Ecology and evolution of light-dependent and light-independent phytogenic volatile organic carbon. *New Phytol.* **2003**, *157*, 199–211. [CrossRef]
211. Sharkey, T.D.; Monson, R.K. Isoprene research—60 years later, the biology is still enigmatic. *Plant Cell Environ.* **2017**, *40*, 1671–1678. [CrossRef]
212. Yu, S.-J.; Seo, Y.-S.; Son, Y.-S.; Kim, J.-C. Characteristics of BVOCs emitted from representative tree species in South Korea according to environmental influences. *Atmos. Environ.* **2024**, *318*, 120238. [CrossRef]
213. Aliniaeifard, S.; van Meeteren, U. Can prolonged exposure to low VPD disturb the ABA signalling in stomatal guard cells? *J. Exp. Bot.* **2013**, *64*, 3551–3566. [CrossRef]
214. Arve, L.E.; Terfa, M.T.; Gislerød, H.R.; Olsen, J.E.; Torre, S. High relative air humidity and continuous light reduce stomata functionality by affecting the ABA regulation in rose leaves. *Plant Cell Environ.* **2013**, *36*, 382–392. [CrossRef] [PubMed]
215. Mu, Z.; Asensio, D.; Llusià, J.; Filella, I.; Ogaya, R.; Yi, Z.; Peñuelas, J. Annual and seasonal variations in soil volatile organic compound concentrations in a Mediterranean shrubland and holm oak forest. *Geoderma* **2022**, *405*, 115401. [CrossRef]
216. Yáñez-Serrano, A.M.; Nölscher, A.C.; Williams, J.; Wolff, S.; Alves, E.; Martins, G.A.; Bourtsoukidis, E.; Brito, J.; Jardine, K.; Artaxo, P.; et al. Diel and seasonal changes of biogenic volatile organic compounds within and above an Amazonian rainforest. *Atmos. Chem. Phys.* **2015**, *15*, 3359–3378. [CrossRef]
217. Fanourakis, D.; Giday, H.; Hyldgaard, B.; Bouranis, D.; Körner, O.; Ottosen, C.-O. Low air humidity during cultivation promotes stomatal closure ability in rose. *Eur. J. Hortic. Sci.* **2019**, *84*, 245–252. [CrossRef]
218. Ning, P.; Guo, X.; Tian, S.; Shi, J.; Sun, C. Emission of main BVOCS for typical landscape trees in Kunming. *Zhongnan Daxue Xuebao (Ziran Kexue Ban)/J. Cent. South Univ. (Sci. Technol.)* **2013**, *44*, 1290–1296.
219. Constable, J.V.; Litvak, M.E.; Greenberg, J.P.; Monson, R.K. Monoterpene emission from coniferous trees in response to elevated CO2 concentration and climate warming. *Glob. Change Biol.* **1999**, *5*, 252–267. [CrossRef]
220. Jasoni, R.; Kane, C.; Green, C.; Peffley, E.; Tissue, D.; Thompson, L.; Payton, P.; Paré, P.W. Altered leaf and root emissions from onion (Allium cepa L.) grown under elevated CO2 conditions. *Environ. Exp. Bot.* **2004**, *51*, 273–280. [CrossRef]
221. Rapparini, F.; Baraldi, R.; Miglietta, F.; Loreto, F. Isoprenoid emission in trees of Quercus pubescens and Quercus ilex with lifetime exposure to naturally high CO2 environment. *Plant Cell Environ.* **2004**, *27*, 381–391. [CrossRef]
222. Heiden, A.C.; Kobel, K.; Langebartels, C.; Schuh-Thomas, G.; Wildt, J. Emissions of Oxygenated Volatile Organic Compounds from Plants Part I: Emissions from Lipoxygenase Activity. *J. Atmos. Chem.* **2003**, *45*, 143–172. [CrossRef]
223. Kim, S.-Y.; Jiang, X.; Lee, M.; Turnipseed, A.; Guenther, A.; Kim, J.-C.; Lee, S.-J.; Kim, S. Impact of biogenic volatile organic compounds on ozone production at the Taehwa Research Forest near Seoul, South Korea. *Atmos. Environ.* **2013**, *70*, 447–453. [CrossRef]
224. Sha, T.; Ma, X.; Zhang, H.; Janechek, N.; Wang, Y.; Wang, Y.; Castro García, L.; Jenerette, G.D.; Wang, J. Impacts of Soil NO_x Emission on O_3 Air Quality in Rural California. *Environ. Sci. Technol.* **2021**, *55*, 7113–7122. [CrossRef]

225. Tiwari, A.; Kumar, P.; Baldauf, R.; Zhang, K.M.; Pilla, F.; Di Sabatino, S.; Brattich, E.; Pulvirenti, B. Considerations for evaluating green infrastructure impacts in microscale and macroscale air pollution dispersion models. *Sci. Total Environ.* **2019**, *672*, 410–426. [CrossRef]
226. Llusià, J.; Peñuelas, J.; Gimeno, B.S. Seasonal and species-specific response of VOC emissions by Mediterranean woody plant to elevated ozone concentrations. *Atmos. Environ.* **2002**, *36*, 3931–3938. [CrossRef]
227. Li, L.; Cao, J.; Hao, Y. Spatial and species-specific responses of biogenic volatile organic compound (BVOC) emissions to elevated ozone from 2014–2020 in China. *Sci. Total Environ.* **2023**, *868*, 161636. [CrossRef]
228. Vo, T.; Faiola, C.L. Acute ozone exposure decreases terpene emissions from Canary Island pines. *Agric. For. Meteorol.* **2023**, *333*, 109416. [CrossRef]
229. Barreira, L.M.F.; Ylisirniö, A.; Pullinen, I.; Buchholz, A.; Li, Z.; Lipp, H.; Junninen, H.; Hõrrak, U.; Noe, S.M.; Krasnova, A.; et al. The importance of sesquiterpene oxidation products for secondary organic aerosol formation in a springtime hemiboreal forest. *Atmos. Chem. Phys.* **2021**, *21*, 11781–11800. [CrossRef]
230. Salvador, C.M.; Chou, C.C.-K.; Ho, T.-T.; Tsai, C.-Y.; Tsao, T.-M.; Tsai, M.-J.; Su, T.-C. Contribution of Terpenes to Ozone Formation and Secondary Organic Aerosols in a Subtropical Forest Impacted by Urban Pollution. *Atmosphere* **2020**, *11*, 1232. [CrossRef]
231. Charan, S.M.; Huang, Y.; Seinfeld, J.H. Computational Simulation of Secondary Organic Aerosol Formation in Laboratory Chambers. *Chem. Rev.* **2019**, *119*, 11912–11944. [CrossRef]
232. Hellén, H.; Praplan, A.P.; Tykkä, T.; Ylivinkka, I.; Vakkari, V.; Bäck, J.; Petäjä, T.; Kulmala, M.; Hakola, H. Long-term measurements of volatile organic compounds highlight the importance of sesquiterpenes for the atmospheric chemistry of a boreal forest. *Atmos. Chem. Phys.* **2018**, *18*, 13839–13863. [CrossRef]
233. Kawamura, K.; Bikkina, S. A review of dicarboxylic acids and related compounds in atmospheric aerosols: Molecular distributions, sources and transformation. *Atmos. Res.* **2016**, *170*, 140–160. [CrossRef]
234. Yue, H.; Zhang, C.; Lin, X.; Wen, Z.; Zhang, W.; Mostafa, S.; Luo, P.-L.; Zhang, Z.; Hemberger, P.; Fittschen, C.; et al. Dimeric Product of Peroxy Radical Self-Reaction Probed with VUV Photoionization Mass Spectrometry and Theoretical Calculations: The Case of C2H5OOC2H5. *Int. J. Mol. Sci.* **2023**, *24*, 3731. [CrossRef] [PubMed]
235. Dunlea, E.J.; Ravishankara, A.R. Measurement of the rate coefficient for the reaction of O(1D) with H2O and re-evaluation of the atmospheric OH production rate. *Phys. Chem. Chem. Phys.* **2004**, *6*, 3333–3340. [CrossRef]
236. Monks, P.S. Gas-phase radical chemistry in the troposphere. *Chem. Soc. Rev.* **2005**, *34*, 376–395. [CrossRef]
237. Wang, X.; Liu, L.; Fang, W.; Chen, X. Theoretical insight towards the photo-dissociation dynamics of O3–H2O complex: Deep understanding the source of atmospheric hydroxyl radical. *Chem. Phys. Lett.* **2014**, *608*, 95–101. [CrossRef]
238. Atkinson, R. Kinetics and mechanisms of the gas-phase reactions of the hydroxyl radical with organic compounds under atmospheric conditions. *Chem. Rev.* **1986**, *86*, 69–201. [CrossRef]
239. Zhao, Y.; Wingen, L.M.; Perraud, V.; Greaves, J.; Finlayson-Pitts, B.J. Role of the reaction of stabilized Criegee intermediates with peroxy radicals in particle formation and growth in air. *Phys. Chem. Chem. Phys.* **2015**, *17*, 12500–12514. [CrossRef] [PubMed]
240. Brown, S.S.; Stutz, J. Nighttime radical observations and chemistry. *Chem. Soc. Rev.* **2012**, *41*, 6405–6447. [CrossRef] [PubMed]
241. Geyer, A.; Alicke, B.; Konrad, S.; Schmitz, T.; Stutz, J.; Platt, U. Chemistry and oxidation capacity of the nitrate radical in the continental boundary layer near Berlin. *J. Geophys. Res.-Atmos.* **2001**, *106*, 8013–8025. [CrossRef]
242. Schindler, M. A QSAR for the prediction of rate constants for the reaction of VOCs with nitrate radicals. *Chemosphere* **2016**, *154*, 23–33. [CrossRef]
243. Mogensen, D.; Gierens, R.; Crowley, J.N.; Keronen, P.; Smolander, S.; Sogachev, A.; Nölscher, A.C.; Zhou, L.; Kulmala, M.; Tang, M.J.; et al. Simulations of atmospheric OH, O_3 and NO_3 reactivities within and above the boreal forest. *Atmos. Chem. Phys.* **2015**, *15*, 3909–3932. [CrossRef]
244. Xu, L.; Tsona, N.T.; You, B.; Zhang, Y.; Wang, S.; Yang, Z.; Xue, L.; Du, L. NOx enhances secondary organic aerosol formation from nighttime γ-terpinene ozonolysis. *Atmos. Environ.* **2020**, *225*, 117375. [CrossRef]
245. Waring, M.S.; Wells, J.R. Volatile organic compound conversion by ozone, hydroxyl radicals, and nitrate radicals in residential indoor air: Magnitudes and impacts of oxidant sources. *Atmos. Environ.* **2015**, *106*, 382–391. [CrossRef]
246. Liu, Y.; Qiu, P.; Li, C.; Li, X.; Ma, W.; Yin, S.; Yu, Q.; Li, J.; Liu, X. Evolution and variations of atmospheric VOCs and O3 photochemistry during a summer O3 event in a county-level city, Southern China. *Atmos. Environ.* **2022**, *272*, 118942. [CrossRef]
247. Sun, J.; Yu, X.; Ling, Z.; Fang, G.; Ming, L.; Zhao, J.; Zou, S.; Guan, H.; Wang, H.; Wang, X.; et al. Roles of photochemical consumption of VOCs on regional background O3 concentration and atmospheric reactivity over the pearl river estuary, Southern China. *Sci. Total Environ.* **2024**, *928*, 172321. [CrossRef]
248. Ng, N.L.; Canagaratna, M.R.; Zhang, Q.; Jimenez, J.L.; Tian, J.; Ulbrich, I.M.; Kroll, J.H.; Docherty, K.S.; Chhabra, P.S.; Bahreini, R.; et al. Organic aerosol components observed in Northern Hemispheric datasets from Aerosol Mass Spectrometry. *Atmos. Chem. Phys.* **2010**, *10*, 4625–4641. [CrossRef]
249. Atkinson, R.; Arey, J. Atmospheric Degradation of Volatile Organic Compounds. *Chem. Rev.* **2003**, *103*, 4605–4638. [CrossRef] [PubMed]

250. Docherty, K.S.; Wu, W.; Lim, Y.B.; Ziemann, P.J. Contributions of Organic Peroxides to Secondary Aerosol Formed from Reactions of Monoterpenes with O3. *Environ. Sci. Technol.* **2005**, *39*, 4049–4059. [CrossRef]
251. Kroll, J.H.; Seinfeld, J.H. Chemistry of secondary organic aerosol: Formation and evolution of low-volatility organics in the atmosphere. *Atmos. Environ.* **2008**, *42*, 3593–3624. [CrossRef]
252. Orlando, J.J.; Tyndall, G.S. Laboratory studies of organic peroxy radical chemistry: An overview with emphasis on recent issues of atmospheric significance. *Chem. Soc. Rev.* **2012**, *41*, 6294–6317. [CrossRef]
253. Berndt, T.; Richters, S.; Kaethner, R.; Voigtländer, J.; Stratmann, F.; Sipilä, M.; Kulmala, M.; Herrmann, H. Gas-Phase Ozonolysis of Cycloalkenes: Formation of Highly Oxidized RO2 Radicals and Their Reactions with NO, NO2, SO2, and Other RO2 Radicals. *J. Phys. Chem. A* **2015**, *119*, 10336–10348. [CrossRef] [PubMed]
254. Calfapietra, C.; Fares, S.; Manes, F.; Morani, A.; Sgrigna, G.; Loreto, F. Role of Biogenic Volatile Organic Compounds (BVOC) emitted by urban trees on ozone concentration in cities: A review. *Environ. Pollut.* **2013**, *183*, 71–80. [CrossRef]
255. Chen, W.; Guenther, A.B.; Jia, S.; Mao, J.; Yan, F.; Wang, X.; Shao, M. Synergistic effects of biogenic volatile organic compounds and soil nitric oxide emissions on summertime ozone formation in China. *Sci. Total Environ.* **2022**, *828*, 154218. [CrossRef]
256. Xie, X.; Shao, M.; Liu, Y.; Lu, S.; Chang, C.-C.; Chen, Z.-M. Estimate of initial isoprene contribution to ozone formation potential in Beijing, China. *Atmos. Environ.* **2008**, *42*, 6000–6010. [CrossRef]
257. Kim, S.; Lee, M.; Kim, S.; Choi, S.; Seok, S.; Kim, S. Photochemical characteristics of high and low ozone episodes observed in the Taehwa Forest observatory (TFO) in June 2011 near Seoul South Korea. *Asia-Pac. J. Atmos. Sci.* **2013**, *49*, 325–331. [CrossRef]
258. Lin, C.; Feng, X.; Heal, M.R. Temporal persistence of intra-urban spatial contrasts in ambient NO2, O3 and Ox in Edinburgh, UK. *Atmos. Pollut. Res.* **2016**, *7*, 734–741. [CrossRef]
259. Pancholi, P.; Kumar, A.; Bikundia, D.S.; Chourasiya, S. An observation of seasonal and diurnal behavior of O3–NOx relationships and local/regional oxidant (OX = O3 + NO2) levels at a semi-arid urban site of western India. *Sustain. Environ. Res.* **2018**, *28*, 79–89. [CrossRef]
260. Simon, H.; Fallmann, J.; Kropp, T.; Tost, H.; Bruse, M. Urban Trees and Their Impact on Local Ozone Concentration—A Microclimate Modeling Study. *Atmosphere* **2019**, *10*, 154. [CrossRef]
261. Zhang, Y.; Dai, J.; Li, Q.; Chen, T.; Mu, J.; Brasseur, G.; Wang, T.; Xue, L. Biogenic volatile organic compounds enhance ozone production and complicate control efforts: Insights from long-term observations in Hong Kong. *Atmos. Environ.* **2023**, *309*, 119917. [CrossRef]
262. Liu, C.; Shi, K. A review on methodology in O3-NOx-VOC sensitivity study. *Environ. Pollut.* **2021**, *291*, 118249. [CrossRef] [PubMed]
263. Ren, J.; Guo, F.; Xie, S. Diagnosing ozone–NO_x–VOC sensitivity and revealing causes of ozone increases in China based on 2013–2021 satellite retrievals. *Atmos. Chem. Phys.* **2022**, *22*, 15035–15047. [CrossRef]
264. Iyer, S.; Reiman, H.; Møller, K.H.; Rissanen, M.P.; Kjaergaard, H.G.; Kurtén, T. Computational Investigation of RO2 + HO2 and RO2 + RO2 Reactions of Monoterpene Derived First-Generation Peroxy Radicals Leading to Radical Recycling. *J. Phys. Chem. A* **2018**, *122*, 9542–9552. [CrossRef]
265. Bianchi, F.; Junninen, H.; Bigi, A.; Sinclair, V.A.; Dada, L.; Hoyle, C.R.; Zha, Q.; Yao, L.; Ahonen, L.R.; Bonasoni, P.; et al. Biogenic particles formed in the Himalaya as an important source of free tropospheric aerosols. *Nat. Geosci.* **2021**, *14*, 4–9. [CrossRef]
266. Friedman, B.; Farmer, D.K. SOA and gas phase organic acid yields from the sequential photooxidation of seven monoterpenes. *Atmos. Environ.* **2018**, *187*, 335–345. [CrossRef]
267. Kulmala, M.; Nieminen, T.; Chellapermal, R.; Makkonen, R.; Bäck, J.; Kerminen, V.-M. Climate Feedbacks Linking the Increasing Atmospheric CO2 Concentration, BVOC Emissions, Aerosols and Clouds in Forest Ecosystems. In *Biology, Controls and Models of Tree Volatile Organic Compound Emissions*; Niinemets, Ü., Monson, R.K., Eds.; Springer: Dordrecht, The Netherlands, 2013; pp. 489–508, ISBN 978-94-007-6606-8.
268. Jimenez, J.L.; Canagaratna, M.R.; Donahue, N.M.; Prevot, A.S.H.; Zhang, Q.; Kroll, J.H.; DeCarlo, P.F.; Allan, J.D.; Coe, H.; Ng, N.L.; et al. Evolution of Organic Aerosols in the Atmosphere. *Science* **2009**, *326*, 1525–1529. [CrossRef]
269. Robinson, A.L.; Donahue, N.M.; Shrivastava, M.K.; Weitkamp, E.A.; Sage, A.M.; Grieshop, A.P.; Lane, T.E.; Pierce, J.R.; Pandis, S.N. Rethinking Organic Aerosols: Semivolatile Emissions and Photochemical Aging. *Science* **2007**, *315*, 1259–1262. [CrossRef]
270. Carlton, A.G.; Wiedinmyer, C.; Kroll, J.H. A review of Secondary Organic Aerosol (SOA) formation from isoprene. *Atmos. Chem. Phys.* **2009**, *9*, 4987–5005. [CrossRef]
271. Dada, L.; Stolzenburg, D.; Simon, M.; Fischer, L.; Heinritzi, M.; Wang, M.; Xiao, M.; Vogel, A.L.; Ahonen, L.; Amorim, A.; et al. Role of sesquiterpenes in biogenic new particle formation. *Sci. Adv.* **2023**, *9*, eadi5297. [CrossRef] [PubMed]
272. Bates, K.H.; Burke, G.J.P.; Cope, J.D.; Nguyen, T.B. Secondary organic aerosol and organic nitrogen yields from the nitrate radical (NO_3) oxidation of alpha-pinene from various RO_2 fates. *Atmos. Chem. Phys.* **2022**, *22*, 1467–1482. [CrossRef]
273. Ziemann, P.J.; Atkinson, R. Kinetics, products, and mechanisms of secondary organic aerosol formation. *Chem. Soc. Rev.* **2012**, *41*, 6582–6605. [CrossRef] [PubMed]

274. Ehn, M.; Thornton, J.A.; Kleist, E.; Sipilä, M.; Junninen, H.; Pullinen, I.; Springer, M.; Rubach, F.; Tillmann, R.; Lee, B.; et al. A large source of low-volatility secondary organic aerosol. *Nature* **2014**, *506*, 476–479. [CrossRef]
275. Perring, A.E.; Pusede, S.E.; Cohen, R.C. An Observational Perspective on the Atmospheric Impacts of Alkyl and Multifunctional Nitrates on Ozone and Secondary Organic Aerosol. *Chem. Rev.* **2013**, *113*, 5848–5870. [CrossRef]
276. Kirkby, J.; Duplissy, J.; Sengupta, K.; Frege, C.; Gordon, H.; Williamson, C.; Heinritzi, M.; Simon, M.; Yan, C.; Almeida, J.; et al. Ion-induced nucleation of pure biogenic particles. *Nature* **2016**, *533*, 521–526. [CrossRef]
277. Tröstl, J.; Chuang, W.K.; Gordon, H.; Heinritzi, M.; Yan, C.; Molteni, U.; Ahlm, L.; Frege, C.; Bianchi, F.; Wagner, R.; et al. The role of low-volatility organic compounds in initial particle growth in the atmosphere. *Nature* **2016**, *533*, 527–531. [CrossRef]
278. Cash, J.M.; Heal, M.R.; Langford, B.; Drewer, J. A review of stereochemical implications in the generation of secondary organic aerosol from isoprene oxidation. *Environ. Sci. Process. Impacts* **2016**, *18*, 1369–1380. [CrossRef]
279. Yang, Z.; Du, L.; Li, Y.; Ge, X. Secondary organic aerosol formation from monocyclic aromatic hydrocarbons: Insights from laboratory studies. *Environ. Sci. Process. Impacts* **2022**, *24*, 351–379. [CrossRef]
280. Gao, Y.; Yan, F.; Ma, M.; Ding, A.; Liao, H.; Wang, S.; Wang, X.; Zhao, B.; Cai, W.; Su, H.; et al. Unveiling the dipole synergic effect of biogenic and anthropogenic emissions on ozone concentrations. *Sci. Total Environ.* **2022**, *818*, 151722. [CrossRef]
281. Shrivastava, M.; Cappa, C.D.; Fan, J.; Goldstein, A.H.; Guenther, A.B.; Jimenez, J.L.; Kuang, C.; Laskin, A.; Martin, S.T.; Ng, N.L.; et al. Recent advances in understanding secondary organic aerosol: Implications for global climate forcing. *Rev. Geophys.* **2017**, *55*, 509–559. [CrossRef]
282. Fierravanti, A.; Fierravanti, E.; Cocozza, C.; Tognetti, R.; Rossi, S. Eligible reference cities in relation to BVOC-derived O3 pollution. *Urban For. Urban Green.* **2017**, *28*, 73–80. [CrossRef]
283. Yenisoy-Karakaş, S.; Dörter, M.; Odabasi, M. Intraday and interday variations of 69 volatile organic compounds (BVOCs and AVOCs) and their source profiles at a semi-urban site. *Sci. Total Environ.* **2020**, *723*, 138028. [CrossRef] [PubMed]
284. Li, N.; He, Q.; Greenberg, J.; Guenther, A.; Li, J.; Cao, J.; Wang, J.; Liao, H.; Wang, Q.; Zhang, Q. Impacts of biogenic and anthropogenic emissions on summertime ozone formation in the Guanzhong Basin, China. *Atmos. Chem. Phys.* **2018**, *18*, 7489–7507. [CrossRef]
285. Choi, J.-S.; Choi, Y.; Ahn, J.; Park, J.; Oh, J.; Lee, G.; Park, T.; Park, G.; Owen, J.S.; Lee, T. Observation of Secondary Organic Aerosol and New Particle Formation at a Remote Site in Baengnyeong Island, Korea. *Asian J. Atmos. Environ.* **2017**, *11*, 300–312. [CrossRef]
286. Ghirardo, A.; Xie, J.; Zheng, X.; Wang, Y.; Grote, R.; Block, K.; Wildt, J.; Mentel, T.; Kiendler-Scharr, A.; Hallquist, M.; et al. Urban stress-induced biogenic VOC emissions and SOA-forming potentials in Beijing. *Atmos. Chem. Phys.* **2016**, *16*, 2901–2920. [CrossRef]
287. Šimpraga, M.; Ghimire, R.P.; Van Der Straeten, D.; Blande, J.D.; Kasurinen, A.; Sorvari, J.; Holopainen, T.; Adriaenssens, S.; Holopainen, J.K.; Kivimäenpää, M. Unravelling the functions of biogenic volatiles in boreal and temperate forest ecosystems. *Eur. J. Forest Res.* **2019**, *138*, 763–787. [CrossRef]
288. Yong, J.; Xie, Y.; Guo, H.; Li, Y.; Sun, S. Unraveling the influence of biogenic volatile organic compounds and their constituents on ozone and SOA formation within the Yellow River Basin, China. *Chemosphere* **2024**, *353*, 141549. [CrossRef]
289. Jia, Y.; Qiao, L.; Xie, W.; Li, L. Spatial-temporal characteristics of ambient isoprene and monoterpene and their ozone and secondary organic aerosol formation potential in China. *Environ. Rev.* **2024**, *32*, 203–213. [CrossRef]
290. Zheng, J.; Shao, M.; Che, W.; Zhang, L.; Zhong, L.; Zhang, Y.; Streets, D. Speciated VOC Emission Inventory and Spatial Patterns of Ozone Formation Potential in the Pearl River Delta, China. *Environ. Sci. Technol.* **2009**, *43*, 8580–8586. [CrossRef]
291. Ryu, Y.-H.; Baik, J.-J.; Kwak, K.-H.; Kim, S.; Moon, N. Impacts of urban land-surface forcing on ozone air quality in the Seoul metropolitan area. *Atmos. Chem. Phys.* **2013**, *13*, 2177–2194. [CrossRef]
292. Li, Y.; Yin, S.; Yu, S.; Bai, L.; Wang, X.; Lu, X.; Ma, S. Characteristics of ozone pollution and the sensitivity to precursors during early summer in central plain, China. *J. Environ. Sci.* **2021**, *99*, 354–368. [CrossRef] [PubMed]
293. Mahilang, M.; Deb, M.K.; Pervez, S. Biogenic secondary organic aerosols: A review on formation mechanism, analytical challenges and environmental impacts. *Chemosphere* **2021**, *262*, 127771. [CrossRef]
294. Yan, F.; Chen, W.; Jia, S.; Zhong, B.; Yang, L.; Mao, J.; Chang, M.; Shao, M.; Yuan, B.; Situ, S.; et al. Stabilization for the secondary species contribution to PM2.5 in the Pearl River Delta (PRD) over the past decade, China: A meta-analysis. *Atmos. Environ.* **2020**, *242*, 117817. [CrossRef]
295. Déméautis, T.; Delles, M.; Tomaz, S.; Monneret, G.; Glehen, O.; Devouassoux, G.; George, C.; Bentaher, A. Pathogenic Mechanisms of Secondary Organic Aerosols. *Chem. Res. Toxicol.* **2022**, *35*, 1146–1161. [CrossRef] [PubMed]
296. Zeng, J.; Zhang, Y.; Zhang, H.; Song, W.; Wu, Z.; Wang, X. Design and characterization of a semi-open dynamic chamber for measuring biogenic volatile organic compound (BVOC) emissions from plants. *Atmos. Meas. Tech.* **2022**, *15*, 79–93. [CrossRef]

297. Xu, Y.; Ling, L.; Xie, P.; Hu, R.; Li, Z.; Chen, H. Low cost and high resolution electrochemical sensing for atmospheric NO2 measurement. *Huanjing Kexue Xuebao/Acta Sci. Circumstantiae* **2018**, *38*, 3488–3495. [CrossRef]
298. Kim, S.; Guenther, A.; Apel, E. Quantitative and qualitative sensing techniques for biogenic volatile organic compounds and their oxidation products. *Environ. Sci. Process. Impacts* **2013**, *15*, 1301–1314. [CrossRef]

toxics

MDPI

Article

Optical Properties and Molecular Composition of Fine Organic Aerosols in Nanjing, China: A Comparison of 2019 and 2023

Binhuang Zhou [1], Yu Huang [1], Liangyu Feng [1], Zihao Zhang [1], Haiwei Li [1], Yun Wu [1], Jianhuai Ye [2] and Xinlei Ge [1,3,*]

[1] Jiangsu Key Laboratory of Atmospheric Environment Monitoring and Pollution Control, Collaborative Innovation Center of Atmospheric Environment and Equipment Technology, School of Environmental Science and Engineering, Nanjing University of Information Science and Technology, Nanjing 210044, China
[2] School of Environmental Science and Engineering, Southern University of Science and Technology, Shenzhen 518055, China
[3] School of Environment and Energy Engineering, Anhui Jianzhu University, Hefei 230601, China
* Correspondence: caxinra@163.com

Abstract: Optical properties and chemical composition of atmospheric fine particles ($PM_{2.5}$) are critical to their environmental and health effects. In this study, we analyzed the organic aerosols (OA) in $PM_{2.5}$ samples in Nanjing, China, collected during the summer and winter of 2019 and 2023. Results show a decline in both concentrations and light-absorbing abilities of methanol—soluble organic carbon (MSOC) and water-soluble OC (WSOC) in OA from 2019 to 2023. Due to increased combustion activities, MSOC and WSOC concentrations, and their corresponding mass absorption efficiencies were all higher in winter than in summer. Furthermore, fluorescence indices suggest that OA in Nanjing was influenced by a mix of microbial/biogenic sources. Fluorescent properties of both WSOC and MSOC were dominated by humic-like components but the remaining contribution from protein-like components was more significant in MSOC. The molecular composition of OA did not show a remarkable difference between 2019 and 2023. Overall, CHON compounds were the most abundant species, followed by CHO and CHN compounds, and aliphatic compounds dominated all molecular types except for CHN (in positive mode) and CHON, CHOS (in negative mode). Regarding the OA sources, the numbers of molecules from fossil fuel combustion and biomass burning (BB) were a bit more in 2023 than in 2019, and signal intensities of BB-related molecules were also higher in winter than in summer; the presence of organosulfates indicate the contribution of aqueous-phase oxidation to OA, especially during high relative humidity conditions. At last, correlations between OA molecules and light absorption efficiencies indicate that the key light-absorbing species in winter and summer were likely quite different despite similar chemical compositions, and in summer, CH and CHN compounds were important to light absorption, whereas CHNS compounds became more important in winter.

Keywords: molecular composition; light absorption; fluorescent properties

1. Introduction

Organic aerosol (OA) is a major and highly complicated component of atmospheric aerosols, accounting for 20% to 90% of their total mass [1], and playing a crucial role in air quality, climate change and human health. OA includes species emitted from multiple primary sources, including biomass burning, biofuel combustion [2,3], fossil fuel combustion (such as traffic activities) [4–7], and species from various secondary chemical reactions [2,8–12]. Many OA species can absorb ultraviolet-visible (UV-Vis) lights, which are collectively termed "brown carbon" (BrC), and are an important driver of aerosol's radiative effect [13–16]; for example, it can occupy 30–50% of the total OA light absorption of in 300–400 nm wavelength range [5,17]; it also affects the cloud condensation nuclei (CCN), therefore cloud formation and climate indirectly [18,19]. BrC mainly includes some highly conjugated organic compounds, such as polycyclic aromatic hydrocarbons (PAHs), humic-like substances (HULIS), etc. [20–22]; a recent study shows that nitrogen-containing organics dominate the light absorption of OA [23]. The light absorption properties of OA are typically measured by the ultraviolet-visible (UV-Vis) spectrophotometry [16,24]. The excitation-emission matrix (EEM) fluorescence spectroscopy is on the other hand used to determine the fluorescence properties of OA, and further parallel factor analysis (PARAFAC) can infer compositions of its key fluorophores [25–27]. The molecular structure of OA has a significant impact on its optical characteristics [28] yet due to the complex composition and diverse sources of OA, our understanding remains poor.

Offline analysis of OA requires proper pre-treatment of the samples at first. Various solvents can be used to extract the OA, and organic solvents (such as methanol) typically offer significantly higher extraction efficiencies than water [29]. Some studies show that water-insoluble organic carbon (WISOC) can contribute substantially to OA's light absorption, and have a mass absorption efficiency potentially larger than that of water-soluble OA [22,30–34]. Three-dimensional fluorescence spectra of OA extracted using different solvents, and following identified fluorophores also exhibit differences [27,33,35]. Therefore, OA extraction by organic solvents such as methanol can achieve a relatively comprehensive analysis of the composition and optical properties of OA. Subsequently, molecular characterization of OA is often fulfilled by using non-targeted high-resolution mass spectrometry coupled with liquid chromatography (LC) or gas chromatography (GC) [36,37]. This approach, together with key parameters like double bond equivalents (DBE) and elemental ratios of the identified molecules, offers valuable insights into the understanding of OA chemical properties. Relatively, some polar compounds and compounds with medium weight are easier to identify due to high ionization efficiency. For instance, Wang et al. (2018) observed notable differences in the chemical composition of OA between Beijing and Mainz, with Beijing showing a higher number of identified compounds particularly combustion-related aromatic hydrocarbons [38]. Wang et al. (2021) found that the molecular compositions of OA in Taiyuan, China, exhibited distinct seasonal variations [39], with more organosulfates in summer, resulting in a more oxidized and saturated OA. Mao et al. (2022) investigated the molecular composition of OA in coal combustion-polluted areas and identified nitrogen-containing organics as an important component [37], and those nitro-PAHs containing one nitrogen as markers of coal combustion emissions.

Nanjing is a densely populated megacity in the Yangtze River Delta (YRD) region in China, therefore fine particulate matter ($PM_{2.5}$) pollution is a critical environmental issue. Although there are some results available that analyzed the chemical composition, optical properties, health risks, and sources of fine particles in Nanjing including a number of earlier studies from our group [40–42], understanding of the optical and chemical properties of ambient OA in Nanjing is still incomplete, especially the molecular level chemical composition. To address this, this work collected and measured the optical properties and molecular composition of $PM_{2.5}$ samples during the summer and winter of 2019 and 2023, respectively. The comparison of 2019 and 2023 also offers insights into the changes in OA properties before and after the three-year COVID-19 influences in China. Elucidation of chemical characteristics of OA serves also as a basis for further toxicological studies.

2. Experimental Methods

2.1. Sample Collection and Pre-Treatment

$PM_{2.5}$ samples were collected from the rooftop of the library building (~21 m above the ground) in Nanjing University of Information Science and Technology (32.20° N, 118.71° E), Nanjing, China. The site locates in suburban Nanjing, near major traffic roads, and is influenced by a mix of industry, traffic, residential and agricultural emissions. A high-flow sampler (KB-1000, Jinstar, Qingdao, China) was used to collect $PM_{2.5}$ samples at a flow rate of 1.05 m^3/min. The samples were collected on pre-baked quartz filters (at 450 °C for 4 h) to remove potential contaminants. Blank samples were processed in the same way as background. The sampler uses a cutter to separate $PM_{2.5}$ by removing particles larger than 2.5 μm based on the principle of inertia. Afterwards, the air carrying $PM_{2.5}$ continues to move forward and reaches the quartz filter with a specific pore size, on which $PM_{2.5}$ are intercepted. Each $PM_{2.5}$ sample was collected for 22 h from 12:00 PM to 10:00 AM of the following day. Samples were not collected during precipitation. In total, 193 samples were obtained: 1 June–31 August 2019 (73 samples), 3 December 2019–20 January 2020 (38 samples), 2 June–26 August 2023 (52 samples), and 5 December 2023–28 February 2024 (30 samples). After collection, the $PM_{2.5}$ mass concentrations on filters were immediately determined using a digital balance (OHAUS DV215CD, Parsippany, NJ, USA; precision 0.01 mg). The filters were then wrapped in clean aluminum foils and stored in the refrigerator (−20 °C) before analysis. Throughout the sampling period, meteorological parameters, including temperature, relative humidity, wind speed, and wind direction, were recorded by a weather station located ~50 m from the sampling site.

Round pieces of the filters, each of 25.13 cm^2 and 9.08 cm^2, were punched from the filters, and extracted using 40 mL of ultrapure water (18.2 MΩ·cm, TOC < 5 ppb) and 20 mL of methanol (MERCK, Rahway, NJ, USA), respectively. Ultrasonic extraction was performed in an ice bath for 30 min, followed by filtration through a 0.45 μm filter, to remove and minimize possible influences of the insoluble substances For methanol-extracted samples, the filters were air-dried overnight in a fume hood, wrapped in clean aluminum foils and stored in a desiccator until further analysis of organic carbon (OC) and elemental carbon (EC). Blank samples were processed in the same manner to ensure consistency.

2.2. Chemical Analysis

The concentrations of water-soluble organic carbon (WSOC) in samples were determined using a total organic carbon (TOC) analyzer (TOC-L, Shimadzu, Kyoto, Japan). OC and EC contents were determined by a thermal-optical OC/EC analyzer (Sunset Laboratory, Portland, OR, USA), using the round piece of 2.01 cm^2 (17 mm diameter) for each filter. OC and EC contents of the filters after methanol extraction were also measured. Methanol-soluble organic carbon (MSOC) was then calculated as the difference of OC concentration before and after extraction. More details are provided in our previous work [43,44]. Measured OC, EC, WSOC and MSOC concentrations are used for calculation of optical properties (Section 2.3).

The molecular composition of OA was measured by a high-performance liquid chromatography (HPLC) coupled with a quadrupole time-of-flight mass spectrometer (Q-ToF-MS). We selected six samples (three with high and three with low $PM_{2.5}$ concentrations) in each period (in total 24 samples) for analysis. In details, five round pieces (3.14 cm^2) of each filter were subjected to extraction with 5 mL methanol (Optima LC/MS grade, Fisher Chemical, Waltham, MA, USA) ultrasonically at room temperature for 30 min. The extraction was repeated three times. Afterwards, the samples were filtered through a 0.22 μm PTFE filter and blew dried by nitrogen gas. The residue was re-dissolved in 300 μL methanol and stored in a 1.5 mL amber vial. Compound separation was performed using a Luna Omega C18 (PHENOMENEX, Torrance, CA, USA) 1.6 μm C18 column (100 mm × 2.1 mm × 1.6 μm) at 40 °C (a sample chromatogram is provide in Figure S1a). The electrospray ionization (ESI) was conducted in both positive and negative ion modes. The mass-to-charge (m/z) scanning range was 50–1100 amu, with an interval of 1 s. The ionization parameters were: drying gas temperature at 320 °C, nitrogen gas flow rate at 8 L/min, sheath gas flow rate at 11 L/min, sheath gas temperature at 350 °C, and capillary voltages at 4000 V (positive mode) and 3500 V (negative mode).

In positive mode, the mobile phase was consisted of solvent A (ultrapure water containing 0.1% v/v formic acid) and solvent B (acetonitrile containing 0.1% v/v formic acid). The eluent was programmed as: 0–1 min: 5% B; 1–3 min: linearly increase to 25% B; 3–20 min: linearly increase to 75% B; 20–50 min: linearly increase to 100% B; 50–53 min: maintain at 100% B; 53–55 min: decrease to 5% B. In negative mode, the mobile phase was consisted of solvent A (ultrapure water containing 0.05% v/v ammonia) and solvent B (acetonitrile containing 0.1% v/v formic acid). The eluent was set as 50% solvent A and 50% solvent B in the first 2 min, after that the sample was directly injected into the mass spectrometer.

The data were processed by the MS-DIAL software (version 4.92) [45], which includes peak extraction, peak alignment, and deconvolution. The ion fragments analyzed included $[M-H^-]$ $[M+H^+]$, $[M+NH_4{}^+]$, and $[M+Na^+]$. All deconvoluted spectra were exported to MS-FINDER (version 3.52) to assign the molecular formulas (the sample mass spectra for both positive and negative modes are provide in Figure S1b,c respectively). Several constraints were then applied to exclude unrealistic formulas: (1) Atomic numbers: C: 1–50, H: 1–100, O: 0–40, N: 0–5, S: 0–2; (2) Elemental ratios: H/C: 0.3–3.0, O/C: 0–1.2, N/C: 0–1.3, S/C: 0–0.5; (3) Double bond equivalence (DBE): 0–25. DBE is used to indicate the unsaturation degree [46] with the following formula:

$$\mathrm{DBE} = \frac{(2 \times \mathrm{C} - \mathrm{H} + \mathrm{N} + 2)}{2} \tag{1}$$

where C, H, and N represent the numbers of carbon, hydrogen, and nitrogen elements in the molecular formula. Additionally, the aromaticity equivalent (Xc) is widely used to distinguish aliphatic compounds from aromatic and polycyclic aromatic compounds [47], as follows:

$$\mathrm{Xc} = \frac{3 \times (\mathrm{DBE} - (p \times O + q \times S)) - 2}{\mathrm{DBE} - (p \times O + q \times S)} \tag{2}$$

where p and q represent the fractions of oxygen and sulfur atoms involved in the π bond structure, respectively. In this study, for substances in ESI− mode, $p = q = 0.5$, and for those in ESI+ mode, $p = q = 1$ [48,49]. Compounds with Xc < 2.5 are considered to be aliphatic, those with Xc ≥ 2.5 are aromatic, and compounds with Xc ≥ 2.7 are classified as polycyclic aromatic compounds [47]. The carbon oxidation state (OSc) [50], defined as 2 × O/C-H/C, can be used to describe the oxidation degree of molecule. Equations (1) and (2) were applied for each individual identified peak in the chromatogram of each sample.

Moreover, based on the relative abundances of identified molecules, the average O/C, H/C, and DBE values of each sample were calculated using the following formula [51]:

$$\frac{O}{C} = \frac{\sum\left(lin_i \times \frac{O}{C}_i\right)}{\sum lin_i} \tag{3}$$

$$\frac{H}{C} = \frac{\left(\sum lin_i \times \frac{H}{C}_i\right)}{\sum lin_i} \tag{4}$$

$$DBE = \frac{(\sum lin_i \times DBE_i)}{\sum lin_i} \quad (5)$$

Here, *lin* represents the relative abundance of molecule *i*, and O/C_i, H/C_i, DBE_i are the corresponding values of molecule *i*.

2.3. Optical Analysis

The light absorption of both water and methanol extracts were measured in 200–800 nm wavelength using a UV-Vis spectrophotometer (UV-3600, Shimadzu, Japan). The light absorption coefficient (Abs) at a specific wavelength (λ) was calculated using the following Equation (6):

$$Abs_\lambda = \frac{(A_\lambda - A_{700})V_1}{V_a \times l} \times \ln(10) \quad (6)$$

Here, A_{700} is used to account for baseline drift. V_1 is volume of the solvent, V_a is volume of the corresponding sampled air, and l is the path length of the quartz cuvette (1 cm here).

The wavelength dependence of the light absorption can be represented by the Ångström Absorption Exponent (AAE) [18], as shown in the following Equation (7):

$$Abs_\lambda = K \times \lambda^{-AAE} \quad (7)$$

where *K* is a dimensionless constant related to light absorption, and the range of λ here is 300–450 nm.

The mass absorption efficiency (MAE, m^2 g^{-1}) at 365 nm (values at this wavelength can effectively eliminate interferences of inorganic salts) [19,27,52,53] can be calculated below (4):

$$MAE = \frac{Abs_{365}}{WSOC(MSOC)} \quad (8)$$

The imaginary part of the particle refractive index, k can be used to represent the light absorption rate of atmospheric aerosols [19], and is expressed as follows (9):

$$k = \frac{MAE_\lambda \times \lambda \times \rho}{4\pi} \quad (9)$$

where ρ (g cm^{-3}) is the particle density, which is set to 1.5 in this study [9].

For the fluorescent properties, the EEM spectra were determined by a fluorescence spectrophotometer (Cary Eclipse, Agilent, Santa Clara, CA, USA). The excitation (Ex) wavelength was 235–500 nm with an interval of 5 nm, and the emission (Em) wavelength was 300–600 nm with an interval of 2 nm. The scanning speed was 1200 nm min^{-1}, with a slit width of 5 nm. Systematic biases, Raman and Rayleigh scattering, as well as blanks from water and methanol, were corrected. Then, the fluorescence data were processed using the PARAFAC model and drEEM toolbox (version 0.6.5, MATLAB R2018a) [54]. The model was run with non-negative constraints, and a half-split analysis.

Additionally, a few indices were calculated to investigate the fluorescence properties of OA. The humification Index (HIX) is the ratio of integrated emission signal of 436–480 nm to that of 300–344 nm at an excitation wavelength of 255 nm. The fluorescence index (FI) is the ratio of emission at 470 nm to 520 nm upon excitation at 370 nm. The biological index (BIX) is the ratio of emission at 380 nm to 430 nm upon excitation at 310 nm [55,56].

3. Results and Discussion

3.1. Light Absorption Properties

Table 1 lists the average concentrations (OC, EC, WSOC and MSOC) and light absorption parameters of OA during different periods. First of all, we need to conduct the significance test on the different pairs of data so that comparison and relevant interpretation are reliable. Before that, the tests of normality distribution were conducted and results are shown in Table S1, which shows that some data (for example, EC concentrations in all periods) do not follow the hypothesis of normality distribution, therefore the Wilcoxon test rather than the two-sample *t*-test is more suitable to our dataset. The corresponding results are presented in Table S2. The p values of four pairs of data (summer 2019 vs. winter 2019, summer 2019 vs. summer 2023, summer 2023 vs. winter 2023, winter 2019 vs. winter 2023) are less than 0.05 (on 95% confidence interval) (many items are actually close to zero). These statistical tests support our discussion and interpretation on the measurement data are reliable.

Table 1. Mass concentrations (±one standard deviation) and optical parameters of carbonaceous components at different sampling periods in Nanjing, China.

Terms	2019		2023	
	Summer (n = 73)	Winter (n = 38)	Summer (n = 52)	Winter (n = 30)
EC (μg m^{-3})	1.17 ± 0.34	1.72 ± 1.08	0.94 ± 0.43	1.02 ± 0.79
OC (μg m^{-3})	8.73 ± 2.49	12.39 ± 5.06	6.70 ± 2.82	10.25 ± 6.11
WSOC (μg m^{-3})	5.34 ± 1.59	6.73 ± 2.74	3.95 ± 1.75	5.69 ± 2.96
MSOC (μg m^{-3})	7.25 ± 2.21	10.94 ± 4.65	6.06 ± 2.71	8.26 ± 4.85
Water-soluble OC (WSOC)				
$Abs_{WSOC\text{-}365}$ (M m^{-1})	2.87 ± 0.86	5.70 ± 3.56	1.76 ± 1.00	3.73 ± 2.46
$MAE_{WSOC\text{-}365}$ (m^2 g^{-1})	0.54 ± 0.14	0.81 ± 0.29	0.45 ±0.15	0.62 ± 0.18
AAE_{WSOC}	6.04 ± 0.40	6.64 ± 1.21	6.28 ± 0.61	6.53 ± 0.71
$k_{WSOC\text{-}365}$	0.0236 ± 0.0060	0.0355 ± 0.0127	0.0195 ± 0.0066	0.0271 ± 0.0078
Methanol-soluble OC (MSOC)				
$Abs_{MSOC\text{-}365}$ (M m^{-1})	3.92 ± 0.95	12.73 ± 4.56	2.97 ± 1.12	5.14 ± 2.65
$MAE_{MSOC\text{-}365}$ (m^2 g^{-1})	0.57 ± 0.19	1.20 ± 0.23	0.52 ± 0.15	0.65 ± 0.15
AAE_{MSOC}	5.53 ± 0.50	4.40 ± 0.32	6.34 ± 0.96	6.09 ± 0.32
$k_{MSOC\text{-}365}$	0.0247 ± 0.0085	0.0525 ± 0.0099	0.0226 ± 0.0065	0.0284 ± 0.0064

As expected, winter levels were higher and that of 2019 winter was the highest. The mean methanol extraction efficiency (MSOC/OC) (0.80~0.90), was significantly higher than that of water (0.56~0.63), affirming that MSOC can better represent OA. Correspondingly, across the measured wavelength range (300–700 nm), the light absorption of MSOC was also noticeably higher than WSOC (Figure 1), consistent with prior findings in Seoul, South Korea [8], Beijing [29], Xi'an [22], and Nanjing [27], etc. For the values at 365 nm, $Abs_{MSOC\text{-}365}$ were generally higher than $Abs_{WSOC\text{-}365}$ during all four periods; and again, those during winter were higher than those in summer, similar to previous observFations [8,27,53], and those of 2023 became smaller than those of 2019. Figure S2 further presents scatter plots among $Abs_{365\text{-}WSOC}$ and MSOC, where the tight correlations suggest a large degree of overlap between MSOC and WSOC.

As shown in Table 1 and Figure 2, the MAE of methanol extracts were higher than those of water extracts as well. This result aligns with previous studies [8,21,27], suggesting that the water insoluble OC (WIOC) exhibits a stronger light absorption capacity and can be efficiently extracted by methanol. The MAE values observed in this study are comparable to those from a South China site in Nanling (summer: $MAE_{WSOC\text{-}365}$ = 0.61 ± 0.16 m^2 g^{-1}, winter: MAE $_{WSOC\text{-}365}$ = 1.0 ± 0.38 m^2 g^{-1}) [57], and from our previous results in Nanjing (summer: MAE $_{WSOC\text{-}365}$/MAE $_{MSOC\text{-}365}$ = 0.67 ± 0.20/0.93 ± 0.21, winter: $MAE_{WSOC\text{-}365}$/$MAE_{MSOC\text{-}365}$ = 1.17 ± 0.30/1.04 ± 0.24) [27], but are lower than those in northern China, such as Beijing (winter: $MAE_{WSOC-365}$ = 1.22 ± 0.11 $m^2 g^{-1}$, $MAE_{MSOC-365}$ = 1.40 ± 0.20 $m^2 g^{-1}$) [29], Lanzhou (winter: $MAE_{WSOC-365}$ = 1.19 ± 0.12 $m^2 g^{-1}$), Xining (winter: $MAE_{WSOC-365}$ = 1.22 ± 0.18 $m^2 g^{-1}$), Yinchuan (winter: $MAE_{WSOC-365}$ = 1.02 ± 0.23 $m^2 g^{-1}$), and Xi'an (summer: $MAE_{WSOC-365}/MAE_{MOC-365}$ = 0.98 ± 0.21/1.65 ± 0.36, winter: $MAE_{WSOC-365}/MAE_{MSOC-365}$ = 0.78 ± 0.23/1.33 ± 0.34) [58]. This result likely suggests that light absorption ability of OA in South China is generally weaker than it in North China.

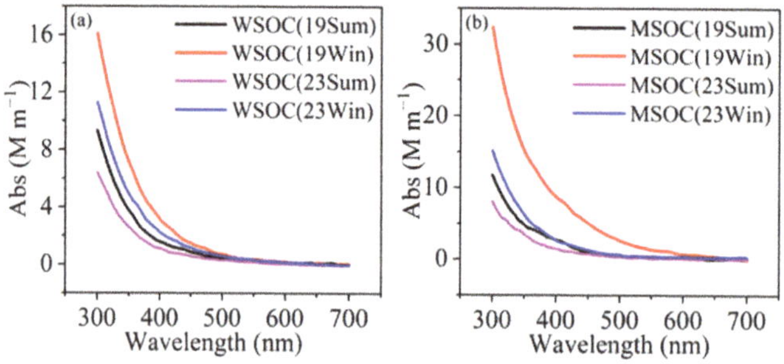

Figure 1. The average light absorption spectra of (**a**) WSOC and (**b**) MSOC during four periods.

Unlike black carbon (BC), BrC has strong wavelength dependence and different species have significantly different average AAE values (Figure 2). As shown in Figure 2, the AAE values for WSOC and MSOC observed in this study (marked in the bottom part) are comparable to those reported in previous studies in Nanjing (4.95–5.97) [27], Seoul (4.09–8.71) [8], etc. However, the values are lower than those reported in winter Beijing (7.1 ± 0.45 for MSOC) [29] and United States (7.63–8.71) [59]. Some studies report AAE values of wood burning particles to be 6.9–11.6 [16], and those from laboratory simulated secondary reactions are 5.2–8.8 [60]. Since the AAE values observed here are generally lower than 7, likely indicating a dominance of secondary sources to both WSOC and MSOC. MSOC appeared to have smaller AAE values than those of WSOC (Table 1), indicating that the WIOC has much weaker wavelength dependence than the WSOC [61]. MSOC AAE values in summer were larger than those in winter for both 2019 and 2023, while the differences of WSOC AAE in different periods were insignificant. In addition, The MSOC AAE values were higher in 2023 than those in 2019, both in summer and winter, suggest a substantial change in sources and compositions of MSOC rather than WSOC before and after the three COVID-19 years.

The imaginary refractive index (k) is another important parameter to assess the direct radiative forcing of aerosols in climate models [18,62]. At 365 nm, average $k_{MSOC-365}$ in Nanjing during the summer (winter) of 2019 were 0.0247 (0.0525), while in 2023, it ranged from 0.0226 (0.0284) (Table 1). These variations follow the same trend as the MAE_{365}, consistent with previous studies [63,64]. The k_{365} value of this study is slightly lower than that in northern Chinese cities such as Jinan (0.035) [65], Lanzhou, Xining, Yinchuan, and Urumqi (0.023–0.034) [58], and in Kanpur, India (0.042) [66], but higher than those in Himalayan cryosphere (0.009–0.026) [67]. In addition, previous studies have shown that photochemical aging may lower the k_{365} values [19,68]. Both MAE_{365} and k_{365} exhibited significant seasonal changes, reflecting the differences regarding primary sources and secondary processes (such as photo-bleaching) of OA across seasons. The higher MAE_{365} and k_{365} values in winter were likely due to increased anthropogenic activities such as biomass and coal burning. Moreover, the overall decreases in MAE_{365} and k_{365} in summer and winter of 2023 compared to 2019 demonstrates a decline of the atmospheric light absorption capacity of OA.

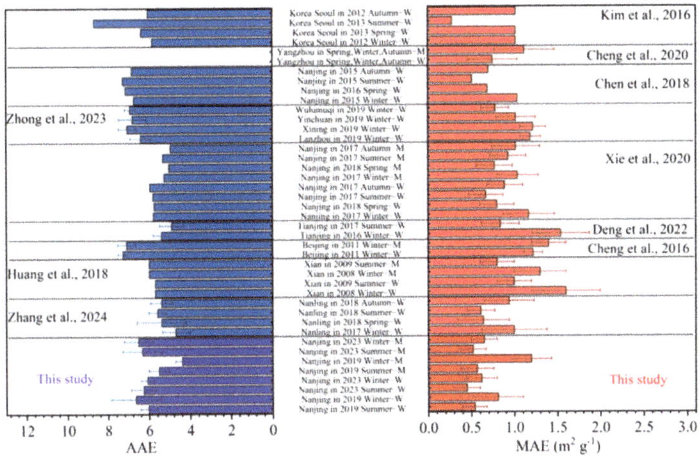

Figure 2. The values of Ångström absorption exponent (AAE) and mass absorption efficiency (MAE) in this study and other key selected studies (Zhong et al., 2023 [58], Huang et al., 2018 [22], Zhang et al., 2024 [57], Kim et al., 2016 [8], Chen et al., 2020 [53], Chen et al., 2018 [52], Xie et al., 2020 [27], Deng et al., 2022 [63], Cheng et al., 2016 [29]) (the name of campaign with a suffix of "-M" means that the data is for MSOC, while that with a "-W" means it is for WSOC) (blue bars represent AAE values while red bars represent MAE values).

3.2. Fluorescent Properties

3.2.1. Fluorescent Indicators

FI is often used as an indicator of organic sources in water and soil environments [69,70], and now it has been applied in aerosol chemistry [63,71,72]. FI < 1.4 indicates the terrestrial origin of OA, while FI > 1.9 indicates a dominance of microorganisms, and in general FI is inversely proportional to the aromaticity of OA. BIX in general reflects the relative contribution of recently emitted organics (such as those from microbial activities), therefore a high BIX indicates a high contribution of fresh organics [73]. HIX can reflect the degree of humification [55,56], and a higher HIX typically refers to a higher OA aromaticity [54,74].

As shown in Figure 3 and Table 2, the mean HIX values for WSOC were high (5.58 ± 1.07 in summer, 7.16 ± 1.06 in winter in 2019; 5.24 ± 0.70 in summer, 6.09 ± 0.77 in winter in 2023), much higher than those in Lanzhou (2.0 in summer and 1.2 in winter) [72] and Tianjin (3.12 in summer and 2.47 in winter) [63], but in agreement with previous results in Nanjing, China (6.97 in summer and 7.20 in winter) [27]. The variations of HIX_{MSOC} was similar to that of WSOC in the four periods, and the average HIX_{MSOC} values were much lower than HIX_{WSOC}, suggesting that water-extracted OA has a much higher aromaticity than its methanol-extracted companion. For both WSOC and MSOC, the mean FI values were higher in summer than in winter, but different from HIX, mean FI_{MSOC} values (2.25–3.35) were all higher than the FI_{WSOC} values (1.87–2.02) in all four periods. This result indicates the WIOC dissolved in methanol rather than water might be associated with biogenic sources. The FI_{WSOC} values in this study are similar to our earlier results in Nanjing (1.82 in summer and 1.91 in winter) [27], and those of MSOC are slightly higher (1.54 in summer and 2.28 in winter for MSOC). For BIX, the differences for both WSOC and MSOC values across all sampling periods were small (0.85–0.87 for WSOC, and 0.85–1.04 for MSOC). As shown further in Figure 3, most BIX values distributed within the range of 0.6 to 1 [69], while the FI values distributed in the range of >1.6 [56], indicating that the OA in summer and winter of both 2019 and 2023 were influenced by microbial/biogenic source.

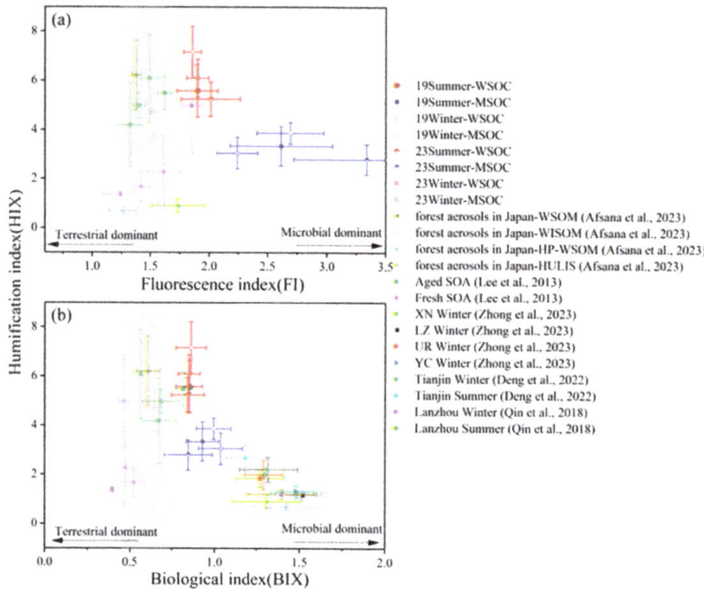

Figure 3. Distribution of Fluorescence Indices in This Study Compared with Other Studies (Afsana et al., 2023 [35]; Lee et al., 2013 [70]; Zhong et al., 2023 [58]; Deng et al., 2022 [63]; Qin et al., 2018 [72]): (**a**) Fluorescence Index (FI) vs. Humification Index (HIX), (**b**) Biological Index (BIX) vs. Humification Index (HIX). The shaded areas represent BIX values from 0.6 to 1 [69] and FI values from 1.6 to 1.9 [56].

Table 2. Average Fluorescence Index Values of Water-Soluble Organic Carbon (WSOC) and Methanol-Soluble Organic Carbon (MSOC).

	2019		**2023**	
	Summer	**Winter**	**Summer**	**Winter**
		WSOC		
FI_{WSOC}	1.91 ± 0.18	1.87 ± 0.07	2.02 ± 0.26	1.91 ± 0.09
BIX_{WSOC}	0.86 ± 0.08	0.87 ± 0.09	0.85 ± 0.09	0.86 ± 0.06
HIX_{WSOC}	5.58 ± 1.07	7.16 ± 1.06	5.24 ± 0.70	6.09 ± 0.77
		MSOC		
FI_{MSOC}	2.62 ± 0.44	2.70 ± 0.28	3.35 ± 0.62	2.25 ± 0.18
BIX_{MSOC}	0.93 ± 0.10	1.00 ± 0.10	0.85 ± 0.14	1.04 ± 0.13
HIX_{MSOC}	3.34 ± 0.81	3.87 ± 0.43	2.80 ± 0.62	3.05 ± 0.64

In addition, the HIX values in 2023 for both WSOC and MSOC appeared to drop significantly from those in 2019, demonstrating a big decrease of OA's aromaticity in Nanjing. Except for MSOC in winter, the FI values were larger in 2023 than in 2019, which are consistent with that FI is inversely related to aromaticity. The 2023 to 2019 difference of BIX was, not significant.

3.2.2. Key Fluorescent Components

The key fluorescent components resolved by the PARAFAC analysis on the 3D-EEM spectra are shown in Figure 4. Four components were identified for WSOC. C1 exhibited two characteristic peaks at Ex/Em = 240/410 nm and Ex/Em = 320/410 nm. Generally, a bi-modal distribution in the fluorescence spectra indicates the presence of HULIS [75–77], and the second peak at relatively long excitation wavelengths refers to condensed aromatic substances, conjugated structures, or non-linear cyclic systems. C3 displayed a similar peak pattern therefore was also linked with HULIS, but with higher excitation and emission wavelengths (Ex/Em = 255/480 nm and Ex/Em = 370/480 nm), suggesting a higher aromaticity. C3 likely contains a richer conjugated unsaturated structure, while C1 is more likely associated with photodegradative decomposition of macromolecules [33]. In contrast to C1 and C3, C2 exhibited a small characteristic peak at Ex/Em $\leq$ 235/400 nm, also classified as a HULIS component, but likely contains more oxygenated structures and may have a lower degree of oxidation [78]. C4 was identified as a protein-like component, exhibiting characteristic peaks at Ex/Em $\leq$ 235/364 nm and Ex/Em = 275/364 nm (shorter wavelengths) [75]. Previous studies have shown that tryptophan-like proteins have excitation wavelengths in the range of 220–290 nm and emission wavelengths in the range of 320–380 nm [79]. Therefore, C4 of WSOC here likely associates with tryptophan-like fluorescent chromophores [54] and may be relevant with anthropogenic activities [33,69,80]. Figure S3a presents the average contributions of these four components in different periods, C2 was the most abundant component especially in winter, followed by C1 whose contribution increased significantly in summer, C3 and C4 were minor components.

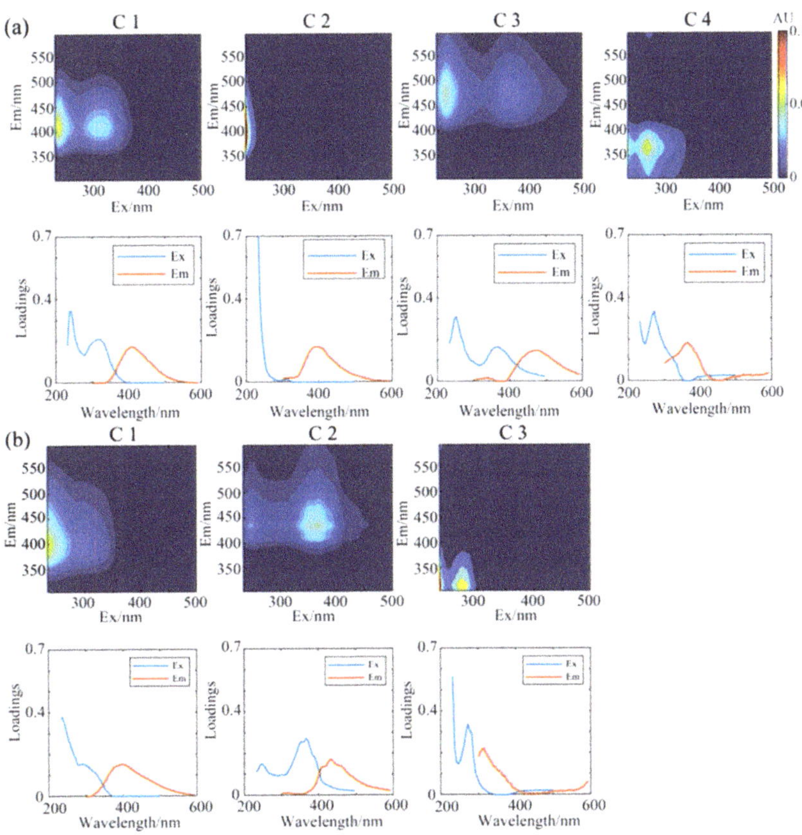

Figure 4. The florescent components resolved by the PARAFAC analyses and its excitation and emission wavelengths: (**a**) WSOC (C1–C4) and (**b**) MSOC (C1–C3).

For MSOC, only three key fluorophores were separated (Figure 4b). C1 and C2 were classified as HULIS while C3 was a protein-like component. C1 exhibited characteristic peaks at Ex/Em = 235/402 nm and Ex/Em = 275/402 nm, similar to C1 of WSOC but with slightly lower excitation and emission wavelengths. C2 had prominent peaks at Ex/Em = 250/436 nm and Ex/Em = 370/436 nm, similar to C3 of WSOC but with a larger Stokes shift. The right-side peak has stronger intensities, indicating possible presence of more conjugated molecules, with larger molecular weights and conjugated bonds, and condensed aromaticity [77]. We showed earlier that MSOC HIX was lower than WSOC HIX in both summer and winter, consistent with the lower emission wavelength of MSOC than WSOC here, reaffirming the higher aromaticity of WSOC than MSOC. C3 had characteristic peaks at Ex/Em $\leq$ 235/316 nm and Ex/Em = 275/316 nm. Its maximum emission wavelength was lower than the protein-like component C4 of WSOC, and was not within the range of aro-matic proteins (330–380 nm), instead lay between those of tryptophan and tyrosine. We postulate that C3 may include not only proteins (such as tyrosine and tryptophan), but also phenolic compounds or PAHs from biomass and/or fossil fuel combustion [53,80–83]. For their contributions in different periods (Figure S3b), C1 became the most abundant component and was slightly more abundant in summer than in winter; the protein-like C3, very differently from that in WSOC, was also rich especially in winter (up to 40.3% in 2019 winter); C2 on the contrary, was the least important component. Overall both MSOC and WSOC fluorescence were dominated by HULIS, but the large difference regarding fluorescent properties between the two, despite their possible large overlap of chemical species, suggests a mismatch of chemical species and fluorophores.

3.3. Molecular Composition

3.3.1. Overview

According to the protocols in Section 2.2, the HPLC-QToF-MS analysis on the methanol extracts identified a total of 729–892 and 821–1662 molecules in negative ion mode (ESI−) and positive ion mode (ESI+), respectively. These molecules were further classified into eight categories: CH, CHO, CHS, CHN, CHON, CHOS, CHNS, and CHONS com-pounds. As shown in Figure 5a, in ESI+ mode, number fraction of CHON species was the largest (41.5~47.0%) in all cases; for samples collected during low-pollution days, its fraction was slightly higher than those during high-pollution days. In terms of relative signal abundance (Figure S4a), CHON remained as the most important group (29.4~44.0%) except for summer 2019. The second most abundant group was CHO, occupying 32.2~37.7% of the total number and 23.4~42.2% of the total signal of identified molecules. Differences in CHO number fractions between summer and winter were insignificant, while those of signal fractions were relatively large, particularly in 2019 (33.9~42.2% in summer vs. 23.4~27.8% in winter). Other relatively important types in number were CHONS (8.5~12.1%) and CHN species (4.4~7.7%); contrastingly, their relatively abundances in signal were only 1.4~4.0% for CHONS but 14.0~35.3% for CHN (Figures 5a and S3a). The rest four compound families were very minor both in number and signal.

In ESI− mode, instead of CHON, CHO species became the most abundant both in number fraction (42.2~51.4%) and signal fraction (38.3~64.6%), and the CHON species became the second (28.6~39.9% in number, and 13.5~43.6% in signal) (Figures 5b and S3b). The high abundance of CHO species in negative mode than in positive ion mode was re-ported before [51]. A possible cause is that CHO species may include carboxyl compounds (such as organic acids) that are more sensitive in ESI− mode, while organic basic com-pounds are more sensitive to ESI+ ionization. Additionally, CHOS and CHONS showed higher abundance in ESI− mode than in ESI+ mode, and the CHONS signal fraction even reached ~39% in high-pollution 2019 summer samples (Figure S4b). CHN and the other three types of compounds were almost negligible in ESI+ mode.

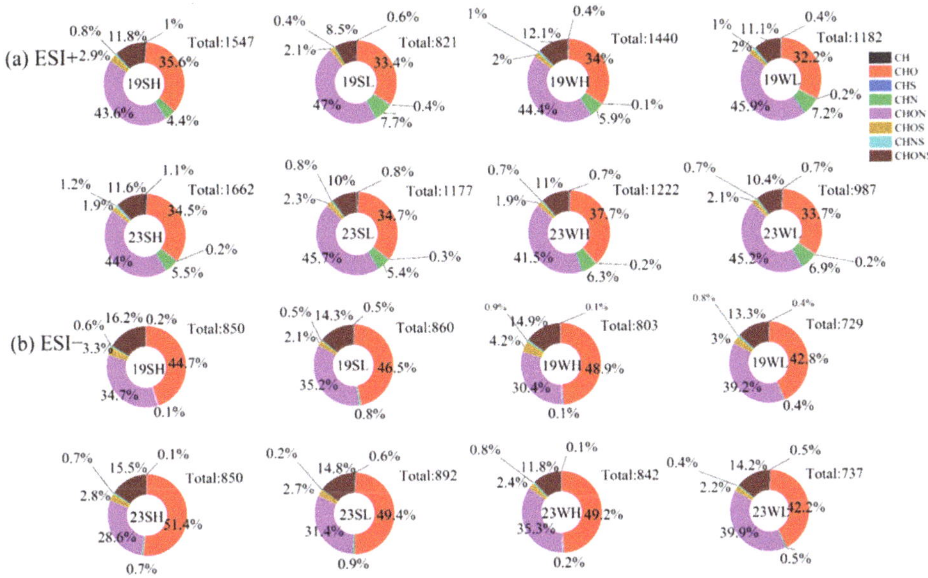

Figure 5. The number fractions of different types of compounds in OA from different periods. 19SH and 19SL represent the selected samples in 2019 summer with relatively high and relatively low $PM_{2.5}$ concentrations, respectively; meanings of 19WH, 19WL, 23SH, 23SL, 23WH and 23WL are similar. (**a**) ESI+ mode; (**b**) ESI− mode.

Overall differences between 2019 and 2023 samples in ESI+ mode were subtle. In ESI− mode, 2023 summer samples appeared to contain a bit more CHO species but less CHON species than the 2019 summer samples, while on the opposite, 2023 winter samples had more CHON species than the 2019 winter samples. The total number of molecules in summer and winter in 2023 was greater than that in 2019 in both ESI− or ESI+ mode; In ESI− mode, the molecular number in summer 2019 or 2023 was greater than those in winter, while in ESI+ mode, the molecular number in summer 2019 was lower than that in winter, while that in summer 2023 was greater than in winter.

3.3.2. CHO Compounds

The Van Krevelen (VK) diagram can distribute the compounds in the H/C vs. O/C space, and the ones for CHO compounds are shown in Figure 6 for both ESI+ and ESI− modes (separate figures for different periods of 2019 and 2023 are presented in Figures S5 and S6). The VK plots can be further divided into five regions [39,84–86]: A—Aliphatic compounds ($H/C \geq 1.5$, $O/C \leq 0.5$); B—Less oxygenated aromatic compounds ($H/C \leq 1.0$, $O/C \leq 0.5$); C—High oxygenated functional group compounds, such as alcohols, esters, and peroxides ($OSc \geq 0$, $O/C \geq 0.6$); D—High reduced functional group compounds, such as carbonyls and organic acids ($OSc < 0$, $O/C \geq 0.6$); E—Moderate oxygenated compounds ($OSc \geq 0$, $O/C \geq 0$, $H/C \leq 1.2$). The Xc value of each compound was also calculated and colored in the VK plot. From Figure 6, we can observe that a majority of molecules fall in the regime of $O/C < 0.5$ (especially regions A and B), but with a wide range of H/C values. Saturated aliphatic CHO species ($Xc < 2.5$) was the most abundant class of compounds (60.6% and 53.5% of total molecules in ESI+ and ESI− modes), which mainly distributed in region A, with small portions in other regions. Unsaturated CHO compounds with $2.5 \leq X < 2.7$ occupied 22.9%

and 23.6%, and those with Xc ≥ 2.7 occupied the remaining 16.5% and 22.5% of the total number of molecules in ESI+ and ESI− modes. These species mainly distributed in region B and that between A and B (Figure 6). Overall, unsaturated compounds were more enriched in ESI− mode than in ESI+ mode. In addition, the aromatic and polycyclic aromatic compounds in both high-pollution and light-pollution samples in winter 2019 were significantly higher than those in summer 2019. However, such difference was not very evident in 2023.

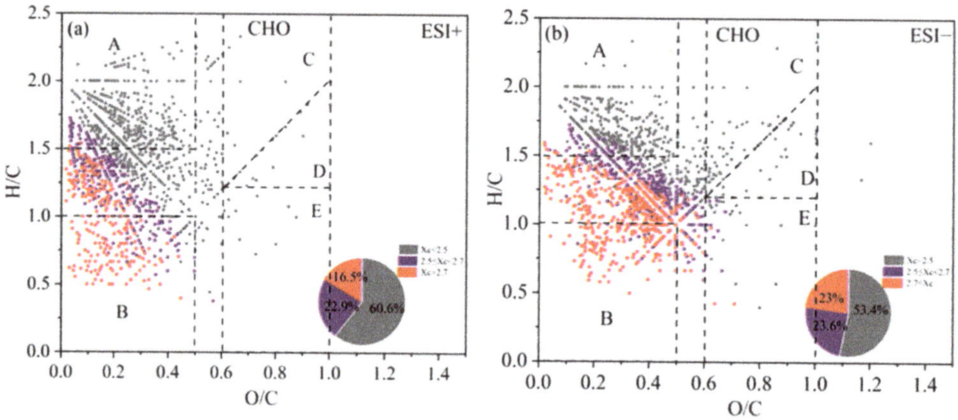

Figure 6. The Van Krevelen (VK) diagram of CHO compounds detected under (**a**) positive ion mode (ESI+) and (**b**) negative ion mode (ESI−) modes. Different colors represent molecules with different aromaticity equivalent (Xc) values, and the pie chart shows corresponding number fractions; dash lines represent the boundaries of difference regimes as explained in the main text.

3.3.3. CHON and CHN Compounds

A similar VK diagram for the CHON compounds in both ESI+ and ESI− modes, with colored Xc, is shown in Figure 7. Furthermore, we also classified some species into different series of compounds. In ESI+ mode, CHON compounds containing one -NO functional group dominate the spectrum. CHON compounds were also predominated by the saturated species (Xc < 2.5) (53.1% of total), and their nitrogen-containing functional groups are likely to be of a reduced nature rather than the nitro- or nitroso- groups. For example, a series of organic homologs, such as $C_6H_{15}NO(CH_2)_n$ (likely N,N-diethyl ethanolamine homologs) and $C_6H_{15}NO_2(CH_2)_n$ (likely diisopropylamine homologs), are likely amines with hydroxyl or ether groups (-O-) and they possess lone pairs of electrons that are prone to protonation [87]. Unsaturated CHON compounds with Xc ≥ 2.5 concentrate in the lower left corner of the plot. Such species include $C_9H_7NO(CH_2)_n$, $C_8H_7NO(CH_2)_n$, $C_7H_7NO(CH_2)_n$, and $C_5H_5NO(CH_2)_n$ (lines 3–6 in Figure 7a), possibly corresponding to homologs of hydroxyquinoline, 4-hydroxyphenylethyl nitrile, benzamide, and hydroxypyridine, respectively [88,89].

In ESI− mode, the CHON compounds showed a broad distribution. Most compounds contained one or two nitrogen atoms, and instead, a majority (69.5%) of the species were categorized as aromatic compounds with Xc ≥ 2.5, including monocyclic and polycyclic structures. Monocyclic compounds such as $C_6H_5NO_3(CH_2)_n$ and $C_6H_5NO_4(CH_2)_n$ (lines 1–2 in Figure 7b) likely indicate nitrophenol and nitroresorcinol homologs, while $C_8H_7NO_3(CH_2)_n$ and $C_8H_7NO_4(CH_2)_n$ (lines 3–4) may correspond to nitroacetophenone and nitrophenylacetic acid homologs, respectively [28,90–92]. Polycyclic CHON compounds, such as $C_{10}H_7NO_3(CH_2)_n$ (line 5 in Figure 7b), probably refers to nitronaphthol homologs. Moreover, in both ESI+ and ESI− modes, the signal and number fractions of CHON compounds in low-pollution samples are nearly equal or even higher than those in high-pollution samples. This finding aligns with previous studies [89,90,93], and together with the fact that CHON species contains many nitro-aromatic species may suggest that the CHON species are related to various secondary reactions such as photochemical aging.

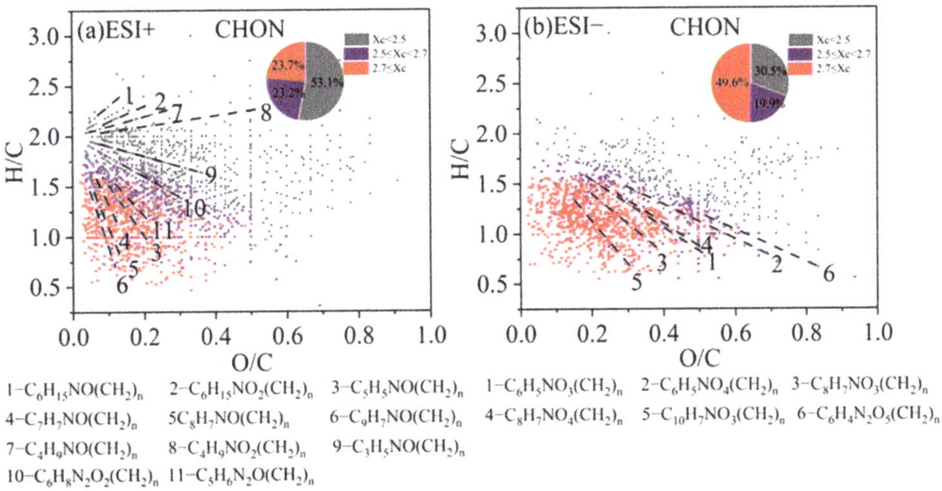

Figure 7. The Van Krevelen (VK) diagram of CHON compounds detected under (**a**) positive ion mode (ESI+) and (**b**) negative ion mode (ESI−) modes. Molecules are also colored according to the aromaticity equivalents (Xc) values and the pie chart shows corresponding number fractions. The different dashed lines represent different series of homologous compounds.

CHN compounds are only enriched in ESI+ mode. The scatter plot in Figure 8 shows the H/C versus N/C ratios for all CHN compounds detected in ESI+ mode. Most of the CHN compounds were amines containing one or two nitrogen atoms. Saturated amines and other monocyclic species (Xc < 2.7) distributed in the upper-middle region of the plot, including $C_6H_{15}N(CH_2)_n$, $C_5H_{11}N(CH_2)_n$, $C_4H_6N_2(CH_2)_n$, $C_5H_6N_2(CH_2)_n$, $C_7H_6N_2(CH_2)_n$, and $C_{11}H_{17}N(CH_2)_n$ (lines 1–6 in Figure 8). Furthermore, many studies have found that species containing two-membered heterocyclic structures are important components of biomass burning OA [94]. The homologs $C_4H_6N_2(CH_2)_n$, $C_5H_6N_2(CH_2)_n$, and $C_7H_6N_2(CH_2)_n$ are believed to be potential markers for biomass burning. In the lower-left corner of the plot, some CHN compounds with Xc > 2.7 were observed, including a series of PAHs) containing one nitrogen atom, such as $C_{10}H_9N(CH_2)_n$ and $C_{15}H_9N(CH_2)_n$. $C_{10}H_9N(CH_2)_n$ is likely an amino-naphthalene homolog [87], which may originate from the combustion of carbonaceous materials. Previous studies have also suggested that 1N-PAHs can be used as markers of coal combustion [37].

3.3.4. CHOS and CHONS Compounds

The VK plots for CHOS and CHONS compounds are illustrated in Figures 9 and 10, respectively. The majority of CHOS compounds were composed of saturated aliphatic compounds (Xc < 2.5) (75.3% and 76.1% in ESI− and ESI+ modes, respectively). Relatively more CHOS compounds resided in the regime of O/C < 0.5, and number fraction of compounds with 2.5 < Xc ≤ 2.7 was also notably higher in ESI+ mode than in ESI− mode. For the CHONS compounds, number fraction of aliphatic species (Xc < 2.5) was still prevailing (56.4%) (contributions of the other two types were on par, 23.0% and 20.6%, respectively) in ESI+ mode, while in ESI− mode, compounds with Xc > 2.7 (polycyclic aromatic species) became the most important portion (41.8%), and fraction of aliphatic species was down to only 29.3%. CHONS species behaved quite distinctly in positive and negative modes and its signal fraction in ESI+ model was quite low relative to the number fraction (Figure S4).

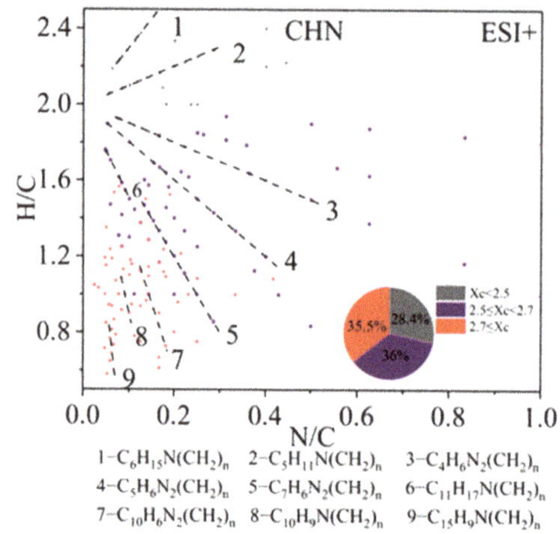

Figure 8. The H/C versus N/C plot of CHN compounds detected under positive ion mode. Molecules are also colored by the aromaticity equivalents (Xc) values and the pie chart shows corresponding number fractions. Different dashed lines represent different series of homologous compounds.

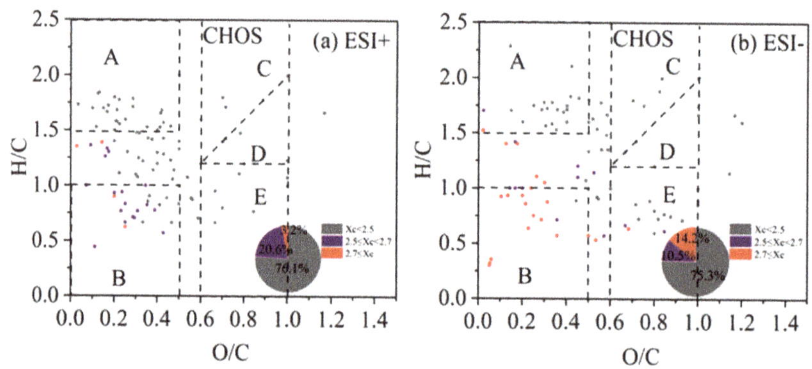

Figure 9. The Van Krevelen (VK) diagram of CHOS compounds detected under (**a**) positive ion mode (ESI+) and (**b**) negative ion mode (ESI−) modes. Molecules are colored according to the aromaticity equivalents (Xc) values and the pie chart shows corresponding number fractions.

3.4. Characteristics of OA Molecules

3.4.1. Tracer Compounds of Different Sources

The ambient OA can contain molecules from multiple primary sources and various secondary processes, which are affected by meteorological conditions, transportation, chemical reactions, deposition, etc. The OSc value can be used to assess the aging or oxidation degree of OA and infer their potential sources [50]. For the most abundant CHO compounds, Figures 11 and S7 demonstrate the relationships between the OSc and the number of carbon atoms for CHO compounds of both ESI+ and ESI− modes, and the data points are roughly assigned to different sources based on results from a previous study [50]: HOA (hydrocarbon-like OA, associated with fossil fuel combustion such as traffic); BBOA (biomass burning OA); SV-OOA (semi-volatile oxygenated OA, secondarily formed generally with a low oxidation degree); LV-OOA (low-volatility oxygenated OA, secondarily formed generally with a high oxidation degree). As presented, a significant number of compounds were categorized as HOA and BBOA in both ESI− and ESI+ modes, reflecting the large impact of anthropogenic combustion activities on OA in Nanjing; the number of BBOA species in ESI− mode were also significantly more than those in ESI+ mode, and the HOA and BBOA species were also a bit more in 2023 than in 2019. A number of species were also identified as potential SV-OOA but much less species belonged to LV-OOA, in agreement with the overall low O/C of CHO compounds shown in Figure 6, indicating that OA in Nanjing was overall relatively fresh.

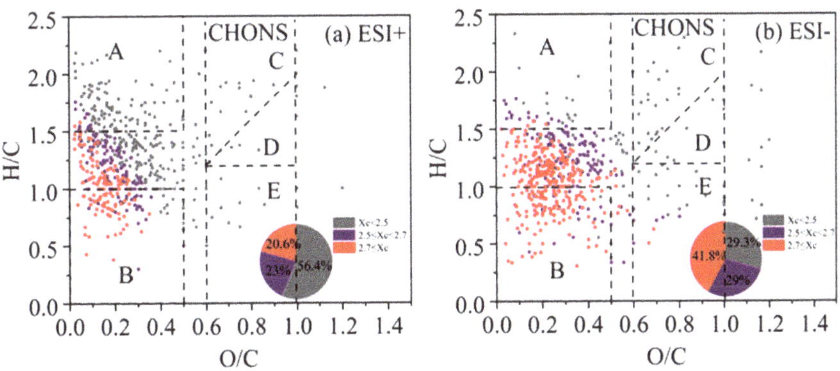

Figure 10. The Van Krevelen (VK) diagram of CHONS compounds detected under (**a**) positive ion mode (ESI+) and (**b**) negative ion mode (ESI−) modes. Molecules are colored according to the aromatici-ty equivalents (Xc) values and the pie chart shows corresponding number fractions.

Since non-targeted molecular characterization is inherently good in identification rather than quantification of the OA compounds, hereby we only selected a few molecules with relatively large signals representing different sources to examine the OA properties in different periods (Table S4). For LV-OOA, SV-OOA and HOA, we chose the compounds locating in the regimes marked in Figure 11. For LV-OOA, we selected $C_5H_8O_5$ and $C_5H_6O_4$, and for SV-OOA, we selected $C_6H_{10}O_5$ and $C_{14}H_{16}O_8$. Signal fractions of these four molecules were slightly higher in summer than in winter, suggesting that their secondary formations were likely more important in summer. Another finding is that their overall in-fluences on OA were also higher in high-pollution samples, indicating that secondary re-actions became even more important during high-pollution episodes. For HOA, we selected $C_{27}H_{44}O_6$, and its portion was higher in winter than in summer, whose value in 2023 decreased from that of 2019. The result is reasonable as fossil fuel combustion emissions are typically more significant in winter, while the contribution might be reduced due to emission control in 2023 relative to that in 2019. For the biomass burning OA (BBOA), we chose $C_6H_{10}O_5$, $C_9H_8O_3$, and $C_6H_5NO_4$, likely levoglucosan, pentose aldehyde (levoglucosan aldehyde) (derived from lignin decomposition), and 4-nitrocatechol, respectively [95–97]. Signal proportions of these BBOA tracers in winter were notably higher than those observed in summer; their proportions were larger in low-pollution samples in 2019, while they were larger in high-pollution samples in 2023, suggesting a change of BBOA's role in the $PM_{2.5}$ pollution in Nanjing. For cooking OA (COA) [98], we selected $C_{16}H_{30}O_3$ and $C_{18}H_{34}O_3$, and their signal proportions were found to be higher in summer and in high pollution samples.

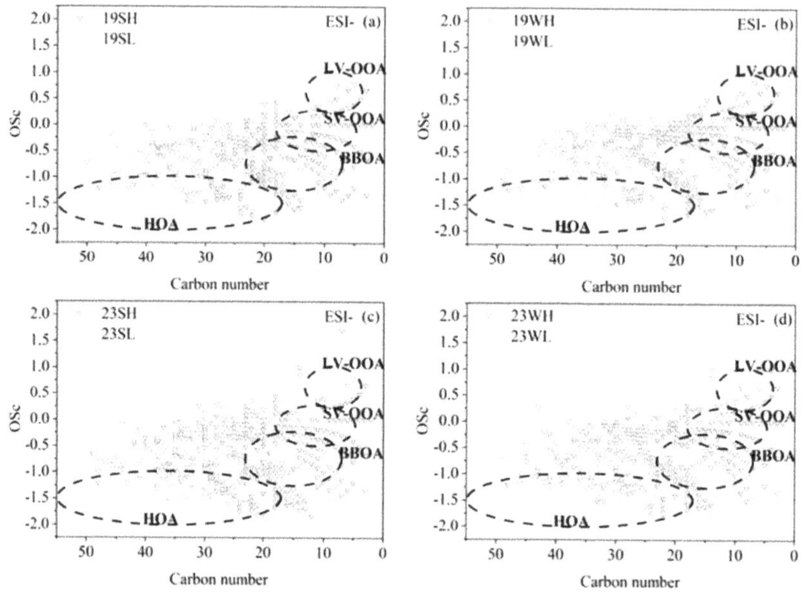

Figure 11. The carbon oxidation states (OSc) of CHO compounds against carbon numbers of the compounds. The black dashed circled regions are labeled as hydrocarbon-like OA (HOA), biomass burning OA (BBOA), semi-volatile oxygenated OA (SV-OOA), and low-volatility oxygenated OA (LV-OOA) (Kroll et al., 2011 [50]). 19SH and 19SL represent the selected samples in 2019 summer with relatively high and relatively low $PM_{2.5}$ concentrations, respectively; meanings of 19WH, 19WL, 23SH, 23SL, 23WH and 23WL are similar. Results of (**a**) 2019 summer, (**b**) 2019 winter, (**c**) 2023 summer, (**d**) 2023 winter.

Furthermore, several secondary acids, including $C_7H_{10}O_4$, $C_8H_{12}O_4$, $C_{10}H_{16}O_3$, $C_{10}H_{16}O_4$, $C_8H_{12}O_6$, and $C_{10}H_{16}O_6$, were identified in our samples. These compounds, previously confirmed as photochemically-derived SOA species from α-pinene oxidation [99], exhibited higher proportions in summer (0.495%) than in winter (0.203%), emphasizing the role of photo-oxidation in summer (Table S5). Their summer average increased from 0.437% in 2019 to 0.558% in 2023, suggesting an enhanced photochemical influence after the pandemic. We also examined three aqueous-phase SOA (AqSOA) markers [20,100], such as pyruvic acid ($C_3H_4O_3$), oxalic acid ($C_2H_2O_4$), and dimethyl nitrobenzoic acid ($C_9H_9NO_4$). Their average proportion was lower in summer (0.087%) than in winter (0.189%), indicating a greater role of aqueous-phase reactions in winter. Furthermore, phenolic derivatives, key contributors to aqueous-phase SOA, originate from lignin pyrolysis and are closely linked to biomass burning [101]. As shown in the Table S6, phenolic-derived oligomers were more abundant in winter, further highlighting the significance of aqueous-phase reactions and BBOA contributions to OA during this season.

At last, organosulfates (OrgSs) are a prominent class of CHOS species, typically having oxygen-to-sulfur (O/S) ratios ≥ 4. As listed in Table 3, in -ESI− mode, the number proportion of OSs (on average 70.8%) was slightly higher than that in ESI+ mode (average 64.6%); and the fraction was higher in 2019 winter than in 2023 winter in ESI+ mode, but differences in other periods were not significant. For CHONS compounds, 5.1–17.4% of the molecules in ESI− mode (signals in ESI+ mode were very low, therefore were not considered here) can be assigned to nitrooxy-OrgSs (with an O/(4S+3N) ratio ≥ 1, and possibly $-OSO_3H$ and $-ONO_2$ functional groups) [38]; there was a relatively large difference between high- and low-pollution samples—higher fraction in low-pollution samples, which is consistent with a previous study [38]. Song et al. [102] point out that coal combustion may be an important sources of S-containing compounds, while biomass burning is a source of atmospheric N-containing compounds, which can further form nitroaromatic compounds upon atmospheric aging [103]. Some previous studies also indicate that biogenic, motor vehicular and ship emissions, etc. might promote the formation of Nitro-OrgSs too [104–107]. However, the nitro-oxy-OrgSs are mostly likel associated with aqueous phase reactions [108]. Since their relative fractions were high in low-pollution samples closely associated with high RH conditions in this work (Table S3), suggesting aqueous oxidation might be the most likely source of such species.

Table 3. The number percentages of the organosulfates (OrgSs) in CHOS compounds under two modes and those of nitrooxy-OrgSs in CHONS compounds under ESI− mode.

		2019 Summer		2019 Winter		2023 Summer		2023 Winter	
		2019SH	**2019SL**	**2019WH**	**2019WL**	**2023SH**	**2023SL**	**2023WH**	**2023WL**
CHOS	OrgSs (ESI−)	75.0%	77.8%	64.7%	86.4%	75.0%	70.8%	60.0%	56.3%
	OrgSs (ESI+)	68.9%	70.6%	62.1%	66.7%	62.5%	63.0%	60.9%	61.9%
CHNOS	Nitrooxy-OrgSs (ESI−)	9.4%	16.3%	15.0%	16.5%	10.6%	17.4%	5.1%	14.3%

3.4.2. Correlations with Light Absorption

To investigate the link between optical properties and molecular composition, we investigated the correlations between the signal fractions of different types of molecules and the $MAE_{MSOC\text{-}365}$ values, as shown in Table 4. Very interestingly, correlations of CH, CHO, CHN, CHON and CHS in ESI− mode with $MAE_{MSOC\text{-}365}$ during summer were positive, while the correlations became negative (except CHO and CHS) during winter; on the contrary, correlations of CHOS, CHNS and CHONS in ESI− mode with $MAE_{MSOC\text{-}365}$ were all negative during summer but all positive during winter. In ESI+ mode, except that correlations of CH and CHN with $MAE_{MSOC\text{-}365}$ were all positive in both summer and winter, the correlation coefficients of CHO, CHS, CHON, CHOS, CHNS and CHONS in summer were all appositive to those in winter. These results likely suggest a big compositional difference of the key light absorbing molecules between summer and winter. Furthermore, in summer, $MAE_{365\text{-}MSOC}$ correlated fairly well with CH, CHS, CHN, and CHON in ESI− mode ($r > 0.5$) and with CH and CHN in ESI+ mode ($r > 0.5$). In winter, it showed relatively good correlations with CHO, CHS, CHOS ($r > 0.5$) and CHNS (r of 0.46) in ESI− mode, and with CHN (r of 0.82) and CHNS (r of 0.97) in ESI+ mode. Overall, CH and CHN compounds contributed significantly to light absorption in summer, while the role of CHNS compounds became a bit more important in winter.

Table 4. The cross-correlation coefficients (Pearson's r) between different types of OA molecules (relative signal intensities) with $MAE_{MSOC\text{-}365}$ in summer and winter (2019 and 2023 samples combined).

Mode	Type	Summer	Winter
		$MAE_{MSOC\text{-}365}$	$MAE_{MSOC\text{-}365}$
ESI−	CH	0.78	−0.15
	CHO	0.23	0.59
	CHS	0.76	0.66
	CHN	0.81	−0.7
	CHON	0.5	−0.68
	CHOS	−0.71	0.83
	CHNS	−0.49	0.46
	CHONS	−0.43	0.35

Table 4. *Cont.*

Mode	Type	Summer	Winter
		$MAE_{MSOC\text{-}365}$	$MAE_{MSOC\text{-}365}$
ESI+	CH	0.66	0.39
	CHO	0.38	−0.87
	CHS	0.42	−0.59
	CHN	0.69	0.82
	CHON	0.32	−0.41
	CHOS	0.48	−0.94
	CHNS	−0.44	0.97
	CHONS	0.49	−0.92

4. Conclusions

In this work, we conducted a comprehensive analysis on the optical properties and molecular chemical composition of OA in Nanjing by using samples collected in summer and winter of 2019 and 2023. The results show a clear decline of EC, OC, MSOC and WSOC concentrations from 2019 to 2023 in both summer and winter; in general, Abs_{365} and MAE_{365} values for both WSOC and MSOC were higher in winter than in summer likely due to increased combustion activities, and those of MSOC were higher than MSOC in all periods. Fluorescent properties of both WSOC and MSOC were dominated by humic-like components but the remaining contribution from protein-like component was more significant in MSOC. Fluorescence indices suggest that OA in Nanjing was influenced by a mix of microbial/biogenic source. In particular, the HIX values in 2023 for both WSOC and MSOC appeared to drop significantly from those in 2019, demonstrating a big decrease of OA's aromaticity in Nanjing.

Results of OA molecular composition did not show remarkable difference between 2019 and 2023. Overall, in term of number of identified molecules, the CHON compounds were the most abundant type of species, followed by the CHO and CHN compounds; aliphatic compounds dominated the major molecular types except for CHN (in ESI+ mode) and CHON, CHOS (in ESI− mode). Qualitative source analysis shows the numbers of molecules from fossil fuel combustion and biomass burning (BB) were a bit more in 2023 than in 2019, and signal intensities of BB-related molecules were also higher in winter than in summer. The presence of organosulfates further indicates the contribution of aqueous-phase oxidation to OA, especially during high relative humidity conditions; identification of a series of phenolic species and their oxidation products also suggests that BB emitted species might be precursors of the aqueous-phase oxidation. At last, cross-correlations between OA molecules (in signal fractions) and light absorption (in MAE) indicate that the key light-absorbing species in winter and summer were likely quite different despite their similar chemical compositions. In summer, CH and CHN compounds were important to light absorption, whereas CHNS compounds became more important in winter. Our findings regarding the optical and chemical properties of ambient OA here are implicate for future PM2.5 reduction and air quality improvement.

Supplementary Materials: The following supporting information can be downloaded at https://www.mdpi.com/article/10.3390/toxics13060443/s1, Table S1: The test results of normality distribution of the mass concentrations and optical parameters of samples during different periods; Table S2: The Wilcoxon significance test results of mass concentrations and relevant optical parameter of different pairs of sampling periods; Table S3: The relative humidity and temperature of the selected 24 samples for OA molecular characterization; Table S4: The relative signal fractions of the selected HOA, BBOA, SV-OOA, and LV-OOA tracer compounds in the samples; Table S5: The relative signal fractions of some selected SOA tracer molecules in the samples; Table S6: The relative signal fractions of some oligomers derived from aqueous-oxidation of phenolic substances (phenol, syringol and eugenol) in the samples; Figure S1: The sample TIC and mass spectra of ESI+ and ESI−; Figure S2: The correlation plots (Pearson's r) of: (**a**) water-soluble organic carbon (WSOC) versus its light absorption at 365 nm ($Abs_{WSOC\text{-}365}$); (**b**) methanol-soluble organic carbon (MSOC) versus its light absorption at 365 nm ($Abs_{MSOC\text{-}365}$); (**c**) WSOC versus MSOC; (**d**) $Abs_{WSOC\text{-}365}$ versus $Abs_{MSOC\text{-}365}$; Figure S3: The relative signal fractions of the resolved fluorescent components derived from parallel factor analysis (PARAFAC) for (**a**) water-soluble organic carbon (WSOC) and (**b**) methanol-soluble organic carbon (MSOC) during different sampling periods; Figure S4: The relative signal fractions of different types of compounds in OA from different periods; Figure S5: The Van Krevelen (VK) diagram of CHO compounds detected under ESI+ mode; Figure S6: The Van Krevelen (VK) diagram of CHO compounds detected under ESI− mode; Figure S7: Carbon oxidation state (OSc) diagram of CHO compounds under ESI+ mode.

Author Contributions: Conceptualization, X.G.; data curation, B.Z., L.F., Z.Z., H.L. and X.G.; formal analysis, B.Z. and Y.H.; funding acquisition, X.G., L.F. and Z.Z.; investigation, B.Z. and J.Y.; methodology, Y.H., H.L., Y.W., J.Y. and X.G.; resources, X.G. and J.Y.; writing—original draft, B.Z.; writing—review & editing, Y.W. and X.G. All authors have read and agreed to the published version of the manuscript.

Funding: National Natural Science Foundation of China (22361162668), and the Postgraduate Research & Practice Innovation Program of Jiangsu Province (SJCX24_0484 and KYCX24_1526).

Institutional Review Board Statement: Not applicable.

Informed Consent Statement: Not applicable.

Data Availability Statement: The data presented in this article are available on request from the corresponding author.

Conflicts of Interest: The authors declare no conflict of interest.

References

1. Jimenez, J.L.; Canagaratna, M.R.; Donahue, N.M.; Prevot, A.S.H.; Zhang, Q.; Kroll, J.H.; DeCarlo, P.F.; Allan, J.D.; Coe, H.; Ng, N.L.; et al. Evolution of Organic Aerosols in the Atmosphere. *Science* **2009**, *326*, 1525–1529. [CrossRef] [PubMed]
2. Hecobian, A.; Zhang, X.; Zheng, M.; Frank, N.; Edgerton, E.S.; Weber, R.J. Water-Soluble Organic Aerosol material and the light-absorption characteristics of aqueous extracts measured over the Southeastern United States. *Atmos. Chem. Phys.* **2010**, *10*, 5965–5977. [CrossRef]
3. Srinivas, B.; Sarin, M.M. Brown carbon in atmospheric outflow from the Indo-Gangetic Plain: Mass absorption efficiency and temporal variability. *Atmos. Environ.* **2014**, *89*, 835–843. [CrossRef]
4. Costabile, F.; Gilardoni, S.; Barnaba, F.; Di Ianni, A.; Di Liberto, L.; Dionisi, D.; Manigrasso, M.; Paglione, M.; Poluzzi, V.; Rinaldi, M.; et al. Characteristics of brown carbon in the urban Po Valley atmosphere. *Atmos. Chem. Phys.* **2017**, *17*, 313–326. [CrossRef]
5. Kirillova, E.N.; Andersson, A.; Tiwari, S.; Srivastava, A.K.; Bisht, D.S.; Gustafsson, Ö. Water-soluble organic carbon aerosols during a full New Delhi winter: Isotope-based source apportionment and optical properties. *J. Geophys. Res. Atmos.* **2014**, *119*, 3476–3485. [CrossRef]
6. Liu, J.; Scheuer, E.; Dibb, J.; Diskin, G.S.; Ziemba, L.D.; Thornhill, K.L.; Anderson, B.E.; Wisthaler, A.; Mikoviny, T.; Devi, J.J.; et al. Brown carbon aerosol in the North American continental troposphere: Sources, abundance, and radiative forcing. *Atmos. Chem. Phys.* **2015**, *15*, 7841–7858. [CrossRef]
7. Xie, M.; Shen, G.; Holder, A.L.; Hays, M.D.; Jetter, J.J. Light absorption of organic carbon emitted from burning wood, charcoal, and kerosene in household cookstoves. *Environ. Pollut.* **2018**, *240*, 60–67. [CrossRef]
8. Kim, H.; Kim, J.Y.; Jin, H.C.; Lee, J.Y.; Lee, S.P. Seasonal variations in the light-absorbing properties of water-soluble and insoluble organic aerosols in Seoul, Korea. *Atmos. Environ.* **2016**, *129*, 234–242. [CrossRef]
9. Liu, J.; Bergin, M.; Guo, H.; King, L.; Kotra, N.; Edgerton, E.; Weber, R.J. Size-resolved measurements of brown carbon in water and methanol extracts and estimates of their contribution to ambient fine-particle light absorption. *Atmos. Chem. Phys.* **2013**, *13*, 12389–12404. [CrossRef]
10. Park, S.; Son, S.-C.; Lee, S. Characterization, sources, and light absorption of fine organic aerosols during summer and winter at an urban site. *Atmos. Res.* **2018**, *213*, 370–380. [CrossRef]

11. Lu, J.; Ge, X.; Liu, Y.; Chen, Y.; Xie, X.; Ou, Y.; Ye, Z.; Chen, M. Significant secondary organic aerosol production from aqueous-phase processing of two intermediate volatility organic compounds. *Atmos. Environ.* **2019**, *211*, 63–68. [CrossRef]
12. Ye, Z.; Qu, Z.; Ma, S.; Luo, S.; Chen, Y.; Chen, H.; Chen, Y.; Zhao, Z.; Chen, M.; Ge, X. A comprehensive investigation of aqueous-phase photochemical oxidation of 4-ethylphenol. *Environ. Sci. Technol.* **2019**, *685*, 976–985. [CrossRef] [PubMed]
13. Feng, Y.; Ramanathan, V.; Kotamarthi, V.R. Brown carbon: A significant atmospheric absorber of solar radiation? *Atmos. Chem. Phys.* **2013**, *13*, 8607–8621. [CrossRef]
14. Jo, D.S.; Park, R.J.; Lee, S.; Kim, S.W.; Zhang, X. A global simulation of brown carbon: Implications for photochemistry and direct radiative effect. *Atmos. Chem. Phys.* **2016**, *16*, 3413–3432. [CrossRef]
15. Park, R.J.; Kim, M.J.; Jeong, J.I.; Youn, D.; Kim, S. A contribution of brown carbon aerosol to the aerosol light absorption and its radiative forcing in East Asia. *Atmos. Environ.* **2010**, *44*, 1414–1421. [CrossRef]
16. Chen, Y.; Bond, T.C. Light absorption by organic carbon from wood combustion. *Atmos. Chem. Phys.* **2010**, *10*, 1773–1787. [CrossRef]
17. Hoffer, A.; Gelencsér, A.; Guyon, P.; Kiss, G.; Schmid, O.; Frank, G.P.; Artaxo, P.; Andreae, M.O. Optical properties of humic-like substances (HULIS) in biomass-burning aerosols. *Atmos. Chem. Phys.* **2006**, *6*, 3563–3570. [CrossRef]
18. Andreae, M.O.; Gelencsér, A. Black carbon or brown carbon? The nature of light-absorbing carbonaceous aerosols. *Atmos. Chem. Phys.* **2006**, *6*, 3131–3148. [CrossRef]
19. Laskin, A.; Laskin, J.; Nizkorodov, S.A. Chemistry of Atmospheric Brown Carbon. *Chem. Rev.* **2015**, *115*, 4335–4382. [CrossRef]
20. Desyaterik, Y.; Sun, Y.; Shen, X.; Lee, T.; Wang, X.; Wang, T.; Collett, J.L., Jr. Speciation of "brown" carbon in cloud water impacted by agricultural biomass burning in eastern China. *J. Geophys. Res. Atmos.* **2013**, *118*, 7389–7399. [CrossRef]
21. Zhang, X.; Lin, Y.-H.; Surratt, J.D.; Weber, R.J. Sources, Composition and Absorption Ångström Exponent of Light-absorbing Organic Components in Aerosol Extracts from the Los Angeles Basin. *Environ. Sci. Technol.* **2013**, *47*, 3685–3693. [CrossRef] [PubMed]
22. Huang, R.-J.; Yang, L.; Cao, J.; Chen, Y.; Chen, Q.; Li, Y.; Duan, J.; Zhu, C.; Dai, W.; Wang, K.; et al. Brown Carbon Aerosol in Urban Xi'an, Northwest China: The Composition and Light Absorption Properties. *Environ. Sci. Technol.* **2018**, *52*, 6825–6833. [CrossRef] [PubMed]
23. Li, Y.; Fu, T.-M.; Yu, J.Z.; Zhang, A.; Yu, X.; Ye, J.; Zhu, L.; Shen, H.; Wang, C.; Yang, X.; et al. Nitrogen dominates global atmospheric organic aerosol absorption. *Science* **2025**, *387*, 989–995. [CrossRef]
24. Liu, J.; Lin, P.; Laskin, A.; Laskin, J.; Kathmann, S.M.; Wise, M.; Caylor, R.; Imholt, F.; Selimovic, V.; Shilling, J.E. Optical properties and aging of light-absorbing secondary organic aerosol. *Atmos. Chem. Phys.* **2016**, *16*, 12815–12827. [CrossRef]
25. Chen, Q.; Li, J.; Hua, X.; Jiang, X.; Mu, Z.; Wang, M.; Wang, J.; Shan, M.; Yang, X.; Fan, X.; et al. Identification of species and sources of atmospheric chromophores by fluorescence excitation-emission matrix with parallel factor analysis. *Environ. Sci. Technol.* **2020**, *718*, 137322. [CrossRef]
26. Li, J.; Chen, Q.; Hua, X.; Chang, T.; Wang, Y. Occurrence and sources of chromophoric organic carbon in fine particulate matter over Xi'an, China. *Environ. Sci. Technol.* **2020**, *725*, 138290. [CrossRef]
27. Xie, X.; Chen, Y.; Nie, D.; Liu, Y.; Liu, Y.; Lei, R.; Zhao, X.; Li, H.; Ge, X. Light-absorbing and fluorescent properties of atmospheric brown carbon: A case study in Nanjing, China. *Chemosphere* **2020**, *251*, 126350. [CrossRef]
28. Lin, P.; Bluvshtein, N.; Rudich, Y.; Nizkorodov, S.A.; Laskin, J.; Laskin, A. Molecular Chemistry of Atmospheric Brown Carbon Inferred from a Nationwide Biomass Burning Event. *Environ. Sci. Technol.* **2017**, *51*, 11561–11570. [CrossRef]
29. Cheng, Y.; He, K.-B.; Du, Z.-Y.; Engling, G.; Liu, J.-M.; Ma, Y.-L.; Zheng, M.; Weber, R.J. The characteristics of brown carbon aerosol during winter in Beijing. *Atmos. Environ.* **2016**, *127*, 355–364. [CrossRef]
30. Chen, Q.; Ikemori, F.; Nakamura, Y.; Vodicka, P.; Kawamura, K.; Mochida, M. Structural and Light-Absorption Characteristics of Complex Water-Insoluble Organic Mixtures in Urban Submicrometer Aerosols. *Environ. Sci. Technol.* **2017**, *51*, 8293–8303. [CrossRef]
31. Corbin, J.C.; Czech, H.; Massabò, D.; de Mongeot, F.B.; Jakobi, G.; Liu, F.; Lobo, P.; Mennucci, C.; Mensah, A.A.; Orasche, J.; et al. Infrared-absorbing carbonaceous tar can dominate light absorption by marine-engine exhaust. *Npj Clim. Atmos. Sci.* **2019**, *2*, 12. [CrossRef]
32. Afsana, S.; Zhou, R.; Miyazaki, Y.; Tachibana, E.; Deshmukh, D.K.; Kawamura, K.; Mochida, M. Abundance, chemical structure, and light absorption properties of humic-like substances (HULIS) and other organic fractions of forest aerosols in Hokkaido. *Sci. Rep.* **2022**, *12*, 14379. [CrossRef] [PubMed]
33. Chen, Q.; Ikemori, F.; Mochida, M. Light Absorption and Excitation–Emission Fluorescence of Urban Organic Aerosol Components and Their Relationship to Chemical Structure. *Environ. Sci. Technol.* **2016**, *50*, 10859–10868. [CrossRef]
34. Huang, R.-J.; Yang, L.; Shen, J.; Yuan, W.; Gong, Y.; Guo, J.; Cao, W.; Duan, J.; Ni, H.; Zhu, C.; et al. Water-Insoluble Organics Dominate Brown Carbon in Wintertime Urban Aerosol of China: Chemical Characteristics and Optical Properties. *Environ. Sci. Technol.* **2020**, *54*, 7836–7847. [CrossRef]

35. Afsana, S.; Zhou, R.; Miyazaki, Y.; Tachibana, E.; Deshmukh, D.K.; Kawamura, K.; Mochida, M. Fluorescence of solvent-extractable organics in sub-micrometer forest aerosols in Hokkaido, Japan. *Atmos. Environ.* **2023**, *303*, 119710. [CrossRef]
36. Kuang, Y.; Shang, J.; Sheng, M.; Shi, X.; Zhu, J.; Qiu, X. Molecular Composition of Beijing $PM_{2.5}$ Brown Carbon Revealed by an Untargeted Approach Based on Gas Chromatography and Time-of-Flight Mass Spectrometry. *Environ. Sci. Technol.* **2023**, *57*, 909–919. [CrossRef]
37. Mao, J.; Cheng, Y.; Bai, Z.; Zhang, W.; Zhang, L.; Chen, H.; Wang, L.; Li, L.; Chen, J. Molecular characterization of nitrogen-containing organic compounds in the winter North China Plain. *Environ. Sci. Technol.* **2022**, *838*, 156189. [CrossRef]
38. Wang, K.; Zhang, Y.; Huang, R.-J.; Cao, J.; Hoffmann, T. UHPLC-Orbitrap mass spectrometric characterization of organic aerosol from a central European city (Mainz, Germany) and a Chinese megacity (Beijing). *Atmos. Environ.* **2018**, *189*, 22–29. [CrossRef]
39. Wang, W.; Zhang, Y.; Jiang, B.; Chen, Y.; Song, Y.; Tang, Y.; Dong, C.; Cai, Z. Molecular characterization of organic aerosols in Taiyuan, China: Seasonal variation and source identification. *Environ. Sci. Technol.* **2021**, *800*, 149419. [CrossRef]
40. Wang, J.; Ge, X.; Chen, Y.; Shen, Y.; Zhang, Q.; Sun, Y.; Xu, J.; Ge, S.; Yu, H.; Chen, M. Highly time-resolved urban aerosol characteristics during springtime in Yangtze River Delta, China: Insights from soot particle aerosol mass spectrometry. *Atmos. Chem. Phys.* **2016**, *16*, 9109–9127. [CrossRef]
41. Wu, Y.; Ge, X.; Wang, J.; Shen, Y.; Ye, Z.; Ge, S.; Wu, Y.; Yu, H.; Chen, M. Responses of secondary aerosols to relative humidity and photochemical activities in an industrialized environment during late winter. *Atmos. Environ.* **2018**, *193*, 66–78. [CrossRef]
42. Li, D.; Cui, S.; Wu, Y.; Wang, J.; Ge, X. Direct Measurement of Aerosol Liquid Water Content: A Case Study in Summer in Nanjing, China. *Toxics* **2024**, *12*, 164. [CrossRef] [PubMed]
43. Ge, X.; Li, L.; Chen, Y.; Chen, H.; Wu, D.; Wang, J.; Xie, X.; Ge, S.; Ye, Z.; Xu, J.; et al. Aerosol characteristics and sources in Yangzhou, China resolved by offline aerosol mass spectrometry and other techniques. *Environ. Pollut.* **2017**, *225*, 74–85. [CrossRef]
44. Chen, Y.; Chen, Y.; Xie, X.; Ye, Z.; Li, Q.; Ge, X.; Chen, M. Chemical Characteristics of $PM_{2.5}$ and Water-Soluble Organic Nitrogen in Yangzhou, China. *Atmosphere* **2019**, *10*, 178. [CrossRef]
45. Tsugawa, H.; Cajka, T.; Kind, T.; Ma, Y.; Higgins, B.; Ikeda, K.; Kanazawa, M.; VanderGheynst, J.; Fiehn, O.; Arita, M. MS-DIAL: Data-independent MS/MS deconvolution for comprehensive metabolome analysis. *Nat. Methods.* **2015**, *12*, 523–526. [CrossRef]
46. Bae, E.; Yeo, I.J.; Jeong, B.; Shin, Y.; Shin, K.-H.; Kim, S. Study of Double Bond Equivalents and the Numbers of Carbon and Oxygen Atom Distribution of Dissolved Organic Matter with Negative-Mode FT-ICR MS. *Anal. Chem.* **2011**, *83*, 4193–4199. [CrossRef]
47. Yassine, M.; Harir, M.; Dabek, E.; Schmitt-Kopplin, P. Structural characterization of organic aerosol using Fourier transform ion cyclotron resonance mass spectrometry: Aromaticity equivalent approach. *Rapid Commun. Mass Spectrom.* **2014**, *28*, 2445–2454. [CrossRef]
48. Kourtchev, I.; Godoi, R.H.M.; Connors, S.; Levine, J.G.; Archibald, A.T.; Godoi, A.F.L.; Paralovo, S.L.; Barbosa, C.G.G.; Souza, R.A.F.; Manzi, A.O.; et al. Molecular composition of organic aerosols in central Amazonia: An ultra-high-resolution mass spectrometry study. *Atmos. Chem. Phys.* **2016**, *16*, 11899–11913. [CrossRef]
49. Tong, H.; Kourtchev, I.; Pant, P.; Keyte, I.J.; O'Connor, I.P.; Wenger, J.C.; Pope, F.D.; Harrison, R.M.; Kalberer, M. Molecular composition of organic aerosols at urban background and road tunnel sites using ultra-high resolution mass spectrometry. *Faraday Discuss.* **2016**, *189*, 51–68. [CrossRef]
50. Kroll, J.H.; Donahue, N.M.; Jimenez, J.L.; Kessler, S.H.; Canagaratna, M.R.; Wilson, K.R.; Altieri, K.E.; Mazzoleni, L.R.; Wozniak, A.S.; Bluhm, H.; et al. Carbon oxidation state as a metric for describing the chemistry of atmospheric organic aerosol. *Nat. Chem.* **2011**, *3*, 133–139. [CrossRef]
51. Lin, P.; Rincon, A.G.; Kalberer, M.; Yu, J.Z. Elemental Composition of HULIS in the Pearl River Delta Region, China: Results Inferred from Positive and Negative Electrospray High Resolution Mass Spectrometric Data. *Environ. Sci. Technol.* **2012**, *46*, 7454–7462. [CrossRef] [PubMed]
52. Chen, Y.; Ge, X.; Chen, H.; Xie, X.; Chen, Y.; Wang, J.; Ye, Z.; Bao, M.; Zhang, Y.; Chen, M. Seasonal light absorption properties of water-soluble brown carbon in atmospheric fine particles in Nanjing, China. *Atmos. Environ.* **2018**, *187*, 230–240. [CrossRef]
53. Chen, Y.; Xie, X.; Shi, Z.; Li, Y.; Gai, X.; Wang, J.; Li, H.; Wu, Y.; Zhao, X.; Chen, M.; et al. Brown carbon in atmospheric fine particles in Yangzhou, China: Light absorption properties and source apportionment. *Atmos. Res.* **2020**, *244*, 105028. [CrossRef]
54. Murphy, K.R.; Stedmon, C.A.; Graeber, D.; Bro, R. Fluorescence spectroscopy and multi-way techniques. PARAFAC. *Anal. Methods* **2013**, *5*, 6557–6566. [CrossRef]
55. Birdwell, J.E.; Engel, A.S. Characterization of dissolved organic matter in cave and spring waters using UV–Vis absorbance and fluorescence spectroscopy. *Org. Geochem.* **2010**, *41*, 270–280. [CrossRef]
56. McKnight, D.M.; Boyer, E.W.; Westerhoff, P.K.; Doran, P.T.; Kulbe, T.; Andersen, D.T. Spectrofluorometric characterization of dissolved organic matter for indication of precursor organic material and aromaticity. *Limnol. Oceanogr.* **2001**, *46*, 38–48. [CrossRef]
57. Zhang, B.; Tang, J.; Geng, X.; Mo, Y.; Zhao, S.; Zhong, G.; Li, J.; Zhang, G. Seasonal changes in water-soluble brown carbon (BrC) at Nanling background station in South China. *Front. Environ. Sci.* **2024**, *12*, 1360453. [CrossRef]

58. Zhong, M.; Xu, J.; Wang, H.; Gao, L.; Zhu, H.; Zhai, L.; Zhang, X.; Zhao, W. Characterizing water-soluble brown carbon in fine particles in four typical cities in northwestern China during wintertime: Integrating optical properties with chemical processes. *Atmos. Chem. Phys.* **2023**, *23*, 12609–12630. [CrossRef]
59. Soleimanian, E.; Mousavi, A.; Taghvaee, S.; Shafer, M.M.; Sioutas, C. Impact of secondary and primary particulate matter (PM) sources on the enhanced light absorption by brown carbon (BrC) particles in central Los Angeles. *Environ. Sci. Technol.* **2020**, *705*, 135902. [CrossRef]
60. Lambe, A.T.; Cappa, C.D.; Massoli, P.; Onasch, T.B.; Forestieri, S.D.; Martin, A.T.; Cummings, M.J.; Croasdale, D.R.; Brune, W.H.; Worsnop, D.R.; et al. Relationship between Oxidation Level and Optical Properties of Secondary Organic Aerosol. *Environ. Sci. Technol.* **2013**, *47*, 6349–6357. [CrossRef]
61. Fan, X.; Li, M.; Cao, T.; Cheng, C.; Li, F.; Xie, Y.; Wei, S.; Song, J.; Peng, P.A. Optical properties and oxidative potential of water- and alkaline-soluble brown carbon in smoke particles emitted from laboratory simulated biomass burning. *Atmos. Environ.* **2018**, *194*, 48–57. [CrossRef]
62. Shamjad, P.M.; Tripathi, S.N.; Thamban, N.M.; Vreeland, H. Refractive Index and Absorption Attribution of Highly Absorbing Brown Carbon Aerosols from an Urban Indian City-Kanpur. *Sci. Rep.* **2016**, *6*, 37735. [CrossRef] [PubMed]
63. Deng, J.; Ma, H.; Wang, X.; Zhong, S.; Zhang, Z.; Zhu, J.; Fan, Y.; Hu, W.; Wu, L.; Li, X.; et al. Measurement report: Optical properties and sources of water-soluble brown carbon in Tianjin, North China—insights from organic molecular compositions. *Atmos. Chem. Phys.* **2022**, *22*, 6449–6470. [CrossRef]
64. Li, J.; Zhang, Q.; Wang, G.; Li, J.; Wu, C.; Liu, L.; Wang, J.; Jiang, W.; Li, L.; Ho, K.F.; et al. Optical properties and molecular compositions of water-soluble and water-insoluble brown carbon (BrC) aerosols in northwest China. *Atmos. Chem. Phys.* **2020**, *20*, 4889–4904. [CrossRef]
65. Wen, H.; Zhou, Y.; Xu, X.; Wang, T.; Chen, Q.; Chen, Q.; Li, W.; Wang, Z.; Huang, Z.; Zhou, T.; et al. Water-soluble brown carbon in atmospheric aerosols along the transport pathway of Asian dust: Optical properties, chemical compositions, and potential sources. *Environ. Sci. Technol.* **2021**, *789*, 147971. [CrossRef]
66. Choudhary, V.; Rajput, P.; Gupta, T. Absorption properties and forcing efficiency of light-absorbing water-soluble organic aerosols: Seasonal and spatial variability. *Environ. Pollut.* **2021**, *272*, 115932. [CrossRef]
67. Choudhary, V.; Gupta, T.; Zhao, R. Evolution of Brown Carbon Aerosols during Atmospheric Long-Range Transport in the South Asian Outflow and Himalayan Cryosphere. *ACS Earth Space Chem.* **2022**, *6*, 2335–2347. [CrossRef]
68. Sumlin, B.J.; Pandey, A.; Walker, M.J.; Pattison, R.S.; Williams, B.J.; Chakrabarty, R.K. Atmospheric Photooxidation Diminishes Light Absorption by Primary Brown Carbon Aerosol from Biomass Burning. *Environ. Sci. Technol. Lett.* **2017**, *4*, 540–545. [CrossRef]
69. Huguet, A.; Vacher, L.; Relexans, S.; Saubusse, S.; Froidefond, J.M.; Parlanti, E. Properties of fluorescent dissolved organic matter in the Gironde Estuary. *Org. Geochem.* **2009**, *40*, 706–719. [CrossRef]
70. Lee, H.J.; Laskin, A.; Laskin, J.; Nizkorodov, S.A. Excitation–Emission Spectra and Fluorescence Quantum Yields for Fresh and Aged Biogenic Secondary Organic Aerosols. *Environ. Sci. Technol.* **2013**, *47*, 5763–5770. [CrossRef]
71. Fu, P.; Kawamura, K.; Chen, J.; Qin, M.; Ren, L.; Sun, Y.; Wang, Z.; Barrie, L.A.; Tachibana, E.; Ding, A.; et al. Fluorescent water-soluble organic aerosols in the High Arctic atmosphere. *Sci. Rep.* **2015**, *5*, 9845. [CrossRef] [PubMed]
72. Qin, J.; Zhang, L.; Zhou, X.; Duan, J.; Mu, S.; Xiao, K.; Hu, J.; Tan, J. Fluorescence fingerprinting properties for exploring water-soluble organic compounds in $PM_{2.5}$ in an industrial city of northwest China. *Atmos. Environ.* **2018**, *184*, 203–211. [CrossRef]
73. Dong, Z.; Pavuluri, C.M.; Li, P.; Xu, Z.; Deng, J.; Zhao, X.; Zhao, X.; Fu, P.; Liu, C.Q. Measurement report: Optical characterization, seasonality, and sources of brown carbon in fine aerosols from Tianjin, North China: Year-round observations. *Atmos. Chem. Phys.* **2024**, *24*, 5887–5905. [CrossRef]
74. Fan, X.; Yu, X.; Wang, Y.; Xiao, X.; Li, F.; Xie, Y.; Wei, S.; Song, J.; Peng, P.A. The aging behaviors of chromophoric biomass burning brown carbon during dark aqueous hydroxyl radical oxidation processes in laboratory studies. *Atmos. Environ.* **2019**, *205*, 9–18. [CrossRef]
75. Coble, P.G. Characterization of marine and terrestrial DOM in seawater using excitation-emission matrix spectroscopy. *Mar. Chem.* **1996**, *51*, 325–346. [CrossRef]
76. Murphy, K.R.; Hambly, A.; Singh, S.; Henderson, R.K.; Baker, A.; Stuetz, R.; Khan, S.J. Organic Matter Fluorescence in Municipal Water Recycling Schemes: Toward a Unified PARAFAC Model. *Environ. Sci. Technol.* **2011**, *45*, 2909–2916. [CrossRef]
77. Yu, H.; Liang, H.; Qu, F.; Han, Z.-S.; Shao, S.; Chang, H.; Li, G. Impact of dataset diversity on accuracy and sensitivity of parallel factor analysis model of dissolved organic matter fluorescence excitation-emission matrix. *Sci. Rep.* **2015**, *5*, 10207. [CrossRef]
78. Chen, Q.; Miyazaki, Y.; Kawamura, K.; Matsumoto, K.; Coburn, S.; Volkamer, R.; Iwamoto, Y.; Kagami, S.; Deng, Y.; Ogawa, S.; et al. Characterization of Chromophoric Water-Soluble Organic Matter in Urban, Forest, and Marine Aerosols by HR-ToF-AMS Analysis and Excitation–Emission Matrix Spectroscopy. *Environ. Sci. Technol.* **2016**, *50*, 10351–10360. [CrossRef]

79. Dubnick, A.; Barker, J.; Sharp, M.; Wadham, J.; Lis, G.; Telling, J.; Fitzsimons, S.; Jackson, M. Characterization of dissolved organic matter (DOM) from glacial environments using total fluorescence spectroscopy and parallel factor analysis. *Ann. Glaciol.* **2010**, *51*, 111–122. [CrossRef]
80. Fu, P.; Mostofa, K.M.G.; Wu, F.; Liu, C.-Q.; Li, W.; Liao, H.; Wang, L.; Wang, J.; Mei, Y.I. Excitation-emission matrix characterization of dissolved organic matter sources in two eutrophic lakes (Southwestern China Plateau). *Geochem. J.* **2010**, *44*, 99–112. [CrossRef]
81. Barsotti, F.; Ghigo, G.; Vione, D. Computational assessment of the fluorescence emission of phenol oligomers: A possible insight into the fluorescence properties of humic-like substances (HULIS). *J. Photochem. Photobiol. A Chem.* **2016**, *315*, 87–93. [CrossRef]
82. Chen, Q.; Hua, X.; Li, J.; Chang, T.; Wang, Y. Diurnal evolutions and sources of water-soluble chromophoric aerosols over Xi'an during haze event, in Northwest China. *Environ. Sci. Technol.* **2021**, *786*, 147412. [CrossRef] [PubMed]
83. Cao, T.; Li, M.; Zou, C.; Fan, X.; Song, J.; Jia, W.; Yu, C.; Yu, Z.; Peng, P. Chemical composition, optical properties, and oxidative potential of water- and methanol-soluble organic compounds emitted from the combustion of biomass materials and coal. *Atmos. Chem. Phys.* **2021**, *21*, 13187–13205. [CrossRef]
84. Cottrell, B.A.; Gonsior, M.; Isabelle, L.M.; Luo, W.; Perraud, V.; McIntire, T.M.; Pankow, J.F.; Schmitt-Kopplin, P.; Cooper, W.J.; Simpson, A.J. A regional study of the seasonal variation in the molecular composition of rainwater. *Atmos. Environ.* **2013**, *77*, 588–597. [CrossRef]
85. Tu, P.; Hall, W.A.I.V.; Johnston, M.V. Characterization of Highly Oxidized Molecules in Fresh and Aged Biogenic Secondary Organic Aerosol. *Anal. Chem.* **2016**, *88*, 4495–4501. [CrossRef]
86. Kourtchev, I.; O'Connor, I.P.; Giorio, C.; Fuller, S.J.; Kristensen, K.; Maenhaut, W.; Wenger, J.C.; Sodeau, J.R.; Glasius, M.; Kalberer, M. Effects of anthropogenic emissions on the molecular composition of urban organic aerosols: An ultrahigh resolution mass spectrometry study. *Atmos. Environ.* **2014**, *89*, 525–532. [CrossRef]
87. Ge, X.; Wexler, A.S.; Clegg, S.L. Atmospheric amines—Part I. A review. *Atmos. Environ.* **2011**, *45*, 524–546. [CrossRef]
88. Ma, Y.; Hays, M.D. Thermal extraction–two-dimensional gas chromatography–mass spectrometry with heart-cutting for nitrogen heterocyclics in biomass burning aerosols. *J. Chromatogr. A* **2008**, *1200*, 228–234. [CrossRef]
89. Wang, Y.; Hu, M.; Wang, Y.; Zheng, J.; Shang, D.; Yang, Y.; Liu, Y.; Li, X.; Tang, R.; Zhu, W.; et al. The formation of nitro-aromatic compounds under high NO_x and anthropogenic VOC conditions in urban Beijing, China. *Atmos. Chem. Phys.* **2019**, *19*, 7649–7665. [CrossRef]
90. Wang, L.; Wang, X.; Gu, R.; Wang, H.; Yao, L.; Wen, L.; Zhu, F.; Wang, W.; Xue, L.; Yang, L.; et al. Observations of fine particulate nitrated phenols in four sites in northern China: Concentrations, source apportionment, and secondary formation. *Atmos. Chem. Phys.* **2018**, *18*, 4349–4359. [CrossRef]
91. Song, J.; Li, M.; Fan, X.; Zou, C.; Zhu, M.; Jiang, B.; Yu, Z.; Jia, W.; Liao, Y.; Peng, P.A. Molecular Characterization of Water- and Methanol-Soluble Organic Compounds Emitted from Residential Coal Combustion Using Ultrahigh-Resolution Electrospray Ionization Fourier Transform Ion Cyclotron Resonance Mass Spectrometry. *Environ. Sci. Technol.* **2019**, *53*, 13607–13617. [CrossRef] [PubMed]
92. Lin, P.; Liu, J.; Shilling, J.E.; Kathmann, S.M.; Laskin, J.; Laskin, A. Molecular characterization of brown carbon (BrC) chromophores in secondary organic aerosol generated from photo-oxidation of toluene. *Phys. Chem. Chem. Phys.* **2015**, *17*, 23312–23325. [CrossRef] [PubMed]
93. Harrison, M.A.J.; Barra, S.; Borghesi, D.; Vione, D.; Arsene, C.; Iulian Olariu, R. Nitrated phenols in the atmosphere: A review. *Atmos. Environ.* **2005**, *39*, 231–248. [CrossRef]
94. Wang, Y.; Hu, M.; Lin, P.; Guo, Q.; Wu, Z.; Li, M.; Zeng, L.; Song, Y.; Zeng, L.; Wu, Y.; et al. Molecular Characterization of Nitrogen-Containing Organic Compounds in Humic-like Substances Emitted from Straw Residue Burning. *Environ. Sci. Technol.* **2017**, *51*, 5951–5961. [CrossRef]
95. Iinuma, Y.; Böge, O.; Gräfe, R.; Herrmann, H. Methyl-Nitrocatechols: Atmospheric Tracer Compounds for Biomass Burning Secondary Organic Aerosols. *Environ. Sci. Technol.* **2010**, *44*, 8453–8459. [CrossRef]
96. Kahnt, A.; Behrouzi, S.; Vermeylen, R.; Safi Shalamzari, M.; Vercauteren, J.; Roekens, E.; Claeys, M.; Maenhaut, W. One-year study of nitro-organic compounds and their relation to wood burning in PM_{10} aerosol from a rural site in Belgium. *Atmos. Environ.* **2013**, *81*, 561–568. [CrossRef]
97. Fleming, L.T.; Lin, P.; Roberts, J.M.; Selimovic, V.; Yokelson, R.; Laskin, J.; Laskin, A.; Nizkorodov, S.A. Molecular composition and photochemical lifetimes of brown carbon chromophores in biomass burning organic aerosol. *Atmos. Chem. Phys.* **2020**, *20*, 1105–1129. [CrossRef]
98. Qi, L.; Chen, M.; Stefenelli, G.; Pospisilova, V.; Tong, Y.; Bertrand, A.; Hueglin, C.; Ge, X.; Baltensperger, U.; Prévôt, A.S.H.; et al. Organic aerosol source apportionment in Zurich using an extractive electrospray ionization time-of-flight mass spectrometer (EESI−TOF-MS)—Part 2: Biomass burning influences in winter. *Atmos. Chem. Phys.* **2019**, *19*, 8037–8062. [CrossRef]
99. Kristensen, K.; Jensen, L.N.; Glasius, M.; Bilde, M. The effect of sub-zero temperature on the formation and composition of secondary organic aerosol from ozonolysis of alpha-pinene. *Environ. Sci. Proc. Imp.* **2017**, *19*, 1220–1234. [CrossRef]

100. Daellenbach, K.R.; Cai, J.; Hakala, S.; Dada, L.; Yan, C.; Du, W.; Yao, L.; Zheng, F.; Ma, J.; Ungeheuer, F.; et al. Substantial contribution of transported emissions to organic aerosol in Beijing. *Nat. Geosci.* **2024**, *17*, 747–754. [CrossRef]
101. Boreddy, S.K.R.; Hegde, P.; Aswini, A.R.; Aryasree, S. Chemical Characteristics, Size Distributions, Molecular Composition, and Brown Carbon in South Asian Outflow to the Indian Ocean. *Earth Space Sci.* **2021**, *8*, e2020EA001615. [CrossRef]
102. Song, J.; Li, M.; Jiang, B.; Wei, S.; Fan, X.; Peng, P.A. Molecular Characterization of Water-Soluble Humic like Substances in Smoke Particles Emitted from Combustion of Biomass Materials and Coal Using Ultrahigh-Resolution Electrospray Ionization Fourier Transform Ion Cyclotron Resonance Mass Spectrometry. *Environ. Sci. Technol.* **2018**, *52*, 2575–2585. [CrossRef] [PubMed]
103. Wang, Y.; Hu, M.; Lin, P.; Tan, T.; Li, M.; Xu, N.; Zheng, J.; Du, Z.; Qin, Y.; Wu, Y.; et al. Enhancement in Particulate Organic Nitrogen and Light Absorption of Humic-Like Substances over Tibetan Plateau Due to Long-Range Transported Biomass Burning Emissions. *Environ. Sci. Technol.* **2019**, *53*, 14222–14232. [CrossRef]
104. Qi, L.; Zhang, Z.; Wang, X.; Deng, F.; Zhao, J.; Liu, H. Molecular characterization of atmospheric particulate organosulfates in a port environment using ultrahigh resolution mass spectrometry: Identification of traffic emissions. *J. Hazard. Mater.* **2021**, *419*, 126431. [CrossRef]
105. Wang, Y.; Hu, M.; Guo, S.; Wang, Y.; Zheng, J.; Yang, Y.; Zhu, W.; Tang, R.; Li, X.; Liu, Y.; et al. The secondary formation of organosulfates under interactions between biogenic emissions and anthropogenic pollutants in summer in Beijing. *Atmos. Chem. Phys.* **2018**, *18*, 10693–10713. [CrossRef]
106. Ye, Y.; Zhan, H.; Yu, X.; Li, J.; Wang, X.; Xie, Z. Detection of organosulfates and nitrooxy-organosulfates in Arctic and Antarctic atmospheric aerosols, using ultra-high resolution FT-ICR mass spectrometry. *Sci. Total Environ.* **2021**, *767*, 144339. [CrossRef]
107. Yang, Z.; Tsona, N.T.; Li, J.; Wang, S.; Xu, L.; You, B.; Du, L. Effects of NO_x and SO_2 on the secondary organic aerosol formation from the photooxidation of 1,3,5-trimethylbenzene: A new source of organosulfates. *Environ. Pollut.* **2020**, *264*, 114742. [CrossRef]
108. McNeill, V.F. Aqueous Organic Chemistry in the Atmosphere: Sources and Chemical Processing of Organic Aerosols. *Environ. Sci. Technol.* **2015**, *49*, 1237–1244. [CrossRef]

toxics

MDPI

Article

Urban-Rural Differences in Indoor and Outdoor Air Quality: A Comparative Study in Bangladesh

Masamitsu Kurata [1,*], Akira Hibiki [2], Kazushi Takahashi [3] and Yutaka Matsumi [4]

1 Department of Economics, Sophia University, Tokyo 102-8554, Japan
2 Graduate School of Economics and Management, Tohoku University, Sendai 980-8576, Japan
3 Graduate School of Policy Studies, National Graduate Institute for Policy Studies (GRIPS), Tokyo 106-8677, Japan
4 Institute for Space-Earth Environmental Research, Nagoya University, Nagoya 464-8601, Japan
* Correspondence: kurata@sophia.ac.jp

Abstract: Health hazards caused by indoor air pollution (IAP) remain a global concern, especially in developing countries. IAP has complex mechanisms related to outdoor air pollution (OAP) and various other factors, and their relationship needs to be clarified to examine effective policies. We conducted an indoor and outdoor air monitoring survey in urban, peri-urban, and rural areas in Bangladesh, one of the countries with the most severe air pollution. The results show that IAP is more severe in urban households than in rural households, with a five-fold difference in daily indoor $PM_{2.5}$ concentration between 117 μg/m^3 and 22 μg/m^3, respectively. Regression analysis reveals that IAP is strongly associated with OAP and is hardly affected by solid fuels used in well-ventilated outside kitchens. Our findings support the view that the mitigation of IAP in developing countries can be achieved not only through a transition to clean fuels, which often entails substantial costs, but also through more practical and accessible alternatives, such as the use of outdoor kitchens, electric fans, and careful management of behaviors such as indoor smoking and mosquito coil use.

Keywords: indoor air pollution; outdoor air pollution; solid fuel

1. Introduction

Health hazards caused by indoor air pollution (IAP) are a major global concern, as reported by various literature [1–4]. Ensuring good indoor air quality is critically important for public health, as people typically spend the majority of their time indoors [5,6]. Major sources of IAP include solid fuels such as firewood, coal, and animal dung [7]. Because solid fuels have been primarily used for cooking and heating in rural areas in the developing world, IAP is often regarded as a problem specific to those regions [8–10]. It is also common to emphasize the need for policies that encourage a shift from traditional solid fuels to cleaner fuels to reduce IAP [11].

However, at least two points should be noted when considering such views. First, the use of solid fuels does not always lead to severe IAP. Even when rural households use solid fuels, IAP can be mitigated through appropriate measures such as good ventilation and a separate kitchen [12]. Conversely, even when urban households use clean fuels, it can cause severe IAP due to unhealthy cooking methods [13]. Second, IAP is affected not only by intra-household activities but also by outdoor air pollution (OAP). If OAP is higher in urban areas than in rural areas, as is generally the case, IAP can also be higher in urban areas due to infiltration of outdoor pollutants [14,15]. However, the focus of research in urban areas tends to be biased toward OAP only, and studies targeting both OAP and IAP are still limited [16]. Therefore, it is essential to clarify the relationship between IAP, OAP, and solid fuel use to consider an effective policy for IAP.

This study focuses on Bangladesh, one of the countries with severe air pollution, and provides a comparative analysis of IAP among urban, peri-urban, and rural areas near the capital city of Dhaka. In Bangladesh, there is no consensus on whether IAP is more severe in rural or urban areas. For example, while Dasgupta et al. [12] indicated that IAP is worse in rural areas, Khalequzzaman et al. [17] reported the opposite. To add new findings, we conducted a 48-h monitoring survey for 50 households equipped with individual sensors for IAP and OAP. Combined with data on household characteristics and members' behavior, we analyzed regional variations of IAP and associated factors.

The results show that IAP is much higher in urban and peri-urban areas than in rural areas. Regression analysis revealed that IAP is strongly associated with OAP and is hardly affected by solid fuels in well-ventilated outside kitchens away from living spaces, which is a common settlement structure in rural Bangladesh and other developing countries. We also confirmed the effectiveness of electric fans and the adverse effects of cigarette smoking and the use of mosquito coils.

2. Materials and Methods

2.1. Survey Area and Sampling of Households

We conducted an air monitoring survey in Dhaka, the capital of Bangladesh, and in nearby peri-urban and rural areas in October 2022. These regions were selected because Dhaka is one of the most polluted cities in the world [18], and the surrounding area shows a large variation in air pollution within a small geographic zone. As shown in Figure 1, three study areas were selected to exploit the variation in OAP, namely, an urban area with high OAP, a peri-urban area with moderate OAP, and a rural area with low OAP. First, we identified Mirpur as a representative urban area, as it has a very high population density in Dhaka. Next, we selected Singair, the closest rural area to Mirpur, as a comparative site with similar climatic and geographical conditions. Finally, we selected Savar, located between the two areas, as the study site representing the peri-urban area.

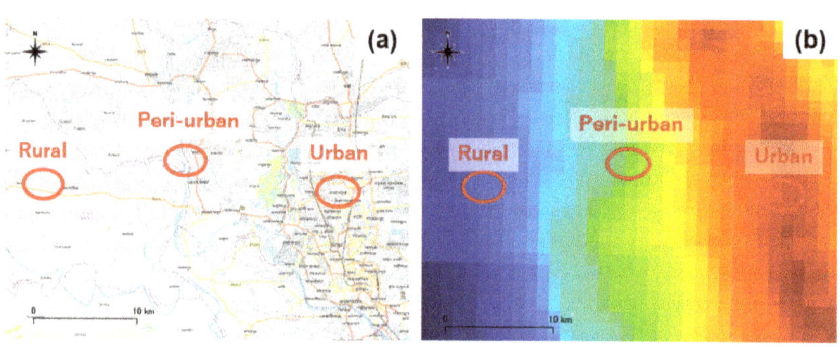

Figure 1. Survey area: (**a**) location of three areas in and nearby Dhaka, and (**b**) the level of ambient air pollution of nitrogen dioxides (NO_2) measured using satellite (Sentinel-5P) in 2022 [19].

In each area, we visited households as randomly as possible to find those willing to cooperate with the installation of sensors and face-to-face interviews. Finally, 50 households consented to the survey, including 10 in urban areas, 20 in peri-urban areas, and 20 in rural areas. The study protocol was approved by the Institutional Review Board of the Institute of Health Economics at the University of Dhaka.

2.2. Air Monitoring and Interview Survey

A palm-sized optical sensor (hereafter referred to as a "P-sensor") was used to monitor $PM_{2.5}$ mass concentration [20]. The P-sensor is designed to estimate $PM_{2.5}$ based on the light scattering intensity distribution and is manufactured by Panasonic Corporation, a prominent Japanese multinational electronics company (https://www.panasonic.com/global/about.html, accessed on 11 June 2025), based on the verification of the high quality of the collected data [20]. According to previous studies using P-sensors [21–23], this study also estimated the mass concentration of $PM_{2.5}$ using a conversion factor of 1.4.

Two P-sensors were installed per household to quantify both IAP and OAP. One unit measuring IAP (indoor sensor) was placed in the living room or bedroom, and the other measuring OAP (outdoor sensor) was placed on the balcony or outside the front door. We installed the indoor sensor in the living room rather than the kitchen because kitchens in rural and peri-urban Bangladesh are often located separately outside the living space [12].

Figure 2 shows typical house configurations in urban and rural areas and the locations where the two sensors are installed. In urban areas, most residences are in multi-story apartments equipped with balconies (Figure 2a), and the kitchen is generally located near the living room. The indoor sensor is installed in the living room, and the outdoor sensor is installed on the outdoor balcony. In rural and peri-urban areas, the main house is used as a living room and/or bedroom, with a toilet and a kitchen usually located outside (Figure 2b). The kitchen often has a roof but no walls on all sides, making it highly ventilated.

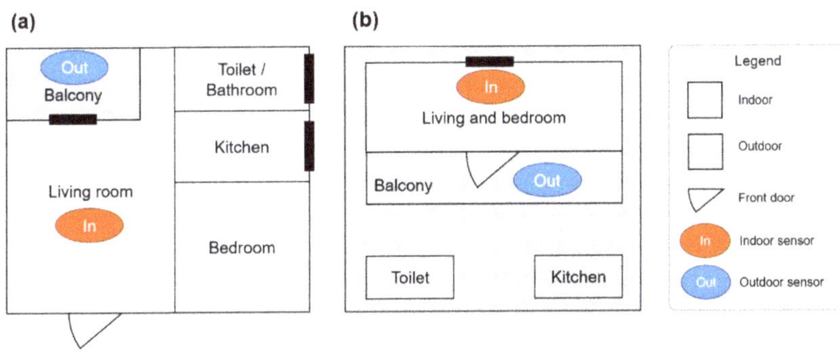

Figure 2. Typical house configurations in (**a**) urban and (**b**) rural areas.

We measured IAP and OAP at 5-min intervals for approximately four days for each household. Due to the limited number of P-sensor equipment, monitoring was staggered by a few days in each area. Generally, sensors were installed in the afternoon of the first day of monitoring and were removed in the afternoon of the fourth day. Because the data from the first and last days contained noise due to the work required for installation and removal, only the data from the middle two full days (48 h) was used for the analysis. Depending on the purpose of the analysis, we aggregate the raw data at 5-min intervals into hourly or daily data.

In addition to air monitoring, interviews were conducted to collect data on various household characteristics. The data can be grouped into four main categories. The first is the type of fuel used for cooking, including solid fuels (such as firewood, agricultural residues, or animal dung), liquefied petroleum gas (LPG), natural gas, and electricity. The second concerns the structure of the house and kitchen, covering the size of the house, the number of rooms, whether the kitchen is located outside, and whether the kitchen is shared with other households. The third relates to household members' behaviors that may affect IAP, such as daily cooking duration, smoking, use of mosquito coils, and use of electric fans or air conditioning. The fourth category includes other household characteristics, such as the gender and age of the household head and the household's income level.

2.3. Data Analysis

We analyze regional differences in IAP and their determinants. The following regression analysis using hourly data are performed to examine the determinants:

$$IAP_{ih} = \beta_1 Urban_i + \beta_2 Periurban_i + \beta_3 OAP_{ih} + \beta_4 Solid_i + \beta_5 OutsideKitchen_i + \beta_6 (Solid_i \times OutsideKitchen_i) + \sum_k \delta_k X_{ik} + \theta_h + \varepsilon_{ih}$$

where, IAP_{ih} and OAP_{ih} are the natural logarithms of indoor and outdoor air pollution measured in terms of $PM_{2.5}$, respectively, for household i and 48-h time h. β_3 indicates the partial correlation between IAP and OAP. $Urban_i$ and $Periurban_i$ are dummy variables for each area with rural as the reference group. The structural differences in IAP between rural and urban or peri-urban areas are reflected in β_1 or β_2.

$Solid_i$ is a dummy variable indicating the use of solid fuels, such as firewood, agricultural residues, and animal dung. $OutsideKitchen_i$ is a dummy variable indicating that the household's kitchen is located outside the living space. Since the impact of solid fuels on IAP would be mitigated if they are used in an outside kitchen, the interaction term is also added to the explanatory variables. In other words, while the effect of the solid fuels on IAP (β_4) would be positive in the case of the inside kitchen, its interaction with $OutsideKitchen_i$ (β_6) is expected to be negative.

X_{ik} indicates other household characteristics (see the next section for details). Associations with various household-level characteristics are captured by δ_k. Moreover, IAP may be subject to time-of-day factors that are common to all households, such as cooking time. θ_h indicates the 24-h time fixed effects controlling these factors. ε_{ih} is a normal error term. Standard errors are clustered at the household level.

3. Results

3.1. Characteristics of Sample Households

The characteristics of the sample households are summarized in Table 1. For cooking fuels, many households use a combination of different types of fuels. In rural areas, most households use solid fuels, while half also use LPG. In peri-urban areas, all households use natural gas, although some also use solid fuel and/or LPG. Only clean fuels are used in urban areas: eight households use natural gas, and four use LPG.

The location of kitchens differed among areas, with 100% of rural areas having outdoor kitchens, 80% in peri-urban areas, and 10% in urban areas. Sharing kitchens with other households is particularly common in peri-urban areas (85%). The cooking time per day was approximately 2.5 to 3 h, with no significant differences among regions.

In the context of Bangladesh, other possible causes of IAP besides cooking are cigarette smoking and the use of mosquito coils. The percentage of smoking was highest among peri-urban households (40%). Mosquito coils are widely used in all areas to prevent infectious diseases such as dengue fever. More than 85% of rural and peri-urban areas use them regularly in their homes.

For cooling, almost all households use fans. Air conditioner use is limited to 10%, even in urban areas. As for housing, the area is larger in rural households, but the number of rooms is similar in all areas. The other notable difference between areas is household income, which is 1.8 times higher in urban areas than in rural areas.

Table 1. Characteristics of sample households.

	Rural (N = 20)	Peri-Urban (N = 20)	Urban (N = 10)
Cooking fuel: solid fuels	19 (95%)	9 (45%)	0 (0%)
Cooking fuel: LPG	11 (55%)	6 (30%)	4 (40%)
Cooking fuel: natural gas	0 (0%)	20 (100%)	8 (80%)
Cooking fuel: electricity	1 (5%)	2 (10%)	1 (10%)
Outside kitchen	20 (100%)	16 (80%)	1 (10%)
Sharing kitchen with other households	1 (5%)	17 (85%)	4 (40%)
Smoking by any household members	4 (20%)	8 (40%)	1 (10%)
Use of mosquito coils	17 (85%)	18 (90%)	6 (60%)
Use of a fan	20 (100%)	19 (95%)	10 (100%)
Use of an air conditioner	0 (0%)	1 (5%)	1 (10%)
Female head of household	2 (10%)	3 (15%)	1 (10%)
Age of the head of household	53.8	38.1	47.9
Number of household members (person)	3.7	3.4	4.1
Size of housing (decimal)	16.4	5.0	1.9
Number of rooms	3.6	3.0	4.3
Average cooking time per day (hour)	2.9	2.5	2.5
Average monthly household income (Taka)	25,950	27,750	47,200

Note: "Cooking fuel: solid fuels" is a dummy variable that indicates the use of firewood, agricultural residues, or animal dung. "Decimal" is a unit of area commonly used in Bangladesh (1 decimal = 40.5 m^2).

3.2. Representative Cases by Region

For illustrative purposes, Figure 3 shows the time paths of $PM_{2.5}$ concentrations in a representative household in each area. The WHO set the air quality guideline level for 24 h at 15 $\mu g/m^3$, which is indicated by the horizontal dashed line in the figure. For the rural household (Figure 3a), IAP generally meets the guideline level despite the use of solid fuels, except for peaks at breakfast and dinner preparation times. However, IAP exceeded the guideline level in the peri-urban and urban regions (Figure 3b,c). This peri-urban household uses natural gas and firewood, while the urban household uses only natural gas. The three cases also showed a clear positive correlation between IAP and OAP. In the rural and peri-urban cases, both indoor and outdoor $PM_{2.5}$ moved in tandem at similar levels. The trend is similar for urban households, but the level is higher for OAP.

3.3. IAP and OAP Variation by Region

Table 2 shows the indoor and outdoor $PM_{2.5}$ concentrations by region using daily and hourly data. For most of the values for IAP and OAP, rural areas had lower levels than urban areas. The mean $PM_{2.5}$ concentration for urban indoor and outdoor environments was about five times higher than that for rural areas. Even at the median, both IAP and OAP were much lower in rural areas than in urban and peri-urban areas. Figure 4 confirms this relationship on a logarithmic scale.

Table 3 shows the variation in indoor and outdoor $PM_{2.5}$ concentrations over a day based on hourly data. Rural IAP and OAP are lower than those for urban and peri-urban areas at all times of the day. The observation is also confirmed by Figure 5, which compares the hourly means on a logarithmic scale.

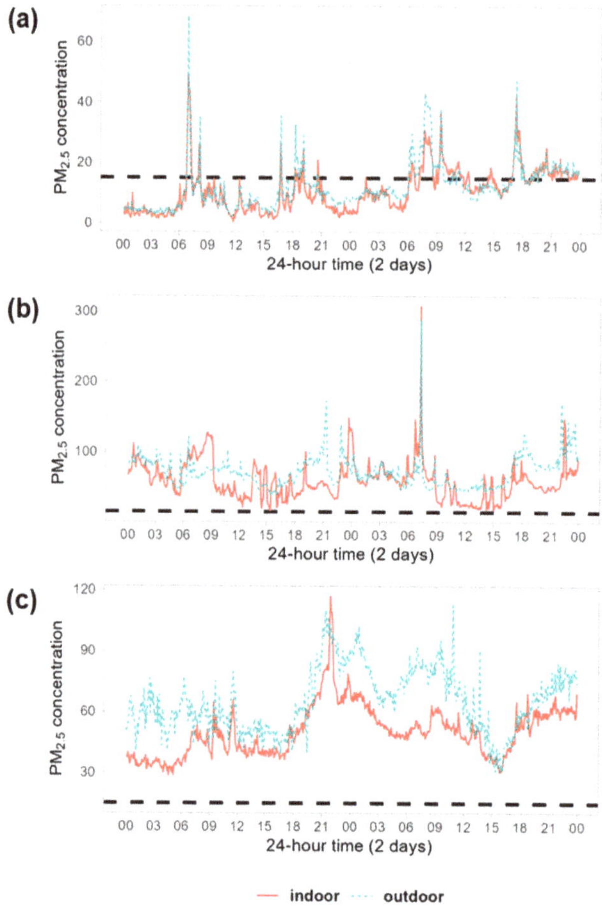

Figure 3. Representative cases of indoor and outdoor $PM_{2.5}$ concentrations (μg/m^3) in (**a**) rural, (**b**) peri-urban, and (**c**) urban areas. The horizontal dashed line indicates 15 μg/m^3, the air quality guideline level of WHO [24].

Table 2. Indoor and outdoor $PM_{2.5}$ concentration (μg/m^3).

		Daily Mean $PM_{2.5}$ Concentration			Hourly Mean $PM_{2.5}$ Concentration		
		Rural	Peri-Urban	Urban	Rural	Peri-Urban	Urban
Indoor	Mean	22.2	95.0	117.0	22.2	95.0	117.0
	Standard deviation	10.2	55.9	124.3	23.2	122.3	229.9
	Minimum	7.2	32.5	18.7	0.1	8.9	3.5
	Median	21.6	76.1	66.5	14.7	62.0	58.9
	Maximum	47.1	260.9	423.5	185.0	1162.8	1398.6
Outdoor	Mean	15.8	64.9	83.6	15.8	64.9	83.6
	Standard deviation	5.3	15.3	27.4	13.0	34.5	48.9
	Minimum	5.2	36.6	40.4	0.7	17.4	24.1
	Median	15.8	63.8	78.8	13.1	58.1	73.7
	Maximum	28.6	124.3	141.2	175.7	436.7	598.9
Num. of observation		40	40	20	960	960	480

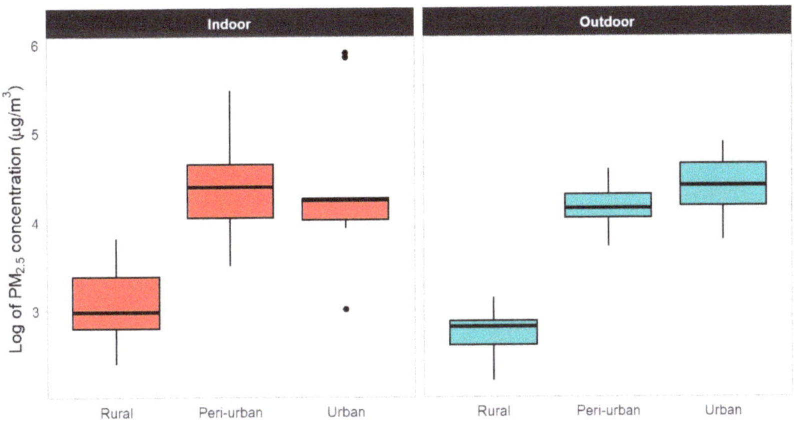

Figure 4. Comparison of indoor and outdoor $PM_{2.5}$ concentration ($\mu g/m^3$) by region.

Table 3. Mean $PM_{2.5}$ concentration ($\mu g/m^3$) by the time of day.

	Time of Day	Rural	Peri-Urban	Urban
Indoor	0:00–6:00	18.3	142.3	252.8
	6:00–12:00	21.7	61.7	74.0
	12:00–18:00	17.1	55.4	49.0
	18:00–24:00	31.7	120.4	92.1
Outdoor	0:00–6:00	8.8	61.5	96.3
	6:00–12:00	22.0	58.0	79.6
	12:00–18:00	15.5	53.0	58.4
	18:00–24:00	16.8	86.9	100.2
Num. of observation		960	960	480

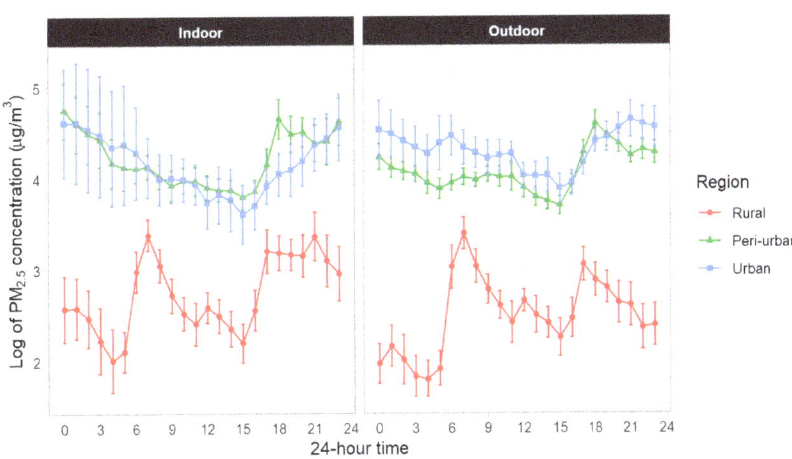

Figure 5. $PM_{2.5}$ concentration by the time of day. The bar indicates a 95% confidence interval.

It should be noted that nighttime IAP (0:00–6:00 and 18:00–24:00) is relatively high in urban and peri-urban areas. There are two possible reasons. First, the daytime natural gas supply was restricted in Dhaka and its suburbs during this survey in October 2022. During the interviews, urban and peri-urban residents reported that they cooked intensively late at night and early in the morning due to the gas crisis caused by high international energy prices. Second, as Table 2 confirms, OAP generally tends to be higher in the evening. This result is consistent with previous studies and may include the influence of the evening rush hour [15,25]. This higher outdoor $PM_{2.5}$ may have eventually contributed to IAP in urban and peri-urban areas.

3.4. Determinants of IAP

The above analysis shows that urban households using only clean fuels are exposed to higher IAP than rural households using mainly solid fuels. Apart from cooking fuels, what factors contribute to this difference? Because we selected three study areas exploiting differences in OAP (Figure 1), OAP could be an important factor. Figure 6 shows the strong correlation between IAP and OAP on a logarithmic scale ($R = 0.73$).

Table 4 shows the regression analysis. Column 1 uses only urban and peri-urban dummies as explanatory variables, with rural households as the reference group. The result again confirms the higher IAP in urban and peri-urban areas. However, the coefficients of regional dummies become much smaller when OAP is added as an explanatory variable in Column 2. This means that most of the regional differences in IAP can be explained by OAP. The coefficient of OAP indicates that a 1 percent increase in OAP is associated with a 0.57 percent increase in IAP.

In Column 3, we added all explanatory variables described in Table 1. The results are summarized in the following three major findings. First, the coefficients for the solid fuels and the outside kitchen are both positive, but their interaction is negative, as expected. This indicates that although solid fuels are an increasing factor in IAP, their impact is significantly reduced when the kitchen is located outside the living space. Second, other factors increasing IAP are cigarette smoking and the use of mosquito coils. Third, the use of fans and larger house sizes contribute to the decrease in IAP. Most other variables omitted from the table were not statistically significant.

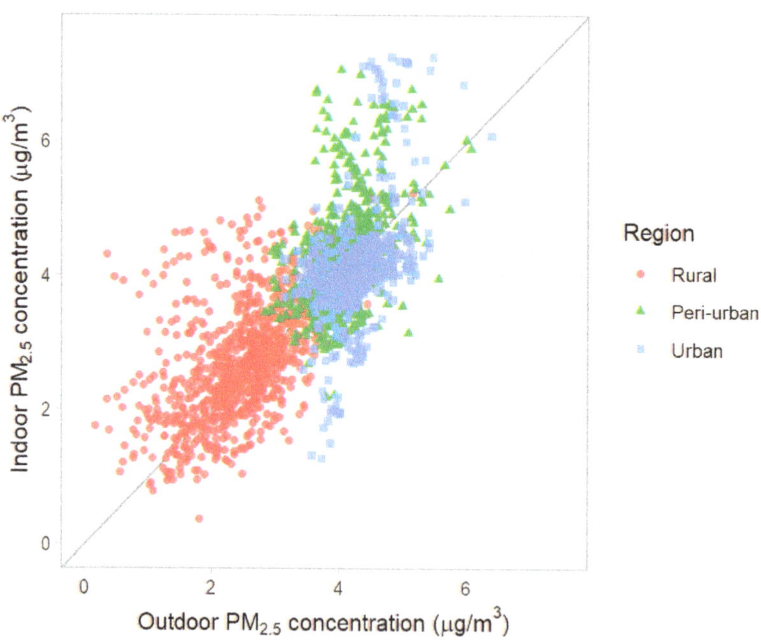

Figure 6. Correlation between indoor and outdoor $PM_{2.5}$ concentration.

Table 4. Regression results: determinants of indoor air pollution ($PM_{2.5}$).

	(1)	(2)	(3)
Urban	1.101 ***	0.265	0.191
	(0.233)	(0.232)	(0.317)
Peri-urban	1.726 ***	0.696 **	0.052
	(0.125)	(0.207)	(0.310)
Log(Outdoor air pollution)		0.565 ***	0.571 ***
		(0.103)	(0.095)
Solid fuel			0.360 *
			(0.173)
Outside kitchen			0.503 *
			(0.189)
Solid fuel × outside kitchen			−0.509 *
			(0.236)
Smoking			0.396*
			(0.179)
Mosquito coil			0.339 *
			(0.156)
Fan			−1.011 **
			(0.368)
House size			−0.030 *
			(0.015)
Controlling for other household characteristics	No	No	Yes
24-h time fixed effects	Yes	Yes	Yes
Num. of observation	2400	2390	2390
Adjusted R2	0.523	0.587	0.637

Notes: Hourly data were used in the regression analysis. The explained variable is the logarithmic scale of indoor $PM_{2.5}$ concentration ($\mu g/m^3$). Standard errors, clustered at the household level, are shown in parentheses. Significance levels: * $p < 0.05$, ** $p < 0.01$, *** $p < 0.001$.

4. Discussion

The results indicate that both IAP and OAP are higher in urban and peri-urban areas than in rural areas. This finding is consistent with Khalequzzaman et al. [17], who compare kitchen IAP and OAP between rural and urban households in Dhaka. However, a key difference between their study and ours is that all urban households in their sample use solid fuels (firewood). In our study, by contrast, most urban and peri-urban households use clean fuels such as LPG and natural gas. Despite this, indoor air quality remains worse than in rural areas. As noted earlier, this may be partly due to limited daytime gas supply during the survey period, which leads to intensive cooking activities at night and in the early morning. Even if this is the case, our findings suggest that the use of clean fuels alone may not be sufficient to achieve low IAP.

Another relevant comparison is with Dasgupta et al. [12], who report that rural households are exposed to higher IAP than urban households. Two main possible reasons explain the divergence from our results. First, their sample includes households with a wide variety of kitchen arrangements. For example, households that use solid fuels and have kitchens located inside the main dwelling exhibit higher IAP. In contrast, all rural households in our sample use well-ventilated outside kitchens. This sampling characteristic may have led to an underestimation of rural IAP in our study. Nevertheless, their finding that IAP is substantially reduced when the kitchen is detached or located outdoors is consistent with our results. This observation is particularly relevant, given that outdoor kitchens are widely used not only in Bangladesh but also in many other developing countries [26].

Second, while the study by Dasgupta et al. [12] was conducted in 2003–2004, OAP in Dhaka has been steadily deteriorating since then. The city's average annual $PM_{2.5}$ concentration increased from approximately 50 $\mu g/m^3$ in 2001 to around 70 $\mu g/m^3$ in

2019 [27]. In Mirpur, our urban study site, it reached as high as 80 $\mu g/m^3$. This worsening OAP has substantially contributed to higher IAP in urban areas, accounting for the different results between their study and ours conducted in 2022.

We acknowledge several limitations in this study. First, it focuses solely on $PM_{2.5}$ and does not include other pollutants such as O_3 or NO_x, which may behave differently under similar conditions. Second, the sample size is relatively small (50 households, including only 10 from urban areas), and the monitoring duration is short-term, limited to 48 h in October 2022. In addition, air quality was measured only in outdoor areas and living rooms, excluding potentially relevant spaces such as kitchens. These factors may introduce biases related to the seasonality of air pollution and household characteristics. Finally, the regression analysis does not control for meteorological conditions. Although the three study sites are geographically close and likely share similar temperatures, other environmental variables such as wind speed and direction may vary by household location.

Beyond these limitations, several issues remain for future research. Our findings confirmed that smoking and the use of mosquito coils contribute to elevated IAP. To better understand the extent to which these effects vary with the frequency and timing of such behaviors, more detailed behavioral data are required. Furthermore, although all rural households in this study had outside kitchens, emissions from these kitchens may affect OAP more than IAP. However, since the outdoor sensors were installed on the balcony of the main dwelling, we were unable to directly capture emissions from the kitchens. Future studies should consider deploying sensors placed both inside and around kitchen areas to more accurately assess their contributions to OAP.

5. Conclusions

This study examined the status and determinants of IAP in Bangladesh, a country known for experiencing severe air pollution. Taking advantage of the substantial variation in ambient air pollution in and around the capital city of Dhaka, the concentrations of $PM_{2.5}$ were measured both indoors and outdoors across 50 households located in three distinct regions: urban, peri-urban, and rural. By combining data on various household characteristics with hourly data on $PM_{2.5}$, the determinants for IAP were examined using regression analysis.

The results indicate that IAP is more pronounced in urban and peri-urban areas than in rural areas. In rural households, the daily concentration of $PM_{2.5}$ in living rooms remained relatively low, averaging 22 $\mu g/m^3$ per day, despite the use of solid fuels. In contrast, urban and peri-urban households recorded substantially higher daily $PM_{2.5}$ concentrations, approximately 100 $\mu g/m^3$, even though clean energy sources such as LPG and natural gas were predominantly used. This result is primarily explained by the infiltration of outdoor air pollutants, which is particularly severe in urban settings. In addition to OAP, smoking and the use of mosquito coils are also associated with high IAP.

The mechanisms underlying IAP are multifaceted, reflecting complex interaction with OAP and various household characteristics. Our findings suggest that high levels of OAP can offset the benefits of clean fuel use in urban areas, thereby limiting the effectiveness of fuel-switching policies alone. To mitigate IAP in developing-country contexts, a comprehensive approach may be needed—one that includes not only transitioning to clean fuels but also practical interventions such as promoting well-ventilated outdoor kitchens, using electric fans to enhance airflow, and encouraging behavioral changes such as avoiding indoor smoking and reducing mosquito coil use. Of course, this comprehensive approach should also include continued efforts to reduce reliance on solid fuels, which remain major contributors to IAP, particularly in poorly ventilated indoor environments.

Author Contributions: Conceptualization, M.K., A.H., K.T. and Y.M.; methodology, M.K., A.H., K.T. and Y.M.; software, M.K.; formal analysis, M.K. and K.T.; investigation, M.K.; resources, Y.M.; writing—original draft preparation, M.K.; writing—review and editing, M.K., A.H., K.T. and Y.M.; visualization, M.K.; supervision, M.K. and A.H.; project administration, M.K. and A.H.; funding acquisition, M.K. and A.H. All authors have read and agreed to the published version of the manuscript.

Funding: This work was financially supported by the Japan Society for the Promotion of Science (JSPS) KAKENHI Grant Number 18H00839, and the Private University Research Branding Project at the Sophia Institute for Human Security.

Institutional Review Board Statement: The study protocol was approved by the Institutional Review Board of the Institute of Health Economics at the University of Dhaka (No. FWA00026031).

Informed Consent Statement: Written informed consent has been obtained from all participants to publish this paper.

Data Availability Statement: The data presented in this study are available on request from the corresponding author.

Acknowledgments: We thank Partha Ghosh for assistance in conducting the survey. During the preparation of this manuscript, the authors used ChatGPT-4o (OpenAI) for the purpose of English language proofreading. The authors have reviewed and edited the output and take full responsibility for the content of this publication.

Conflicts of Interest: The authors declare no conflicts of interest.

Abbreviations

The following abbreviations are used in this manuscript:

IAP	Indoor air pollution
OAP	Outdoor air pollution
$PM_{2.5}$	Particulate matter (diameter $\leq$ 2.5 μm)

References

1. Van Tran, V.; Park, D.; Lee, Y.C. Indoor air pollution related human diseases recent trends in the control improvement of indoor air quality. *Int. J. Environ. Res. Publ. Health* **2020**, *17*, 2927. [CrossRef] [PubMed]
2. Pratiti, R.; Vadala, D.; Kalynych, Z.; Sud, P. Health effects of household air pollution related to biomass cook stoves in resource limited countries and its mitigation by improved cookstoves. *Environ. Res.* **2020**, *186*, 109574. [CrossRef] [PubMed]
3. Dominski, F.H.; Branco, J.H.L.; Buonanno, G.; Stabile, L.; da Silva, M.G.; Andrade, A. Effects of air pollution on health: A mapping review of systematic reviews and meta-analyses. *Environ. Res.* **2021**, *201*, 111487. [CrossRef]
4. Pillarisetti, A.; Ye, W.; Chowdhury, S. Indoor air pollution and health: Bridging perspectives from developing and developed countries. *Annu. Rev. Environ. Resour.* **2022**, *47*, 197–229. [CrossRef]
5. Klepeis, N.E.; Nelson, W.C.; Ott, W.R.; Robinson, J.P.; Tsang, A.M.; Switzer, P.; Behar, J.V.; Hern, S.C.; Engelmann, W.H. The National Human Activity Pattern Survey (NHAPS): A resource for assessing exposure to environmental pollutants. *J. Expo. Sci. Environ. Epidemiol.* **2001**, *11*, 231–252. [CrossRef]
6. Yun, H.; Seo, J.H.; Kim, Y.G.; Yang, J. Impact of scented candle use on indoor air quality and airborne microbiome. *Sci. Rep.* **2025**, *15*, 10181. [CrossRef] [PubMed]
7. Chen, B.H.; Hong, C.J.; Pandey, M.R.; Smith, K.R. Indoor air pollution in developing countries. *World Health Stat. Q.* **1990**, *43*, 127–138.
8. Smith, K.R. Indoor air pollution in developing countries: Recommendations for research. *Indoor Air.* **2002**, *12*, 198–207. [CrossRef]
9. Shen, G. Mutagenicity of particle emissions from solid fuel cookstoves: A literature review and research perspective. *Environ. Res.* **2017**, *156*, 761–769. [CrossRef]
10. Arku, R.E.; Ezzati, M.; Baumgartner, J.; Fink, G.; Zhou, B.; Hystad, P.; Brauer, M. Elevated blood pressure and household solid fuel use in premenopausal women: Analysis of 12 Demographic and Health Surveys (DHS) from 10 countries. *Environ. Res.* **2018**, *160*, 499–505. [CrossRef]
11. Amegah, A.K.; Jaakkola, J.J. Household air pollution and sustainable development goals. *Bull. World Health Organ.* **2016**, *94*, 215–221. [CrossRef]

12. Dasgupta, S.; Huq, M.; Khaliquzzaman, M.; Pandey, K.; Wheeler, D. Indoor air quality for poor families: New evidence from Bangladesh. *Indoor Air.* **2006**, *16*, 426–444. [CrossRef] [PubMed]
13. Kang, K.; Kim, H.; Kim, D.D.; Lee, Y.G.; Kim, T. Characteristics of cooking-generated PM_{10} and $PM_{2.5}$ in residential buildings with different cooking and ventilation types. *Sci. Total Environ.* **2019**, *668*, 56–66. [CrossRef] [PubMed]
14. Shupler, M.; Godwin, W.; Frostad, J.; Gustafson, P.; Arku, R.E.; Brauer, M. Global estimation of exposure to fine particulate matter ($PM_{2.5}$) from household air pollution. *Environ. Int.* **2018**, *120*, 354–363. [CrossRef]
15. Nahian, S.; Roy, S.; Choudhury, T.R.; Begum, B.A.; Salam, A. Indoor particulate matter exposure and correlation of $PM_{2.5}$ with lung efficacy and SpO_2 level of Dhaka City Dwellers. *Air Qual. Atmos. Health* **2024**, *17*, 2515–2528. [CrossRef]
16. Dyer, G.M.; Khomenko, S.; Adlakha, D.; Anenberg, S.; Behnisch, M.; Boeing, G.; Esperon-Rodriguez, M.; Gasparrini, A.; Khreis, H.; Kondo, M.C.; et al. Exploring the nexus of urban form, transport, environment and health in large-scale urban studies: A state-of-the-art scoping review. *Environ. Res.* **2024**, *257*, 119324. [CrossRef]
17. Khalequzzaman, M.; Kamijima, M.; Sakai, K.; Ebara, T.; Hoque, B.A.; Nakajima, T. Indoor air pollution and health of children in biomass fuel-using households of Bangladesh: Comparison between urban and rural areas. *Environ. Health Prev. Med.* **2011**, *16*, 375–383. [CrossRef] [PubMed]
18. IQAir. World Air Quality Report 2023: Region and City PM2.5 Ranking. 2023. Available online: https://www.iqair.com/dl/2023_World_Air_Quality_Report.pdf (accessed on 11 June 2025).
19. European Space Agency. Copernicus Sentinel-5P (Processed by ESA), TROPOMI Level 2 Nitrogen Dioxide Total Column Products. Version 02. 2021. Available online: https://sentinels.copernicus.eu/web/sentinel/data-products/-/asset_publisher/fp37fc19FN8F/content/sentinel-5-precursor-level-2-nitrogen-dioxide (accessed on 11 June 2025).
20. Nakayama, T.; Matsumi, Y.; Kawahito, K.; Watabe, Y. Development and evaluation of a palm-sized optical $PM_{2.5}$ sensor. *Aerosol Sci. Technol.* **2018**, *52*, 2–12. [CrossRef]
21. Dhaka, S.K.; Chetna, V.; Kumar, V.; Panwar, V.; Dimri, A.P.; Singh, N.; Patra, P.K.; Matsumi, Y.; Takigawa, M.; Nakayama, T.; et al. $PM_{2.5}$ diminution and haze events over Delhi during the COVID-19 lockdown period: An interplay between the baseline pollution and meteorology. *Sci. Rep.* **2020**, *10*, 13442. [CrossRef]
22. Ly, B.T.; Matsumi, Y.; Nakayama, T.; Sakamoto, Y.; Kajii, Y.; Nghiem, T.D. Characterizing $PM_{2.5}$ in Hanoi with new high temporal resolution sensor. *Aerosol Air Qual. Res.* **2018**, *18*, 2487–2497. [CrossRef]
23. Othman, M.; Latif, M.T.; Yee, C.Z.; Norshariffudin, L.K.; Azhari, A.; Halim, N.D.A.; Alias, A.; Sofwan, N.M.; Hamid, H.H.A.; Matsumi, Y. $PM_{2.5}$ and ozone in office environments and their potential impact on human health. *Ecotoxicol. Environ. Saf.* **2020**, *194*, 110432. [CrossRef]
24. World Health Organization. WHO global air quality guidelines: Particulate matter (PM2.5 and PM10). Ozone, Nitrogen Dioxide, Sulfur Dioxide and Carbon Monoxide. World Health Organization. 2021. Available online: https://apps.who.int/iris/handle/10665/345329 (accessed on 11 June 2025).
25. Mahmood, A.; Hu, Y.; Nasreen, S.; Hopke, P.K. Airborne particulate pollution measured in Bangladesh from 2014 to 2017. *Aerosol Air Qual. Res.* **2019**, *19*, 272–281. [CrossRef]
26. Langbein, J.; Peters, J.; Vance, C. Outdoor cooking prevalence in developing countries and its implication for clean cooking policies. *Environ. Res. Lett.* **2017**, *12*, 115008. [CrossRef]
27. Rahman, M.; Meng, L. Examining the Spatial and Temporal Variation of $PM_{2.5}$ and Its Linkage with Meteorological Conditions in Dhaka, Bangladesh. *Atmosphere* **2024**, *15*, 1426. [CrossRef]

toxics

MDPI

Article

Characterization and Source Analysis of Water-Soluble Ions in $PM_{2.5}$ at Hainan: Temporal Variation and Long-Range Transport

Xinghong Xu [1], Wenshuai Xu [1,*], Xinxin Meng [1], Xiaocong Cao [2], Biwu Chu [3,*], Chuandong Du [1], Rongfu Xie [2,4], Zhaohe Zeng [1], Hui Sheng [1], Youjing Lin [1], Weijun Yan [1] and Hong He [3]

1 Hainan Provincial Ecological and Environmental Monitoring Center, Haikou 571126, China; sinhom_xu@foxmail.com (X.X.)
2 Hainan Research Academy of Environmental Sciences, Haikou 571126, China
3 Research Center for Eco-Environmental Sciences, Chinese Academy of Sciences, Beijing 100085, China
4 School of Ecology, Hainan University, Haikou 570228, China
* Correspondence: xuwenshuai@hotmail.com (W.X.); bwchu@rcees.ac.cn (B.C.)

Abstract: We explored the mass concentrations of water-soluble ions in $PM_{2.5}$ and their variations across different time scales and concentration levels. Using the Positive Matrix Factorization (PMF) model and backward trajectory analysis, we focused on identifying the sources of $PM_{2.5}$ and its water-soluble ion fractions, with particular emphasis on regional transport. The findings reveal that the average mass concentration of total water-soluble ions in Hainan between 1 August 2021 and 31 July 2022 was 7.0 ± 4.4 μg m^{-3}, constituting 73.5% ± 24.4% of $PM_{2.5}$. Secondary ions (SO_4^{2-}, NO_3^-, NH_4^+) were dominant, accounting for 84.0% ± 12.4% of the total water-soluble ions, followed by sea-salt particles. Seasonal variations were pronounced, with the highest concentrations observed in winter and the lowest in summer. The results of the PMF analysis showed that secondary sources, combustion sources, dust sources, and oceanic sources are the main sources of $PM_{2.5}$ at the monitoring site. The potential sources and transport pathways of water-soluble ions exhibit distinct seasonal characteristics, with the land-based outflows from the YRD–PRD–Fujian corridor controlling Hainan's $PM_{2.5}$ maxima, while southerly marine air delivers the annual minimum; seasonal alternation between dust/secondary aerosols (winter–spring), combustion (autumn), and oceanic dilution (summer) dictates the island's air-quality rhythm.

Keywords: regional transport; $PM_{2.5}$; water-soluble ions; source apportionment

1. Introduction

As a major component of atmospheric pollutants, $PM_{2.5}$ not only reduces atmospheric visibility and triggers hazy weather, but also significantly affects the climate [1–4]. Due to its small size, $PM_{2.5}$ is not only capable of penetrating deep into the respiratory system, posing increased health risks [5–7], but is also easily influenced by atmospheric circulation, enabling it to be transported over long distances to downstream regions. In particular, fine particles can be transported from their sources to leeward regions due to the influence of the Asian Northeast Monsoon, which is prevalent in East Asia. This transport mechanism leads to elevated $PM_{2.5}$ levels in the affected areas [8,9].

$PM_{2.5}$ has a complex composition, including water-soluble ions (WSI), carbon components, trace elements, and crustal elements. WSI, particularly sulfate (SO_4^{2-}), nitrate (NO_3^-), and ammonium (NH_4^+) (collectively termed SNA), dominate $PM_{2.5}$ mass, contributing 45–65% of its total composition [10]. These secondary inorganic aerosols form

 https://doi.org/10.3390/toxics13090804

through complex atmospheric reactions influenced by precursor gases (SO_2, NO_x, and NH_3) and meteorological conditions [11]. To date, many scholars have conducted a series of in-depth studies on the composition, existence form, and transformation characteristics of water-soluble ions in $PM_{2.5}$. However, most of the studies focused on the relatively heavily polluted urban areas such as Beijing-Tianjin-Hebei [12,13], the Yangtze River Delta [14], and the Pearl River Delta [15]. In contrast, there is a notable lack of studies on $PM_{2.5}$ in areas with lower concentrations, particularly regarding the transport mechanisms of $PM_{2.5}$ in these regions.

Hainan Province, located at the southernmost tip of China, is characterized by a relative lack of large-scale industries and enterprises. This scarcity of local emission sources has historically contributed to its reputation for excellent air quality over the past decades [16]. Local pollutants have a negligible influence at Hainan, so long-range transport is the primary route for air pollutants, especially transboundary pollution via monsoonal flows and biomass-burning plumes [17,18]. According to reports, 41.6% of $PM_{2.5}$ in Hainan originates from regional transport from mainland China and neighboring Southeast Asian countries [19]. These transboundary inputs complicate air quality management, underscoring the need for source apportionment.

Accordingly, continuous online monitoring of water-soluble ions in $PM_{2.5}$ at the regional transport site of Hainan was carried out for one year, from August 2021 to July 2022, to analyze the temporal distribution characteristics of $PM_{2.5}$ and its water-soluble ions. The sources of the water-soluble ions in $PM_{2.5}$ were analyzed by combining the Positive Matrix Factorization (PMF) model and the backward trajectory model. The impacts of air masses in different directions on the air quality of the regional station during different seasons were illustrated. The findings of this research are intended to contribute to the scientific basis for the refined prevention and control of atmospheric pollution, specifically tailored to the unique environmental context of tropical islands close to the mainland.

2. Materials and Methods

2.1. Monitoring Locations and Monitoring Time

All data in this study were monitored using online monitoring at the Hainan Wenchang Puqian Regional Atmospheric Super Station (hereafter referred to as the "Regional station") located at the top of Daling Mountain (20°05′ N, 110°36′ E), which is the regional transport point for Hainan Province. The location is depicted in Figure 1. Surrounded by the sea on three sides, with no obvious pollution sources, and high vegetation cover and broad-leaved forests, the regional station is an important station on the atmospheric transport corridor from the Guangdong–Hong Kong–Macao Greater Bay Area to the Hainan Free Trade Port. The monitoring period for this study was from 1 August 2021 to 31 July 2022 at hourly intervals.

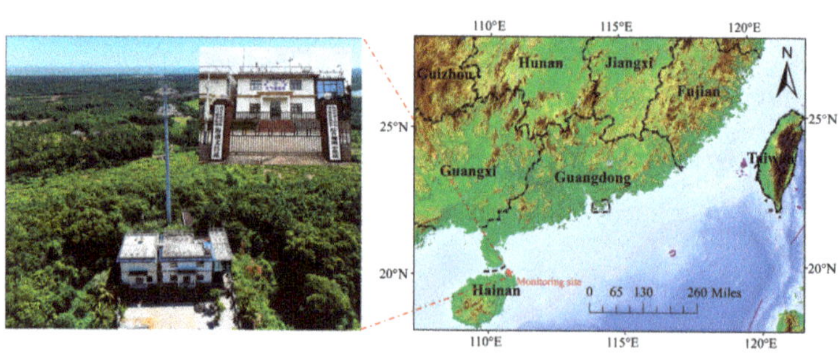

Figure 1. The location of the regional station of Hainan (https://www.91weitu.com/, accessed on 31 August 2025).

2.2. Monitoring Instruments

$PM_{2.5}$ samples were monitored using the Thermo Fisher 5030i instrument by Thermo Fisher Scientific Inc., Franklin, MA, USA. The equipment used for monitoring water-soluble ions was the gas–aerosol water-soluble ion online analysis system (MARGA/ADI2080) by Metrohm, Herisau, Switzerland, and a $PM_{2.5}$ cyclone was employed to measure the mass concentration of water-soluble inorganic ions (Cl^-, NO_3^-, SO_4^{2-}, Na^+, K^+, NH_4^+, Mg^{2+}, and Ca^{2+}) in the aerosol phase at a time resolution of one hour. The system includes a sampling system, an analysis system, a control system, and an auxiliary system. The sampling system utilizes the different diffusion properties of gas and aerosol. Soluble gas is quantitatively absorbed by the rotary liquid cavitator (WRD), while aerosol is collected by the vapor jet aerosol collector (SJAC) through the WRD without any loss. The liquid stream from the WRD and SJAC is collected by the burette, degassed, mixed with the internal standard, and then injected quantitatively into anion and cation chromatography for analysis. The analytical system is a dual-channel ion chromatography system with two built-in independent analytical modules, one for anion and organic acid analysis and one for cation analysis. The Cl^-, NO_3^-, SO_4^{2-}, Na^+, K^+, NH_4^+, Mg^{2+}, and Ca^{2+} detection limits were 0.01, 0.05, 0.04, 0.05, 0.09, 0.05, 0.06, and 0.09 μg m^{-3}, respectively.

Quality assurance and quality control of the MARGA system include the following steps. Firstly, the inlets and sampling lines were cleaned weekly with ultrapure deionized water and dried with zero-grade air. Secondly, a primary standard flowmeter traceable to the National Institute of Metrology, China, was used to measure and verify the atmospheric inlet flow rate every month. Third, a liquid blank was obtained by operating the MARGA with the air pumps disconnected and the denuder inlets sealed, so that only the absorption solution was sampled. With the air pumps off and the denuder inlets sealed, an external-standard test was conducted by substituting the absorption solution with a known liquid standard containing SO_4^{2-}, NH_4^+, and NO_3^-, thereby verifying the analytical accuracy established by the internal lithium bromide (LiBr) standard.

2.3. Methods of Analysis

2.3.1. Positive Definite Matrix Factorization (PMF)

The positive definite matrix factorization (PMF) model was developed by the U.S. Environmental Protection Agency (EPA) and has been widely used for $PM_{2.5}$ component source resolution [20]. In this study, PMF 5.0 software was used for $PM_{2.5}$ source resolution, the basic principle of which is to calculate the error of each chemical component in particulate matter using weights, and then determine its main pollution sources and contribution by the least squares method [21], with the following formula:

$$X_{ij} = \sum_{k=1}^{p} g_{ik} f_{kj} + e_{ij} \tag{1}$$

where X_{ij} is the concentration of the ijth sample, g_{ik} denotes the contribution of the ikth sample, f_{kj} denotes the mass fraction of the kjth sample, and e_{ij} denotes the residual between the measured mass concentration of the ijth sample and its analyzed value:

$$Q_{ij} = \sum_{i=1}^{n} \sum_{j=1}^{m} \left[\frac{X_{ij} - \sum_{k=1}^{p} g_{ik} f_{kj}}{u_{ij}} \right] \tag{2}$$

where Q_{ij} is the sum of all sample residuals and their uncertainty u:

$$u = \frac{5}{6} \times MDL \tag{3}$$

$$u = \sqrt{(\text{error and fraction} \times \text{concentration})^2 + (0.5 \times MDL)^2}$$

where u is the uncertainty. MDL is the method detection limit. If the concentration is less than or equal to the MDL provided, Unc is calculated using a fixed fraction of the MDL. If the concentration is greater than the MDL, the calculation is based on the concentration fraction and MDL.

The principles of PMF have been described in detail in the PMF 5.0 User Guide (USEPA, 2014). In total, eight chemical components were used for the PMF model, which included Cl^-, NO_3^-, SO_4^{2-}, Na^+, K^+, NH_4^+, Mg^{2+}, and Ca^{2+}. First, we calculated the uncertainty (u) for each species based on its concentration fraction and MDL. Second, different numbers of factors were tested with random seeds in 100 iterations of each run. When four factors were specified, with most species having a correlation coefficient above 0.85, the species followed a normal distribution (Table 1). Third, bootstrap (BS), displacement (DISP), and bootstrapping with displacement (BS-DISP) diagnostics were applied to assess the reliability of the resolved factor profiles and contributions. BS detects and quantifies the undue influence of a small subset of observations on the solution and, to a lesser extent, rotational ambiguity. The value of the f peak strength was ensured to be 0.5 to eliminate the rotation ambiguity. BS mapping exceeded 90% for every factor, and there were no swaps with DISP and BS-DISP. Therefore, it is suggested that four factors were appropriate in this study.

Table 1. Summary of linear correlation between the measured and predicted values.

Species	Category	R^2	Intercept	Intercept SE	Slope	Slope SE	SE
Cl^-	Strong	0.986	0.017	0.004	0.954	0.006	0.038
NO_3^-	Strong	0.438	0.636	0.038	0.295	0.019	0.440
SO_4^{2-}	Strong	0.502	0.675	0.119	0.609	0.034	1.047
Na^+	Strong	0.994	−0.005	0.002	1.019	0.005	0.017
K^+	Strong	0.990	0.012	0.001	0.933	0.005	0.015
NH_4^+	Strong	0.610	0.322	0.045	0.569	0.026	0.455
Mg^{2+}	Strong	0.851	0.011	0.001	0.810	0.019	0.014
Ca^{2+}	Strong	0.994	0.000	0.001	0.997	0.004	0.006

2.3.2. HYSPLIT Model

The Hybrid Single-Particle Lagrangian Integrated Trajectory (HYSPLIT) model was first developed by the National Oceanic and Atmospheric Administration of the United States of America and the Australian Bureau of Meteorology to analyze pollutant sources, transport, and dispersion trajectories in conjunction with air-mass trajectories and monitoring data [22]. The model system is a hybrid Lagrangian and Eulerian diffusion model, which uses the Lagrangian method to deal with advection and diffusion processes, while the Eulerian method is used to calculate the pollutant concentrations [23].

To investigate the pollution characteristics and sources of $PM_{2.5}$ and its water-soluble ionic constituents along different air-mass trajectories, we simulated backward trajectories with the TrajStat plug-in (developed from the HYSPLIT model) in MeteoInfo 3.5.10. The Global Data Assimilation System (GDAS) meteorological fields at $1^\circ \times 1^\circ$ resolution were used, with a starting height of 500 m, a backward scale of 72 h, and a temporal resolution of 1 h. Representative months—January, April, July, and October—were selected to capture the four seasons during the monitoring period. A K-means clustering algorithm was then applied to classify the trajectories into distinct clusters. Due to the fact that a single backward trajectory is highly uncertain, we computed multiple trajectories and subjected them to cluster analysis. This multivariate statistical technique, based on the angle-distance algorithm, quantifies the pollution–concentration relationships within each source region.

3. Results and Discussion

3.1. Pollution Characteristics of $PM_{2.5}$ and Water-Soluble Ions

3.1.1. Characteristics of Seasonal Changes

During the monitoring period, the mean mass concentration of $PM_{2.5}$ at the regional station was 9.5 μg m^{-3}, in which all the daily concentration data were lower than the average daily Class I standard for air quality in China (35 μg m^{-3}), which shows that the study area belongs to a cleaner area. The concentrations of the eight water-soluble ions are shown in Table 2, and the proportion of each water-soluble ion to the total water-soluble ions (TWSI) is shown in Figure 2. The annual average concentration of TWSI in $PM_{2.5}$ was 7.0 μg m^{-3}, which accounted for 73.5% of the mass concentration of $PM_{2.5}$, indicating that water-soluble ions are an important component of $PM_{2.5}$. The mass concentrations of the eight water-soluble ions were in the following order: SO_4^{2-} > NO_3^- > NH_4^+ > Cl^- > Na^+ > K^+ > Ca^{2+} > Mg^{2+}, of which the mass concentrations of SO_4^{2-}, NO_3^-, and NH_4^+ were 2.9 μg m^{-3}, 1.5 μg m^{-3}, 1.4 μg m^{-3}, accounting for 41.4%, 21.4%, and 20.0%, and totaling 82.8%, of the TWSI. These were higher than those in Beijing (75.2%) [24], Nanjing (76.8%) [25], and Sanya (57.8%) [26], reflecting that the study area had a strong secondary pollution characteristic. In addition to SNA, the mass concentrations of sea-salt ions such as Cl^- and Na+ totaled 0.8 μg m^{-3}, accounting for 11.4% of the TWSI, which is significantly higher than that of non-coastal cities such as Beijing and Nanjing [24,25], and is clearly influenced by the ocean.

Table 2. Concentration of $PM_{2.5}$ and water-soluble inorganic ion components (μg m^{-3}).

Season	Spring	Summer	Autumn	Winter	Mean
SO_4^{2-}	3.3 ± 1.7	1.7 ± 0.8	2.9 ± 1.7	3.9 ± 1.9	2.9 ± 1.8
NO^{3-}	1.4 ± 0.7	0.79 ± 0.2	1.1 ± 0.6	2.8 ± 2.2	1.5 ± 1.4
NH^{4+}	1.4 ± 0.7	0.73 ± 0.34	1.2 ± 0.7	2.2 ± 0.75	1.4 ± 1.1
Cl^-	0.56 ± 0.31	0.32 ± 0.16	0.44 ± 0.31	0.70 ± 0.41	0.50 ± 0.34
Na^+	0.32 ± 0.21	0.19 ± 0.10	0.29 ± 0.22	0.31 ± 0.23	0.28 ± 0.20
K^+	0.19 ± 0.21	0.11 ± 0.07	0.15 ± 0.15	0.29 ± 0.14	0.18 ± 0.16
Ca^{2+}	0.15 ± 0.086	0.10 ± 0.056	0.052 ± 0.047	0.11 ± 0.081	0.10 ± 0.077
Mg^{2+}	0.065 ± 0.043	0.032 ± 0.028	0.041 ± 0.039	0.058 ± 0.049	0.049 ± 0.042
SNA	6.1 ± 3.2	3.2 ± 1.3	5.3 ± 3.3	9.0 ± 5.5	5.9 ± 4.0
TWSI	7.4 ± 3.7	3.9 ± 1.3	6.3 ± 3.3	10.4 ± 5.5	7.0 ± 4.4
$PM_{2.5}$	9.0 ± 2.9	6.9 ± 1.1	9.2 ± 4.1	13.0 ± 5.8	9.5 ± 4.4

In this study, March–May 2022 was defined as spring, August 2021 and June–July 2022 as summer, September–November 2021 as autumn, and December 2021–February 2022 as winter. The mass concentrations of TWSI during the monitoring period showed a clear seasonal variation, from highest to lowest: winter (10.4 μg m^{-3}) > spring (7.4 μg m^{-3}) > autumn (6.3 μg m^{-3}) > summer (3.9 μg m^{-3}), with the highest concentration season (winter) being 2.67 times higher than the lowest season (summer). The seasonal variations in SNA are influenced by different sources of ions, meteorological factors, and chemical reaction conditions [27]. The mass concentrations of SNA are consistent with a change in TWSI concentrations, and the proportions of SNA in TWSI (80.8–85.9%) were relatively stable across seasons, indicating that SNA is the main driver of TWSI concentration changes. The higher concentrations of SO_4^{2-}, NO_3^-, and NH_4^+ in winter than in summer could be attributed to the following: (1) the lower temperature and higher humidity in winter, which favors the conversion of NH_3 and HNO_3 to particulate states $(NH_4)_2SO_4$ and NH_4NO_3; (2) the higher temperatures and more frequent precipitation in summer, which reduces SO_4^{2-}, NO_3^-, and NH_4^+ concentrations [28]; and (3) the influences from anthropogenic emissions and regional transportations, which will be discussed in later sections. Among the other ions, the concentrations of Na^+ and Mg^{2+} were the highest in spring and were contributed by surrounding marine sources. Ca^{2+}, as an important tracer of soil and sand and dust sources [29], showed high concentrations in spring and winter, which should be explained by the concurrent high-speed northeasterly winds and low precipitation. The concentrations of Cl^- and K^+ were the highest in winter and the lowest in summer, with different seasonal variation characteristics from those of Na^+ and Mg^{2+} (highest in spring and lowest in summer), suggesting that Cl^- is not only affected by marine sources. The unfavorable meteorological conditions and the increase in biomass burning might have led to the increase in the concentrations of K^+ and Cl^- [30].

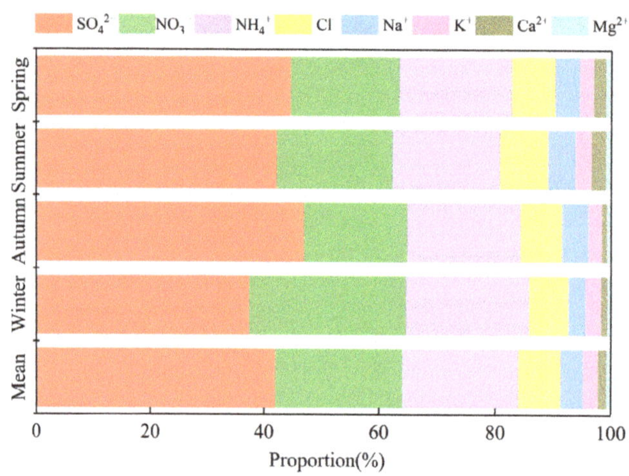

Figure 2. Seasonal proportion of water-soluble inorganic ions.

3.1.2. Characterization of Daily Changes

Figure 3 presents a detailed analysis of the daily fluctuation patterns of water-soluble ions throughout different seasons. The SNA had little fluctuation of the daily change in spring, summer, and autumn, but in winter, when SNA concentrations were high, daily changes were more pronounced. SO_4^{2-} and NH_4^+ showed a single-peak pattern, peaking around 6:00 PM, while NO_3^- showed a double-peak pattern, peaking around 11:00 AM and 3:00 PM. Na^+, Mg^{2+}, and Cl^- display a single-peak daily pattern influenced by marine sources in spring, summer, and autumn. The maximum and minimum values of Na^+ and Mg^{2+} appeared at the same time, but there was a significant fluctuation of Cl^- in the evening of winter, which may be related to the biomass burning and unfavorable diffusion conditions in winter. K^+ concentrations fluctuate throughout the day, with winter levels significantly higher than other seasons, peaking mainly in the evening and night-time. Ca^{2+} shows minor daily variations during summer, autumn, and winter, with spring experiencing greater amplitude in daily fluctuations and high concentrations concentrated between 8:00 AM and 7:00 PM. This analysis underscores the distinct daily variation patterns among water-soluble ions, indicating that their primary influencing factors differ significantly.

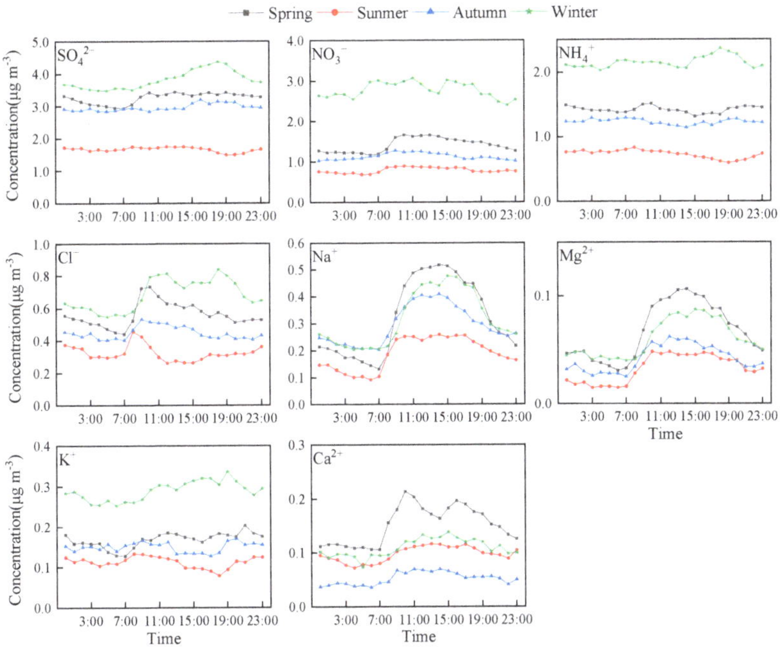

Figure 3. Diurnal variation characteristics of water-soluble inorganic ions in different seasons.

3.1.3. Compositional Characteristics at Different $PM_{2.5}$ Concentrations

With the change of $PM_{2.5}$ concentration, its composition will also change [31]. Therefore, analyzing the component characteristics of water-soluble ions at different $PM_{2.5}$ concentrations in low-concentration areas is of great significance for the future precision control of the atmosphere at low $PM_{2.5}$ concentrations in China. In this study, according to the World Health Organization 2021 Air Quality Guidelines as the basis of division, the concentration of $PM_{2.5}$ was divided into four value domains: I (0 μg m^{-3} < $\rho(PM_{2.5})$ ≤ 5 μg m^{-3}), II (5 μg m^{-3} < $\rho(PM_{2.5})$ ≤ 10 μg m^{-3}), III (10 μg m^{-3} < $\rho(PM_{2.5})$ ≤ 15 μg m^{-3}), and IV ($\rho(PM_{2.5})$ ≥ 15 μg m^{-3}) are discussed. Figure 4 illustrates the proportion of each water-soluble ion in TWSI at varying $PM_{2.5}$ concentrations. As pollution intensifies, the percentages of Na^+, Mg^{2+}, and Cl^- decrease, suggesting a significant marine influence on $PM_{2.5}$ when concentrations are low. Conversely, the respective percentages of NH_4^+ and NO_3^- across the four domains increase from 15.9% to 23.1%, and from 21.4% to 28.4%, indicating that nitrogen pollution plays a crucial role in the increase in $PM_{2.5}$ concentration. SO_4^{2-} consistently represents the highest proportion of water-soluble ions across different seasons and pollution levels. The monitoring site is situated in a regional background area of Hainan Province, characterized by minimal coal combustion emissions. Aside from transportation of emissions from coal combustion, the high proportion of SO_4^{2-} may also originate from natural sources such as dimethylsulfide (DMS) produced by marine phytoplankton [32].

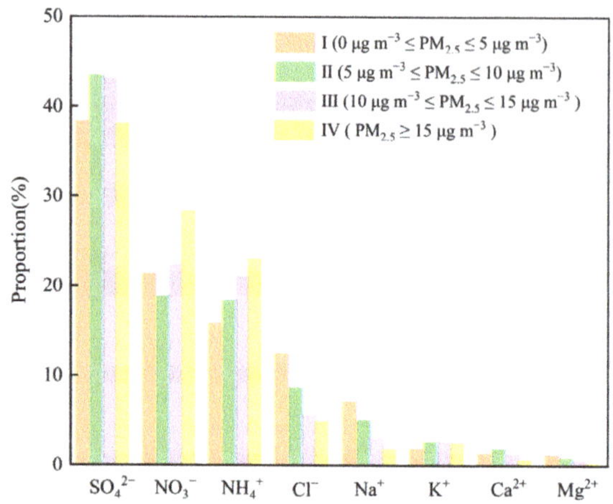

Figure 4. Contribution proportion of $PM_{2.5}$ components under different pollution levels.

3.2. Secondary Conversion Characteristics

Sulfur oxidation rate (SOR) and nitrogen oxidation rate (NOR) can effectively reflect the extent to which gaseous precursors SO_2 and NO_2 are converted into secondary components SO_4^{2-} and NO_3^- [33], calculated as follow: SOR = $n(SO_4^{2-})/(n(SO_2) + n(SO_4^{2-}))$ and NOR = $n(NO_3^-)/(n(NO_2) + n(NO_3^-))$.

When SOR and NOR are less than 0.10, it indicates that the atmosphere is primarily composed of primary pollutants. In this study, as shown in Figure 5, both the SOR and NOR values are significantly greater than 0.1, with annual average values of 0.54 and 0.24, respectively. This indicates that secondary transformation reactions at the monitoring site were intense throughout the observation period. The monitoring site is located in a coastal area, where high salinity and humidity significantly promote the secondary transformation of SO_2 and NO_2, particularly enhancing the formation of SO_4^{2-} [34].

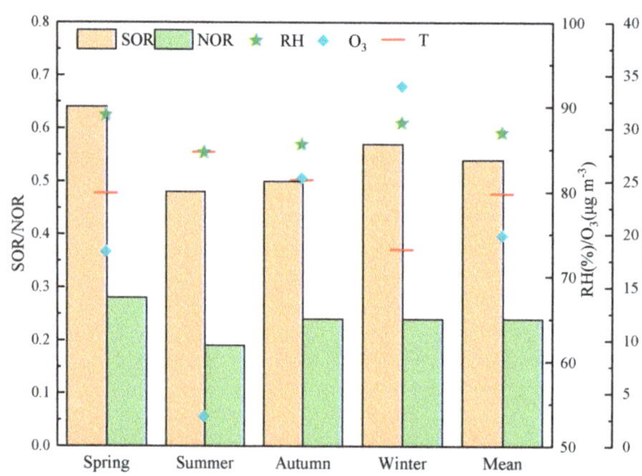

Figure 5. Seasonal distribution of SOR, NOR, O_3, Relative Humidity, and Temperature. Note: "RH" means Relative Humidity, "T" means Temperature.

The transformation of SO_4^{2-} primarily involves gas-phase and aqueous-phase oxidation. Gas-phase oxidation occurs through the reaction of SO_2 with atmospheric O_3 and OH radicals, which is significantly correlated with temperature and O_3 concentration. Aqueous-phase oxidation of SO_2 with oxidants such as HONO, H_2O_2, NO_2, and O_2 (catalyzed by

Mn and Fe) on the surfaces of water droplets and aerosols, on the other hand, is highly correlated with atmospheric humidity [35]. As depicted in Figure 5, the SOR values are higher in spring and winter compared to summer and autumn, indicating that higher humidity effectively promotes the oxidation of SO_2. NOR were 0.28, 0.19, 0.24, and 0.24 in spring, summer, autumn, and winter, respectively, which showed little seasonal difference, except in summer. This is related to the formation mechanism of NO_3^-. Higher temperatures are not conducive to the reaction of NH_3 with HNO_3 to form particulate NH_4NO_3 and reduce its stability [36]. During summer, the higher temperatures could accelerate the dissociation of ammonium nitrate particles to gas-phase ammonia and nitric acid.

Even though elevated SOR and NOR values may not fully indicate the extent of precursor conversion in Hainan, this is due to the influence of regional transport on the urban concentrations of SO_4^{2-} and NO_3^-. As depicted in Figure 6, the wind rosette for the monitoring site varies with the seasons. During summer, the prevailing winds are southerly, with the southern aspect of the site facing the South China Sea. The influx of clean maritime air currents aids in the dilution of pollutants, which likely contributes to the lower ion concentrations observed in this season. In contrast, the spring, autumn, and winter seasons are characterized by predominantly northeasterly winds. These winds are significantly stronger than those in summer. The northeastern region of the monitoring site is proximal to the northern parts of China and the Pearl River Delta. This geographic positioning facilitates the transportation of pollutants from upstream areas to Hainan, potentially leading to higher ion concentrations during these seasons.

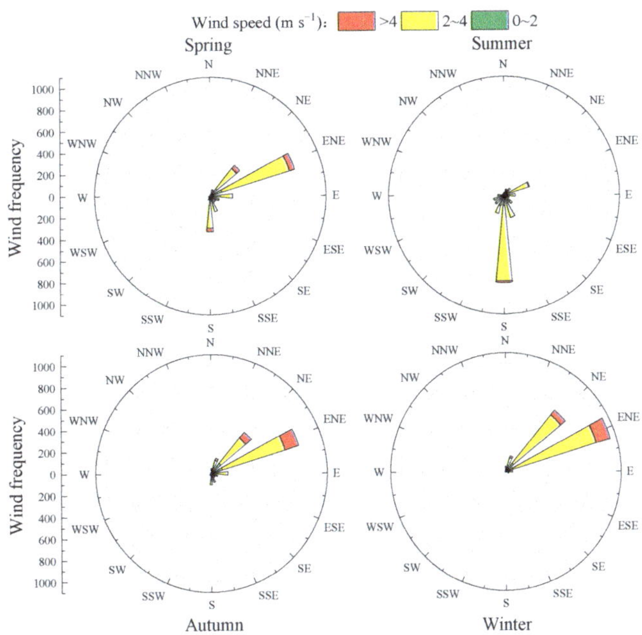

Figure 6. Wind rose in different seasons.

3.3. Correlation Analysis of $PM_{2.5}$ with Water-Soluble Ions and Meteorological Factors

The correlation between $PM_{2.5}$ and its water-soluble ions, as well as among the ions themselves, can shed light on their potential sources and binding mechanisms within particulate matter [37]. Figure 7 illustrates these correlations along with the influence of meteorological factors across different seasons. $PM_{2.5}$ is strongly correlated with SNA components in spring, autumn, and winter, and shows a weaker—though still significant—correlation in summer, indicating that secondary inorganic ions make a substantial contribution to atmospheric $PM_{2.5}$ and should not be overlooked. The high correlations among NH_4^+, SO_4^{2-}, and NO_3^- suggest that NH_4^+ is probably present as $(NH_4)_2SO_4$ and NH_4NO_3. In addition, the high correlation between Na^+ and Cl^- indicates that these ions share a common origin; given that the site is surrounded by the sea on three sides, marine sources exert a considerable influence on $PM_{2.5}$ composition.

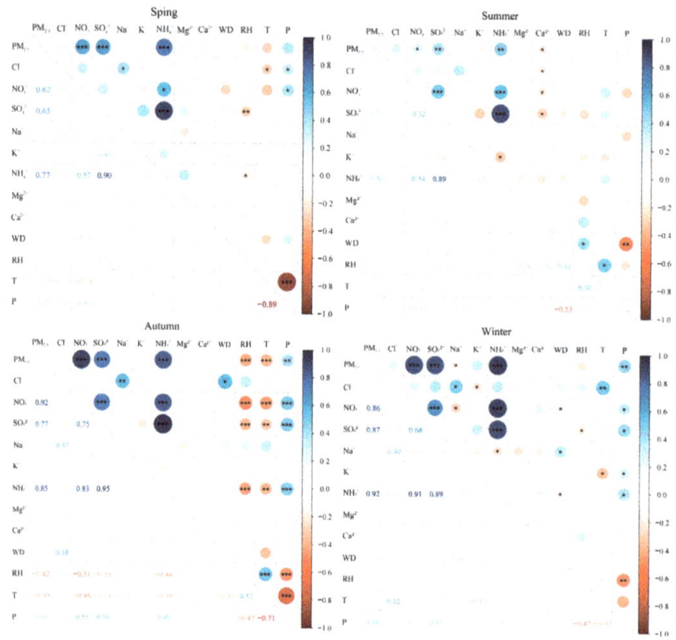

Figure 7. The correlation between $PM_{2.5}$, water-soluble inorganic ions, and meteorological factors. Note: "WD" means Wind Speed, "RH" means Relative Humidity, "T" means Temperature, "P" means Barometric Pressure; *** denotes significant correlation at $p < 0.001$, ** denotes significant correlation at $p < 0.01$, and * denotes significant correlation at $p < 0.05$.

In terms of meteorological influences, the correlation between $PM_{2.5}$ and its water-soluble ion concentrations with meteorological parameters was strongest in autumn, indicating that the monitoring sites are especially sensitive to weather variability during this season. The effects of seasonal change and meteorological factors on ion concentrations also differ. In autumn, SO_4^{2-}, NO_3^-, and NH_4^+ were positively correlated with barometric pressure and negatively correlated with temperature and humidity. This pattern can be ascribed to the southward advance of the continental cold high-pressure system, which transports large amounts of highly concentrated ionic species to the sites. In summer, by contrast, the concentrations of SO_4^{2-}, NO_3^-, and NH_4^+ were positively correlated with temperature, implying that elevated temperatures promote secondary $PM_{2.5}$ formation. Additionally, Na^+ and Cl^- were positively correlated with wind speed in summer, autumn, and winter, and most significantly in autumn, suggesting that stronger winds enhance the dispersion of sea-salt ions from the ocean surface.

3.4. Source Analysis

In this paper, we conducted a factor analysis of water-soluble ion fractions in $PM_{2.5}$ during the monitoring period using the EPA PMF 5.0 software. The analysis resolved a total of four factors, and the results are presented in Figure 8. Factor 1 is characterized by high percentages of Na^+ (89.9%), Mg^{2+} (70.8%), and Cl^- (46.1%). Na+ and Cl^- are typically indicative of sea-salt sources. The Mg^{2+}/Na^+ ratio is used to determine whether these ions in particulate matter primarily originate from the ocean or soil. A ratio less than 1.2 suggests an oceanic source [38]. The annual average Mg^{2+}/Na^+ ratio during the monitoring period was 1.1, indicating a predominant oceanic influence. Therefore, Factor 1 is identified as the oceanic source, with a contribution rate of 10.8%. Factor 2 contains relatively high percentages of K^+ (90.1%), which is an excellent tracer of biomass-burning aerosols [39], followed by SO_4^{2-} (54.8%), and NH_4^+ (54.3%). SO_4^{2-} is formed by the photochemical oxidation of sulfur-containing precursors (SO_2 and H_2S) released by coal combustion and NH_4^+ likely derived from secondary transformations of precursors released by biomass combustion (NH_3) [40]. Consequently, Factor 2 is recognized as a combustion source, with a contribution rate of 33.9%. Factor 3 has a high percentage of Ca^{2+} (79.8%) and Mg^{2+} (24.5%), which are typically associated with building construction, soil and road dust, and long-distance transported sand and dust [41]. Therefore, Factor 3 is identified as a dust source, with a contribution rate of 11.7%. Factor 4 features high percentages of Cl^- (53.8%), NO_3^- (51.3%), NH4$^+$ (45.7%), and SO_4^{2-} (29.8%), which are significant characteristics of secondary sources. NO_3^- and SO_4^{2-} are mainly from the oxidation of NO_x and SO_2, while NH_4^+ probably comes from the conversion processes between ammonia and sulfuric and nitric acid [23]. Thus, Factor 4 is identified as a secondary source, with a contribution rate of 43.6%.

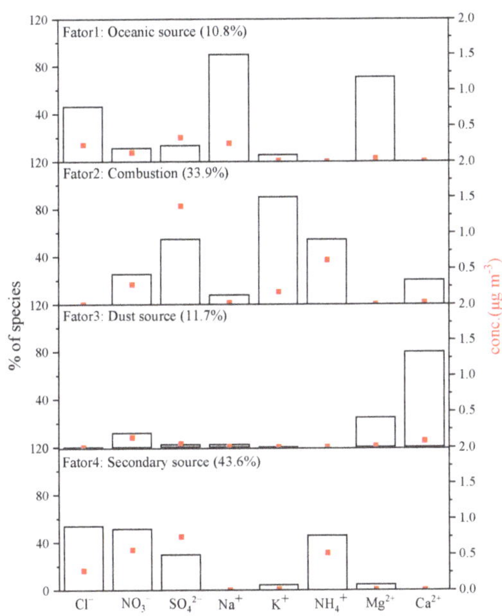

Figure 8. Contribution of pollution sources to water-soluble ions in $PM_{2.5}$.

3.5. Backward Trajectory Analysis

Regional transport of atmospheric pollutants significantly contributes to local pollutant concentrations, particularly in areas with inherently low pollutant levels [42]. This study analyzed the pollution trajectories of different air masses in four seasons, as shown in Figure 9. Each backward trajectory in Figure 9 indicates a 72 h period transportation clustering result, using the regional background site as the simulation starting point. The clustering results show distinct seasonal and spatial differences in air-mass transport paths and result in different characteristics of the pollutants.

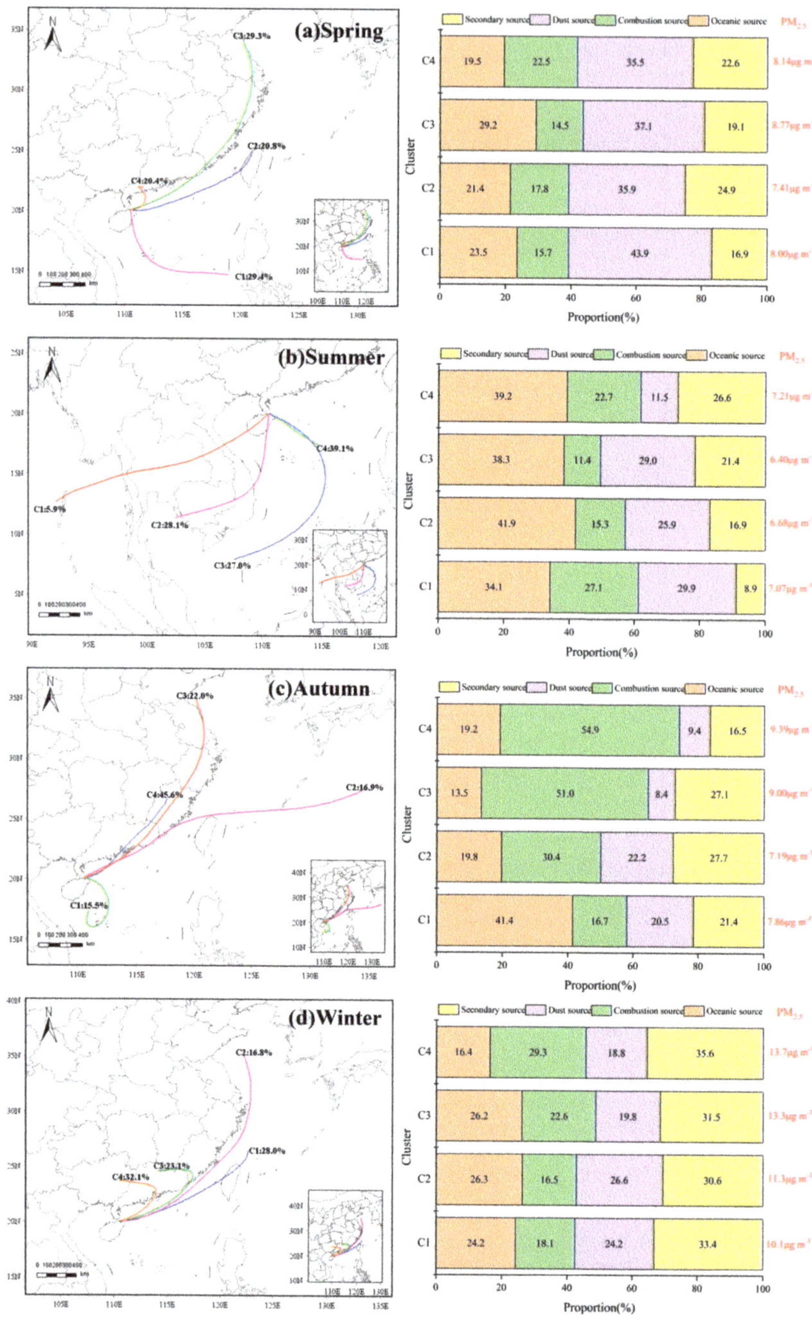

Figure 9. Results for the back-trajectory cluster.

In spring (Figure 9a), 70.5% of the trajectories arriving at the sampling site originated from the northeast of Hainan Island; of these, 29.3% (C3) traveled over the Yangtze River

Delta (YRD) and crossed Fujian Province before reaching the receptor, leading to the highest $PM_{2.5}$ and chemical component concentrations under this cluster influence (8.77 μg m^{-3}). A further 20.8% (C2) and 20.4% (C4) of the trajectories were sourced from Taiwan and the Pearl River Delta (PRD), respectively, yielding $PM_{2.5}$ levels of 7.41 and 8.14 μg m^{-3}. The remaining 29.4% of the trajectories were advected from the coastal Philippines, with a corresponding $PM_{2.5}$ concentration of 8.00 μg m^{-3}. Notably, dust sources constituted the dominant fraction of spring aerosols, accounting for >30% of the total trajectories and even exceeding 40%. The enhanced contribution of fugitive dust can be attributed to industrial and construction, as well as the long-range dust transport [43,44]. All four clusters traverse industrially developed regions; consequently, dust advected from these source areas exerts a substantial influence on the aerosol burden over Hainan.

In summer (Figure 9b), regional transport is overwhelmingly oceanic: all trajectories originate from the seas south of Hainan. The corresponding marine contribution to aerosol mass is markedly higher than in other seasons, and surface $PM_{2.5}$ reaches its annual minimum. The pronounced dilution by clean maritime air masses is a primary driver of the improved ambient air quality observed over Hainan during summer.

In autumn (Figure 9c), the clusters C3 and C4, which originated from the YRD and Fujian Province, collectively accounted for 67.6% of the total trajectories and yielded the highest $PM_{2.5}$ concentrations of 9.00 and 9.39 μg m^{-3}, respectively. Relative to other clusters, these two clusters were characterized by a pronounced enhancement in combustion-related emissions, contributing 51.0% and 54.9% of the total aerosol mass. This increase is likely attributable to industrial activities in the YRD and Fujian regions, as well as to post-harvest biomass burning during the autumn season.

In winter (Figure 9d), all clusters are from the northeast, with typical winter airflow characteristics. The most obvious feature is the increase in secondary sources, accounting for more than 30%, indicating that the clusters in winter are significantly affected by anthropogenic influences. However, Hainan has a sparse distribution of industries, accompanied by a large number of pollutants carried by air masses gradually moving southward in the cold air, which is one of the main reasons for the increase in pollutant concentrations in Hainan in winter. Combining the trajectories of the four seasons, it can be found that the trajectories from the YRD, PRD, and Fujian Province carry larger concentrations of pollutants, while the marine air masses from the south are cleaner.

4. Conclusions

In regions with minimal local emissions, regional transport is a significant pathway contributing to the elevation of pollutant concentrations; determining and quantitatively evaluating the composition and sources of water-soluble ions in $PM_{2.5}$ plays a significant role in the development of policies aimed at controlling particulate pollution. We have generated a high-quality dataset for $PM_{2.5}$ composition, derived from the continuous monitoring of chemical constituents at a regional background site in Hainan from August 2021 to July 2022. The results from the Positive Matrix Factorization (PMF) and backward trajectory analysis were utilized to elucidate the chemical characteristics and assess the regional transport patterns of the identified sources.

During the monitoring period, the average mass of total water-soluble ions at the regional station was 7.0 ± 4.4 μg m^{-3}, constituting 73.5% ± 24.4% of $PM_{2.5}$ mass. The concentration hierarchy was dominated by secondary inorganic aerosols (SNA), particularly SO_4^{2-}, NO_3^-, and NH_4^+, indicating significant secondary pollution. The sharp increase in NH_4^+ and NO_3^- concentrations with rising $PM_{2.5}$ levels highlights their role in pollution formation. SO_4^{2-}, despite low SO_2 levels and no coal heating in winter, remained a major component, suggesting external source contributions and a possible link to marine-derived dimethyl sulfide (DMS). The contribution of identified sources, including secondary sources (43.6%), combustion sources (33.9%), dust sources (11.7%), and oceanic sources (10.8%), had different spatial distributions and seasonal variations. Land-based air masses originating from the YRD, PRD, and Fujian are the dominant pathway driving $PM_{2.5}$ elevations over Hainan, whereas southerly marine air masses consistently supply clean background conditions. Specifically, long-range dust and secondary aerosols prevail in winter–spring, combustion emissions peak in autumn, and the summer minimum in $PM_{2.5}$ is governed by marine dilution—seasonal modulation of regional transport thereby governs Hainan's air-quality evolution. This study offers a scientific foundation for the formulation of more robust air pollution prevention and control measures for Hainan Province and China in the future.

Author Contributions: Project administration, X.X., W.X. and B.C.; writing—original draft preparation, X.X. and Y.L.; writing—review and editing, W.X., X.M., X.C., B.C., R.X., W.Y. and H.H.; data curation, X.M., C.D. and Z.Z.; investigation, C.D. and R.X.; software, Z.Z.; validation, H.S.; methodology, H.S. and Y.L. All authors have read and agreed to the published version of the manuscript.

Funding: This work was supported by the Academician Innovation Platform Scientific Research Special Project of Hainan Province, China (No. YSPTZX202205), and the Major Program of Science and Technology of Hainan Province, China (No. ZDKJ202007).

Institutional Review Board Statement: Not applicable.

Informed Consent Statement: Not applicable.

Data Availability Statement: All data supporting this article have been included as part of the main text.

Conflicts of Interest: The authors declare no conflicts of interest.

Abbreviations

The following abbreviations are used in this manuscript:

PMF	Positive Matrix Factorization
WSI	Water-soluble Ions
SNA	Sulfate (SO_4^{2-}), Nitrate (NO^{3-}), and Ammonium (NH_4^+)
EPA	Environmental Protection Agency
TWSI	Total Water-soluble Ions
DMS	Dimethylsulfide
SOR	Sulfur Oxidation Rate
NOR	Nitrogen Oxidation Rate

References

1. Jiang, Q.; Zhang, B.H.; Zhao, Y.L.; Wang, F.; Sun, Y.L. Growth mechanism of $PM_{2.5}$ and its chemical components in Beijing's urban area from 2013 to 2020. *Chin. J. Atmos. Sci.* **2023**, *47*, 373–386. Available online: http://www.iapjournals.ac.cn/dqkx/article/doi/10.3878/j.issn.1006-9895.2110.21142 (accessed on 18 September 2025). (In Chinese).
2. Wang, X.Q.; Wei, W.; Cheng, S.Y.; Yao, S.; Zhang, H.Y.; Zhang, C. Characteristics of $PM_{2.5}$ and SNA components and meteorological factors impact on air pollution through 2013–2017 in Beijing, China. *Atmos. Pollut. Res.* **2019**, *10*, 1976–1984. [CrossRef]
3. Qi, L.; Zhang, Y.; Ma, Y.; Chen, M.; Ge, X.; Ma, Y.; Zheng, J.; Wang, Z.; Li, S. Source identification of trace elements in the atmosphere during the second Asian Youth Games in Nanjing, China: Influence of control measures on air quality. *Atmos. Pollut. Res.* **2016**, *7*, 547–556. [CrossRef]
4. Kong, S.F.; Li, L.; Li, X.X.; Yin, Y.; Chen, K.; Liu, D.T.; Yuan, L.; Zhang, Y.J.; Shan, Y.P.; Ji, Y.Q. The impacts of firework burning at the Chinese Spring Festival on air quality: Insights of tracers, source evolution and aging processes. *Atmos. Chem. Phys.* **2015**, *14*, 2167–2184. [CrossRef]
5. Kumari, P.; Toshniwal, D. Impact of lockdown measures during COVID-19 on air quality-A case study of India. *Int. J. Environ. Health Res.* **2022**, *32*, 503–510. [CrossRef] [PubMed]

6. Sulaymon, I.D.; Zhang, Y.; Hopke, P.K.; Zhang, Y.; Hua, J.; Mei, X. COVID-19 pandemic in Wuhan: Ambient air quality and the relationships between criteria air pollutants and meteorological variables before, during, and after lockdown. *Atmos. Res.* **2021**, *250*, 105362. [CrossRef] [PubMed]
7. Zheng, H.; Kong, S.F.; Yan, Q.; Wu, F.Q.; Cheng, Y.; Zheng, S.R.; Wu, J.; Yang, G.; Zheng, M.; Tang, L.; et al. The impacts of pollution control measures on $PM_{2.5}$ reduction: Insights of chemical composition, source variation and health risk. *Atmos. Environ.* **2018**, *197*, 103–117. [CrossRef]
8. Yuan, C.S.; Hung, C.M.; Hung, K.N.; Yang, Z.M.; Cheng, P.H.; Soong, K.Y. Route-based chemical significance and source origin of marine $PM_{2.5}$ at three remote islands in East Asia: Spatiotemporal variation and long-range transport. *Atmos. Pollut. Res.* **2023**, *14*, 101762. [CrossRef]
9. Li, T.C.; Yuan, C.S.; Huang, H.C.; Lee, C.L.; Wu, S.P.; Tong, C. Inter-comparison of seasonal variation, chemical characteristics, and source identification of atmospheric fine particles on both sides of the Taiwan Strait. *Sci. Rep.* **2016**, *6*, 22956. [CrossRef]
10. Yue, D.; Zhong, L.; Zhang, T.; Shen, J.; Zhou, Y.; Zeng, L.; Dong, H.; Ye, S. Pollution properties of water-soluble secondary inorganic ions in atmospheric pm_(2.5) in the pearl river delta region. *Aerosol Air Qual. Res.* **2015**, *15*, 1737–1747. [CrossRef]
11. Meng, C.C.; Wang, L.T.; Zhang, F.F.; Wei, Z.; Ma, S.M.; Ma, X.; Yang, J. Characteristics of concentrations and water-soluble inorganic ions in $PM_{2.5}$ in Handan City, Hebei province, China. *Atmos. Res.* **2016**, *171*, 133–146. [CrossRef]
12. Hong, X.P.; Yang, K.; Liang, H.D.; Shi, Y.Y. Characteristics of water-soluble inorganic ions in $PM_{2.5}$ in typical urban areas of Beijing, China. *ACS Omega*, 2022; 7, 35575–35585. [CrossRef]
13. Wang, B.; Tang, Z.; Cai, N.; Niu, H. The characteristics and sources apportionment of water-soluble ions of $PM_{2.5}$ in suburb Tangshan, China. *Urban Clim.* **2021**, *35*, 100742. [CrossRef]
14. Zhao, M.; Qiao, T.; Huang, Z.; Zhu, M.; Xu, W.; Xiu, G.; Tao, J.; Lee, S. Comparison of ionic and carbonaceous compositions of $PM_{2.5}$ in 2009 and 2012 in Shanghai, China. *Sci. Total Environ.* **2015**, *536*, 695–703. [CrossRef]
15. Wu, L.; Wang, Y.; Li, L.; Zhang, G. Acidity and inorganic ion formation in $PM_{2.5}$ based on continuous online observations in a South China megacity. *Atmos. Pollut. Res.* **2020**, *11*, 1339–1350. [CrossRef]
16. Cao, X.C.; Xing, Q.; Hu, S.H.; Xu, W.S.; Xie, R.F.; Xian, A.D.; Xie, W.; Yang, Z.; Wu, X. Characterization, reactivity, source apportionment, and potential source areas of ambient volatile organic compounds in a typical tropical city. *J. Environ. Sci.* **2023**, *123*, 417–429. [CrossRef]
17. Xu, W.S.; Xing, Q.; Pan, L.B.; Wang, Z.S.; Cao, X.C.; Yan, W.J.; Xie, W.; Meng, X.; Wu, X. Characterization, source apportionment, and risk assessment of ambient volatile organic compounds in urban and background regions of Hainan Island, China. *Atmos. Environ.* **2024**, *316*, 120167. [CrossRef]
18. Xu, X.H.; Xu, W.S.; Meng, X.X.; Xing, B.C.; Zeng, Z.H.; Sheng, H.; Yan, W. Analysis of impacts of biomass burning in southeast Asia on air quality in Hainan island. *Environ. Monit. China* **2025**, *41*, 42–52. (In Chinese)
19. Yu, Y.Y.; Zhou, H.Y.; Zhao, Z.Z.; Chang, Y.H.; Wu, D.; Li, Z.Q.; Wang, F.; Fang, M.; Zhou, X. Spatiotemporal distribution, meteorological influence, and potential sources of air pollution over Hainan island, China. *Atmosphere* **2024**, *15*, 1336. [CrossRef]
20. Zong, Z.; Wang, X.P.; Tian, C.G.; Chen, Y.J.; Qu, L.; Ji, L.; Zhi, G.; Li, J.; Zhang, G. Source apportionment of $PM_{2.5}$ at a regional background site in North China using PMF linked with radiocarbon analysis: Insight into the contribution of biomass burning. *Atmos. Chem. Phys.* **2016**, *16*, 11249–11265. [CrossRef]
21. Paatero, P.; Tapper, U. Analysis of different modes of factor analysis as least squares fit problems. *Chemom. Intell. Lab. Syst.* **1993**, *18*, 183–194. [CrossRef]
22. Wang, Y.Q.; Zhang, X.Y.; Draxler, R.R. TrajStat: GIS-based software that uses various trajectory statistical analysis methods to identify potential sources from long-term air pollution measurement data. *Environ. Model. Softw.* **2009**, *24*, 938–939. [CrossRef]
23. Zhan, Y.Z.H.; Xie, M.; Zhao, W.; Wang, T.J.; Gao, D.; Chen, P.L.; Tian, J.; Zhu, K.; Li, S.; Zhuang, B.; et al. Quantifying the seasonal variations in and regional transport of $PM_{2.5}$ in the Yangtze River Delta region. *Atmos. Chem. Phys.* **2023**, *23*, 9837–9852. [CrossRef]
24. Gao, H.; Wei, J.; Wang, Y. Seasonal variation and source analysis of water-soluble inorganic salts in $PM_{2.5}$ in the southern suburbs of Beijing. *Environ. Sci.* **2018**, *39*, 1987–1993. (In Chinese)
25. Wang, X.; An, J.L.; Su, X.Q.; Liang, J.S.; Liu, J.D. Characteristics and optical properties of water-soluble ion pollution in the northern suburbs of Nanjing. *China Environ. Sci.* **2020**, *40*, 506–512. (In Chinese)
26. Hu, S.H.; Wu, X.C.; Xu, W.S.; Xie, R.F.; Xian, A.D.; Yang, Z.H. Characteristics and source analysis of water-soluble ions in $PM_{2.5}$ in Sanya. *Acta Sci. Circumst.* **2021**, *41*, 2540–2549. (In Chinese)
27. Xie, Y.; Lu, H.; Yi, A.; Zhang, Z.; Zheng, N.; Fang, X.; Xiao, H. Characterization and source analysis of water-soluble ions in $PM_{2.5}$ at a background site in Central China. *Atmos. Res.* **2019**, *239*, 104881. [CrossRef]
28. Shi, G.; Xu, J.; Peng, X.; Xiao, Z.; Chen, K.; Tian, Y.; Guan, X.; Feng, Y.; Yu, H.; Nenes, A.; et al. pH of Aerosols in a Polluted Atmosphere: Source Contributions to Highly Acidic Aerosol. *Environ. Sci. Technol.* **2017**, *51*, 4289–4296. [CrossRef]
29. Liu, J.; Wu, D.; Fan, S.; Mao, X.; Chen, H. A one-year, on-line, multi-site observational study on water-soluble inorganic ions in $PM_{2.5}$ over the Pearl River Delta region, China. *Sci. Total Environ.* **2017**, *601–602*, 1720–1732. [CrossRef] [PubMed]

30. Sun, Y.L.; Wang, Z.F.; Fu, P.Q.; Yang, T.; Jiang, Q.; Dong, H.B.; Li, J.; Jia, J.J. Aerosol composition, sources and processes during wintertime in Beijing, China. *Atmos. Chem. Phys.* **2013**, *13*, 4577–4592. [CrossRef]
31. Li, J.-M.; Zhao, S.-M.; Xiao, S.-H.; Li, X.; Wu, S.-P.; Zhang, J.; Schwab, J.J. Source apportionment of water-soluble oxidative potential of $PM_{2.5}$ in a port city of Xiamen, Southeast China. *Atmos. Environ.* **2023**, *314*, 120122. [CrossRef]
32. Fang, Y.; Cao, F.; Fan, M.Y.; Zhang, Y.L. Chemical characteristics and source apportionment of water-soluble ions in atmosphere aerosols over the East China sea island during winter and summer. *Environ. Sci.* **2020**, *41*, 1025–1035. (In Chinese)
33. Zhang, C.; Wang, L.L.; Qi, M.; Ma, X.; Zhao, L.; Ji, S.; Wang, Y.; Lu, X.; Wang, Q.; Xu, R.; et al. Evolution of key chemical components in $PM_{2.5}$ and potential formation mechanisms of serious haze events in Handan, China. *Aerosol Air Qual. Res.* **2018**, *18*, 1545–1557. [CrossRef]
34. Cao, Y.; Liu, J.; Ma, Q.; Zhang, C.; Zhang, P.; Chen, T.; Wang, Y.; Chu, B.; Zhang, X.; Francisco, J.S.; et al. Photoactivation of chlorine and its catalytic role in the formation of sulfate aerosols. *J. Am. Chem. Soc.* **2024**, *146*, 1467–1475. [CrossRef]
35. Cheng, C.; Shi, M.; Liu, W.; Mao, Y.; Hu, J.; Tian, Q.; Chen, Z.; Hu, T.; Xing, X.; Qi, S. Characteristics and source apportionment of water-soluble inorganic ions in $PM_{2.5}$ during a wintertime haze event in Huanggang, central China. *Atmos. Pollut. Res.* **2021**, *12*, 111–123. [CrossRef]
36. Galindo, N.; Nicolás, J.; Yubero, E.; Caballero, S.; Pastor, C.; Crespo, J. Factors affecting levels of aerosol sulfate and nitrate on the Western Mediterranean coast. *Atmos. Res.* **2008**, *88*, 305–313. [CrossRef]
37. Huang, D.; Xu, J.; Zhang, S. Valuing the health risks of particulate air pollution in the Pearl River Delta, China. *Environ. Sci. Policy* **2012**, *15*, 38–47. [CrossRef]
38. Xue, G.Q.; Zhu, B.; Wang, H.L. Size distributions and source apportionment of soluble ions in aerosol in Nanjing. *Environ. Sci.* **2014**, *35*, 1633–1643. Available online: https://www.hjkx.ac.cn/ch/reader/view_abstract.aspx?file_no=20140502&flag=1 (accessed on 18 September 2025). (In Chinese).
39. Zhang, F.; Cheng, H.-R.; Wang, Z.-W.; Lv, X.-P.; Zhu, Z.-M.; Zhang, G.; Wang, X.-M. Fine particles ($PM_{2.5}$) at a CAWNET background site in Central China: Chemical compositions, seasonal variations and regional pollution events. *Atmos. Environ.* **2014**, *86*, 193–202. [CrossRef]
40. Li, H.; Wang, Q.; Yang, M.; Li, F.; Wang, J.; Sun, Y.; Wang, C.; Wu, H.; Qian, X. Chemical characterization and source apportionment of $PM_{2.5}$ aerosols in a megacity of Southeast China. *Atmos. Res.* **2016**, *181*, 288–299. [CrossRef]
41. Pant, P.; Harrison, R.M. Critical review of receptor modelling for particulate matter: A case study of India. *Atmos. Environ.* **2012**, *49*, 1–12. [CrossRef]
42. Shu, L.; Xie, M.; Gao, D.; Wang, T.; Peng, L. Regional severe particle pollution and its association with synoptic weather patterns in the Yangtze River Delta region, China. *Atmos. Chem. Phys.* **2017**, *21*, 12871–12891. [CrossRef]
43. Sheng, H.; Yan, W.J.; Meng, X.X.; Xu, X.H.; Mai, X.Q.; Xu, W.S. Analysis of the impact of a typical dust storm event on air quality in tropical regions of China. *Acta Sci. Circumst.* **2024**, *44*, 310–319. (In Chinese)
44. Xu, H.; Cao, J.; Chow, J.C.; Huang, R.-J.; Shen, Z.; Chen, L.A.; Ho, K.F.; Watson, J.G. Inter-annual variability of wintertime $PM_{2.5}$ chemical composition in Xi'an, China: Evidences of changing source emissions. *Sci. Total Environ.* **2016**, *545*, 546–555. [CrossRef] [PubMed]

toxics

MDPI

Article

Air Pollution in a Northwest Chinese Valley City (2020–2024): Integrated WRF-HYSPLIT Modeling of Pollution Characteristics, Meteorological Drivers, and Transport Pathways in Yining

Xiaoqi Liu [1], Wei Wen [1,*], Xin Ma [2,*], Dayi Qian [1,3], Weiqing Zhang [4] and Shaorui Wang [1]

1 School of Energy and Environmental Engineering, University of Science and Technology Beijing, Beijing 100083, China; 18513216926@163.com (X.L.); qday@ustb.edu.cn (D.Q.); m202520245@xs.ustb.edu.cn (S.W.)
2 CMA Earth System Modeling and Prediction Centre (CEMC), Beijing 100081, China
3 Xinjiang Key Laboratory of Clean Conversion and High Value Utilization of Biomass Resources, Yili Normal University, Yining 835000, China
4 Yining Environmental Monitoring Station, Yining 835000, China; 15909997016@163.com
* Correspondence: wenwei@ustb.edu.cn (W.W.); marsin9955@163.com (X.M.)

Abstract: This study investigates the characteristics, meteorological drivers, and transport pathways of air pollution in Yining City from 2020 to 2024 based on meteorological records and air pollutant monitoring data. An integrated modeling approach combining the Weather Research and Forecasting (WRF) model and the Hybrid Single-Particle Lagrangian Integrated Trajectory (HYSPLIT) model was employed. Results reveal an overall annual decrease in ambient pollutant concentrations in Yining, with $PM_{2.5}$ and PM_{10} consistently below the national secondary standards, In contrast, the O_3 concentration shows a marked yearly increase. Pronounced seasonal variations were identified: the elevated O_3 concentrations in summer were driven by high temperatures and intense solar radiation. The significant increase in $PM_{2.5}$ and PM_{10} concentrations during winter was predominantly attributed to coal-based heating emissions and temperature inversion conditions. Pollutant concentrations were strongly associated with gaseous precursors (e.g., CO and NO_2) and meteorological factors. Higher temperatures and lower relative humidity aggravated O_3 formation, whereas lower temperatures and higher relative humidity favored $PM_{2.5}$ pollution. Correlation analysis revealed that NO_2 and CO showed the strongest correlations with $PM_{2.5}$ ($r = 0.84$) and O_3 ($r = -0.62$), respectively. Backward trajectory analysis revealed that higher pollution levels were associated with air masses originating from the southwest and southeast.

Keywords: air pollution; valley city; meteorological drivers; transport pathways

1. Introduction

With rapid industrialization and urbanization, cities have experienced a substantial increase in energy consumption and air pollutant emissions, leading to increasingly serious environmental challenges. Consequently, air pollution prevention and control have become a crucial aspect of China's ecological civilization initiatives. Air quality exerts a profound influence on public health; for instance, $PM_{2.5}$ can penetrate deeply into the respiratory tract and enter the bloodstream, elevating the risks of cardiovascular diseases, diabetes, lung cancer, and stroke [1]. Similarly, exposure to ozone is associated with respiratory symptoms such as chest tightness, shortness of breath, and throat irritation, while chronic exposure may have also contributed to ocular damage and visual impairment [2].

https://doi.org/10.3390/toxics13100868

In recent years, extensive research has been conducted on the drivers of air pollution and corresponding mitigation strategies across China. However, the majority of these studies have concentrated on economically developed areas, such as the Beijing–Tianjin–Hebei region [3], the Yangtze River Delta [4], and the Pearl River Delta [5]. As a result of industrial relocation policies, many pollution-intensive industries have shifted to western China, such as the Xinjiang Uygur Autonomous Region, where environmental standards are more relaxed [6]. This transition has spurred rapid urban population growth and industrial agglomerations, significantly raising local emissions [7]. Although these developments have stimulated employment and economic growth, they have also exacerbated air pollution.

Yining City, the administrative capital of the Ili Kazakh Autonomous Prefecture, is situated in the Xinjiang Uygur Autonomous Region of northwestern China. It lies within Ili River Valley, a north–south trending basin flanked by the Tianshan Mountains to the north and south, forming a semi-enclosed topography setting [8]. With an average elevation of 1080 m, Yining's urban area occupies the transitional zone between the mountain foothills and the river valley plain. This geographical layout significantly restricts the atmospheric diffusion, particularly during the heating season [9]. Yining City has a moderate continental climate, with an annual precipitation of 245.1 mm and a mean temperature of 10.5 °C, marked by considerable diurnal temperature variation. Temperature inversions are frequently (83% probability) during the heating season, with an average thickness of 511 m. Ground-level and multi-layer inversions dominate [10]. These conditions—combined with generally low wind speeds—aggravate the poor natural dispersal of air pollutants. Furthermore, recent industrial development in Yining has emphasized energy-intensive sectors, including coal-fired power generation, coal chemical processing, biopharmaceuticals, and advanced construction materials [11], resulting in significant deterioration of air quality [12]. Monitoring data from 2015 to 2020 show variable pollution levels in Yining City: after an initial rise, concentrations of PM_{10} and $PM_{2.5}$ began to decline, yet remained consistently above China's Grade II air quality standards. By 2020, PM_{10} and $PM_{2.5}$ concentrations reached 71 μg/m^3 and 43 μg/m^3, respectively, up by 4.4% and 16.2% from 2015, respectively. In the same period, O_3 concentrations rose consistently from 70 μg/m^3 to 83 μg/m^3, an increase of 18.6%. Although SO_2 and CO levels showed a downward trend, Yining City still recorded the highest regional concentrations in Xinjiang, with SO_2 at 14 μg/m^3 and CO concentrations persistently elevated in recent years [13]. Pollution exceedance days in Yining City are classified as light, moderate, or severe, with moderate to severe episodes displaying clear seasonality concentrated mainly in the heating season. The city's unique topography results in distinctly regional pollution patterns. Unlike flat terrains, the valley's restricted airflow greatly diminishes the atmosphere's pollutant carrying capacity, especially in winter, intensifying dispersion challenges. Therefore, identifying the drivers of air pollution in Yining City, analyzing valley-specific pollution characteristics, clarifying transport pathways, and pinpointing potential source regions are of considerable importance for formulating effective control strategies.

Research on air pollution has largely focused on economically developed regions characterized by intensive industrial activity. For instance, Zhang et al. [14] systematically investigated the regional heterogeneity of summertime O_3 pollution in Beijing in 2018 by integrating the Community Multiscale Air Quality modeling system (CMAQ) with process analysis. Their work quantified source contributions and formation mechanisms, revealing that non-local emissions from surrounding regions had a significantly greater impact on O_3 concentrations than urban and suburban sources. Similarly, Ge et al. [15] employed the Comprehensive Air Quality Model with Extensions with the Ozone Source Apportionment Technology (CAMx-OSAT) to simulate ozone reduction scenarios in Texas, concluding that controlling high-altitude point source emissions was more effective for ozone abatement than reducing road emissions. Liang et al. [16] employed the WRF-CAMx with Particulate Source Apportionment Technology (PSAT) to assess the impact of vehicle emission on air quality in the Beijing–Tianjin–Hebei region, demonstrating the significant role of non-local pollution transport. In a related study, Wen et al. [17] utilized the CAMx model to conduct the source apportionment of atmospheric pollutants in Beijing. Regarding the complex topography characteristic of northwestern China, Yang et al. [18] investigated air pollution patterns in the valley city of Lanzhou. They identified particulate matter as the dominant pollutant in Xigu District, with emissions showing strong seasonality—highest in winter due to increased emissions and poor dispersion, and lower in summer. O_3 levels varied seasonally, with higher levels observed in spring compared to summer. This pattern was attributed to higher NOx precursor concentrations and weaker atmospheric diffusion during spring. However, research on air pollution sources in Xinjiang remains limited. The transport mechanisms of pollutants in the Tianshan region is considerably more complex than in plain areas due to mountainous terrain and more dynamic meteorological conditions [19].

As a typical valley city, Yining's unique topography and meteorology leads to the accumulation of air pollutants. However, systematic studies on its pollution sources and key drivers are still lacking. With the city undergoing rapid development, improving regional air quality has become an urgent priority. This study investigates air pollution in Yining through an integrated analysis of meteorological and air quality monitoring data, combined with a customized atmospheric numerical model for the region. It systematically explores the characteristics, formation mechanisms, and influencing factors of air pollution in valley cities, aiming to provide scientific support for Yining's pollution control strategies. The results are also expected to offer valuable references for air quality management in other topographically similar regions.

2. Materials and Methods

2.1. Study Area Description

Yining City, the administrative center of Ili Kazakh Autonomous Prefecture, is situated in the northwestern part of Xinjiang Uygur Autonomous Region, China (43°50′–44°09′ N, 81°04′–81°29′ E). As the western terminus of China National Highway 312, the city serves as a crucial transportation junction. The city is bordered by Yining County to the east, Huocheng County to the west, Qapqal Xibe Autonomous County to the south-separated by a river boundary, and is backed by the Kogurqin Mountains, a branch range of the Tianshan Mountains, to the north. It lies within the Ili River Valley of the northern Tianshan Mountains, characterized by a distinct valley topography with low-lying terrain enclosed by mountains on three sides (as shown in Figure 1). The city covers an administrative area of 524.94 km^2. In 2024, the city's economy grew by 6.1% year-on-year, with its gross domestic product (GDP) reaching CNY 42.691 billion. Sectoral analysis reveals a tertiary-sector-dominated industrial structure: the tertiary sector contributed CNY 27.314 billion (64.0%), followed by the secondary sector at CNY 14.248 billion (33.4%), and the primary sector at CNY 1.129 billion (2.6%). Pollutant source census data from 2020 show that emissions of SO_2 and NO_x in Yining are 18,533.43 t and 42,480.28 t, which were dominated by stationary combustion sources (accounting for 82.0% and 71.5%, respectively), industrial process sources (16.1% and 5.9%), and mobile sources (1.1% and 21.1%). The emission of atmospheric particulate matter is 7197.84 t, which was dominated by stationary combustion sources (62.9%), cooking fume sources (11.9%), and fugitive dust sources (9.9%). The emission of volatile organic compounds (VOCs) is 11,657.36 t, which was mainly dominated by stationary combustion sources (48.2%), mobile sources (15.8%), and natural sources (12.7%).

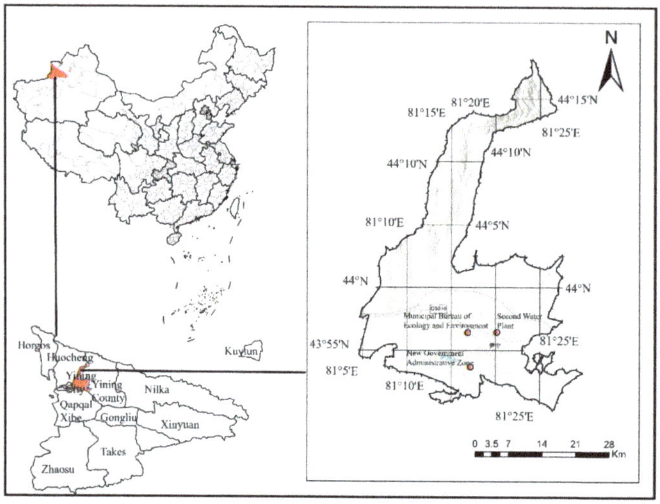

Figure 1. Map showing the geographical location of the study area. The three red dots indicate the air quality monitoring stations in Yining City.

2.2. Data Source

This research analyzed the daily average air pollutant concentrations ($PM_{2.5}$, PM_{10}, SO_2, NO_2, CO, and O_3) in Yining City. Pollution data were obtained from the Environmental Meteorological Service Platform (http://eia-data.com (accessed on 1 May 2025)), which aggregates measurements from three national-level automatic air quality monitoring stations: the Municipal Environmental Protection Bureau station (43.94° N, 81.28° E), the Second Water Treatment Plant station (43.94° N, 81.34° E), and the New Administrative District station (43.89° N, 81.29° E). The spatial distribution of these stations is shown in Figure 1. All stations operate in strict accordance with China's Ambient Air Quality Standards (GB 3095-2012) [20]. This study collected daily average concentrations of $PM_{2.5}$, PM_{10}, SO_2, NO_2, CO, and O_3 from 2020 to 2024, yielding 1825 initial samples. Following quality control screening to remove records with missing values, 1461 complete daily records were retained for analysis.

Meteorological data were primarily sourced from the China Meteorological Administration (CMA) "TianQing" Big Data Cloud Platform, with supplementary observations from the Yining City Meteorological Monitoring Station (43.9406° N, 81.3264° E). The meteorological dataset includes 2 m temperature, relative humidity, precipitation, and 10 m wind vector components (direction and speed). The number of valid meteorological records aligns with the air quality dataset, ensuring temporal consistency for subsequent correlation analysis.

2.3. Model Setup

To analyze pollution sources and transport pathways, this study established a coupled modeling system integrating the Weather Research and Forecasting (WRF v4.1) model and HYSPLIT-WEB trajectory model (NOAA HYSPLIT Model, https://www.ready.noaa.gov/hypub-bin/trajtype.pl?runtype=archive, (accessed on 1 May 2025)). The WRF simulation was configured with a Lambert conformal conic projection and a single-domain configuration centered at 44.1° N, 86.45° E. The modeling domain covered a 525 × 275 grid at 3 km horizontal resolution, encompassing Yining City and adjacent regions (Figure 2). Meteorological field simulation settings are summarized in Table 1. The WRF model was initialized using the Final Operational Global Analysis (FNL) dataset from the National Centers for

Environmental Prediction (NCEP; http://rda.ucar.edu/datasets/ds083.2/, (accessed on 4 May 2025)) as boundary conditions. For pollution trajectory analysis, the HYSPLIT model was driven by NCEP reanalysis data. Given the potential for long-range transboundary transport affecting Yining City, the study selected the NCEP reanalysis dataset for its global coverage, long-term continuity, and direct compatibility with the HYSPLIT model.

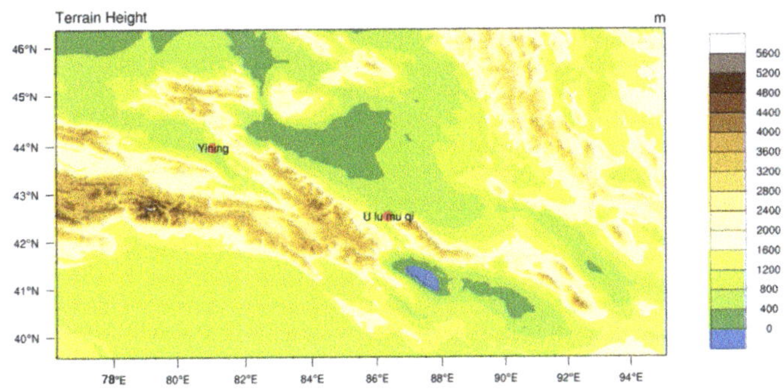

Figure 2. Modeling domain with terrain height profile.

Table 1. WRF model parameter configurations.

Parameter	Configuration/Scheme
Map projection	Lambert
Standard parallels	30° N and 60° N
Boundary conditions	NCEP FNL 1° × 1° reanalysis data (ds083.2)
Microphysics	WSM6 (6-class graupel scheme)
Longwave radiation	RRTM
Shortwave radiation	Goddard
Surface layer	Monin–Obukhov
Land surface	Thermal diffusion
Planetary boundary layer	Yonsei University (YSU)
Cumulus parameterization	Kain–Fritsch (new Eta) scheme for shallow convection

This study employed the HYSPLIT backward trajectory model, an atmospheric dispersion simulation system co-developed by NOAA's Air Resources Laboratory and the Australian Bureau of Meteorology [21]. Based on a Lagrangian framework for simulating advection and dispersion, the system incorporates multi-scale physical processes and heterogeneous emission sources within multi-parameter meteorological fields, enabling a comprehensive simulation of pollutant transport and transformation. As a well-validated numerical tool, HYSPLIT has been widely applied in regional to global studies of atmospheric transport patterns and transboundary pollution assessments [22–24].

2.4. Data Analysis Methodology

This study utilized Spearman correlation analysis to assess bivariate relationships between air pollutants and meteorological parameters. Based on a complete dataset of 1461 valid daily records, this study employed Spearman's rank correlation analysis ($p < 0.01$) using SPSS Statistics 27.0.1 to quantify the relationships between atmospheric contaminants and meteorological factors. This analytical approach established empirical evidence to support pollution source attribution. As a symmetric statistical measure, correlation analysis identifies systemic relationships between variables without implying causation, distinguishing it from directional causal inference methods [25].

Additionally, this study integrated hourly meteorological observations (wind direction/speed) and concurrent pollution concentration data from Yining City (2020–2024) using polar coordinate overlay analysis. By mapping pollutant spatial distributions through color-gradient visualization superimposed on wind direction sector, the method enabled the quantitative evaluation of dispersion patterns as a function of wind speed and directions. This approach provided an effective technical basis for identifying regional pollution transport pathways by combining wind fields and contaminant distributions visualization.

3. Results and Discussion

3.1. Characteristics of Air Pollution in Yining City

This study assessed the air quality status in Yining City from 2020 to 2024 using the Air Quality Index (AQI) methodology outlined in China's HJ633-2012 technical standard [26]. The AQI is a dimensionless indicator used to quantitatively describe ambient air quality conditions, which is determined solely by the highest sub-index among the constituent pollutants. The AQI classification system defines six air quality levels: Excellent (0–50), Good (51–100), Lightly Polluted (101–150), Moderately Polluted (151–200), Heavily Polluted (201–300), and Severely Polluted (>300). The analysis revealed significant annual variations in air quality category distribution, as illustrated in Figure 3a.

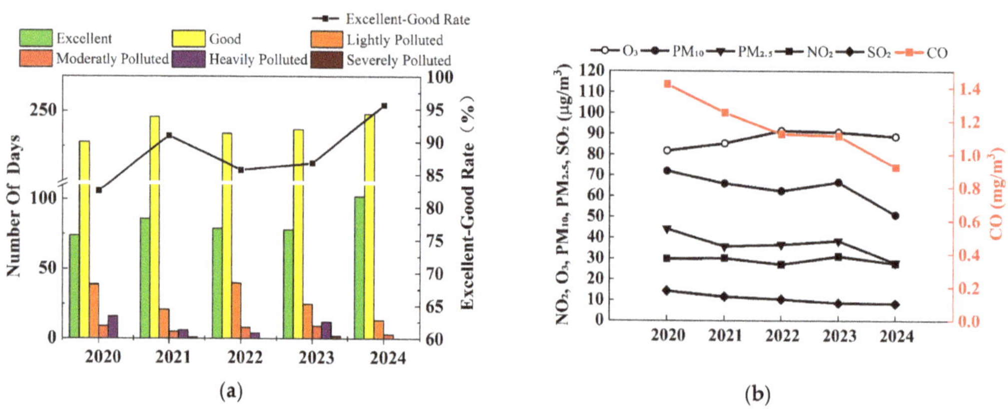

Figure 3. (**a**) Variation in air quality grade days in Yining City (2020–2024); (**b**) Interannual variations in six criteria pollutant concentrations in Yining City.

During the study period, Yining City exhibited a fluctuating yet overall improving air quality. The annual proportion of days with excellent-to-good air quality ($AQI \leq 100$) was 82.51% (302 days) in 2020, 90.96% (332 days) in 2021, 85.75% (313 days) in 2022, 86.30% (315 days) in 2023, and 95.63% (350 days) in 2024. The peak observed in 2020–2021 is likely attributable to COVID-19 control measures, such as industrial suspensions and transportation restrictions, while the subsequent rebound in 2022 coincided with the resumption of socioeconomic activities. Data from 2023 to 2024 indicate that air quality levels have been sustained beyond the levels observed during the pandemic period.

Overall, the excellent-to-good rate showed a gradual upward trend from 2020 to 2024, reflecting measurable enhancements in Yining City's ambient air quality following the continuous implementation of national and local air pollution prevention and control measures.

Interannual variations in pollutant concentration were analyzed using 2020–2024 monitoring data collected from three air quality stations in Yining City (located at the Municipal Environmental Protection Bureau, Second Water Treatment Plant, and New Administrative District). As summarized in Table 2 and Figure 3b, SO_2 and CO concentrations showed marked decreasing trends, while $PM_{2.5}$ and PM_{10} exhibited fluctuating but overall declining patterns. NO_2 levels remained relatively stable. These results reflect measurable improvements Yining's ambient air quality during the five-year study period.

Table 2. Annual mean concentrations of major pollutants in Yining City (2020–2024).

Year	O_3 (μg/m^3)	PM_{10} (μg/m^3)	$PM_{2.5}$ (μg/m^3)	NO_2 (μg/m^3)	SO_2 (μg/m^3)	CO (mg/m^3)
2020	82	72	44	30	14	1.4
2021	85	66	36	30	12	1.3
2022	91	62	37	27	10	1.1
2023	91	67	39	31	9	1.1
2024	89	51	28	28	8	0.9

The annual average $PM_{2.5}$ concentration demonstrated an overall decreasing trend, peaking at 44 μg/m^3 in 2020 before declining to 36 μg/m^3 in 2021. Subsequent rebounds elevated concentrations to 39 μg/m^3 in 2023, followed by a further reduction to 28 μg/m^3 in 2024, resulting in a cumulative decrease of 16 μg/m^3 over the five-year period. Nevertheless, these levels consistently exceeded the 35 μg/m^3 limit (except 2024) set in China's Ambient Air Quality Standards (GB 3095-2012) [20], hereafter referred to as the Standards).

The annual average PM_{10} concentration followed a pattern similar to that of $PM_{2.5}$, reaching a peak of 72 μg/m^3 in 2020 and declining to a low of 51 μg/m^3 in 2024, representing an overall decrease of 21 μg/m^3. Concentrations remained above the 40 μg/m^3 standard limit throughout the period. These results indicate that Yining City continues to face considerable particulate matter pollution, requiring further emission reductions to meet the primary annual air quality standard.

In contrast, SO_2 pollution levels remain relatively low, with the five-year maximum concentration reaching only 14 μg/m^3 in 2020—well below the 60 μg/m^3 standard limit. SO_2 concentrations declined consistently, reaching a minimum of 8 μg/m^3 in 2024, a 43% reduction from 2020. Meanwhile, NO_2 concentrations remained stable and below the 40 μg/m^3 Standard threshold over the five-year period.

The annual average O_3 concentration exhibited a trend opposite to that of particulate matter, showing an overall increase during the study period. Levels rose from 82 μg/m^3 in 2020 to 91 μg/m^3 in 2022, then declined gradually to 89 μg/m^3 in 2024. Although still below the Standard's 160 μg/m^3 limit, the 2024 concentration was 7 μg/m^3 higher than the 2020 baseline. Concurrently, CO concentrations decreased steadily, with the maximum recorded level of 1.4 mg/m^3 in 2020 already below the 4 mg/m^3 regulatory limit. By 2024, CO concentrations reached a minimum of 0.9 mg/m^3, a reduction of 0.5 mg/m^3 from 2020 levels.

The analysis of 2020–2024 monitoring data from Yining City's three national air quality monitoring stations, evaluated against the Standards, reveals the percentage of exceedance days for each pollutant (Table 3). $PM_{2.5}$ was the predominant pollutant, with an average of 12.039% of days exceeding the Standard—significantly higher than other pollutants. PM_{10} ranked second, with 5.144% of days exceeding the Standard over the five-year period. CO and NO_2 showed lower exceedance rates of 3.502% and 1.916%, respectively, and no exceedances were recorded for either pollutant in 2024. This notable improvement confirms the effective implementation of Yining City's Air Quality Attainment Plan (2021–2025), particularly through comprehensive industrial emission controls and multi-sector ultra-low emission initiatives. O_3 showed an emerging trend of exceedance days during the five-year period, transitioning from zero occurrences to measurable levels, though overall frequency remained relatively low. Its formation is influenced by both meteorological conditions (particularly temperature and solar radiation) and gaseous precursor concentrations (notably VOCs, CO, and NO_x). Therefore, future air pollution mitigation strategies should emphasize emission controls from stationary combustion and industrial process sources.

Table 3. Percentage of days exceeding the air quality standards for each pollutant.

Year	CO	NO_2	O_3_8 h	PM_{10}	$PM_{2.5}$	SO_2
2020	6.552%	2.184%	0.000%	9.009%	18.018%	0.000%
2021	3.014%	2.466%	0.000%	4.110%	9.590%	0.000%
2022	3.562%	1.370%	0.822%	5.206%	14.522%	0.000%
2023	4.384%	3.562%	0.274%	6.576%	13.152%	0.000%
2024	0.000%	0.000%	0.273%	0.819%	4.914%	0.000%
Average	3.502%	1.916%	0.274%	5.144%	12.039%	0.000%

O_3_8 h means daily maximum 8 h average O_3 concentration.

$PM_{2.5}$ concentrations exhibit distinct seasonal variability. Following the Northern Hemisphere climate patterns, the annual cycle is divided into winter (December–February), spring (March–May), summer (June–August), and autumn (September–November). Monthly averages from 2020 to 2024 reveal a consistent seasonal pattern (Figure 4a), with concentrations ranked as follows: winter (76 μg/m^3) > autumn (31 μg/m^3) > spring (23 μg/m^3) > summer (16 μg/m^3). Winter concentrations exceed other seasons by factors of 2.5–4.7, reflecting the highest pollution levels across the year. Similarly, PM_{10} follows a comparable seasons patterns (Figure 4b), with elevated concentrations in autumn and winter, ordered as winter (96 μg/m^3) > autumn (62 μg/m^3) > spring (54 μg/m^3) > summer (41 μg/m^3).

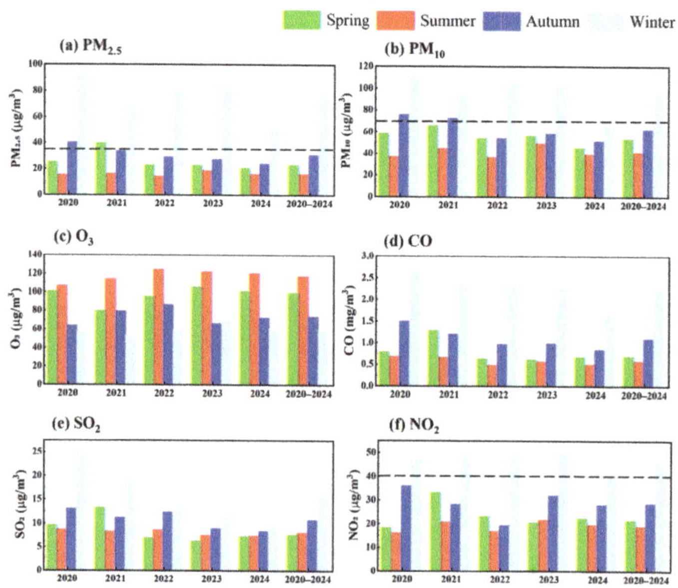

Figure 4. Seasonal variations in six criteria pollutants (**a–f**) in Yining City (2020–2024). The black dashed line denotes the secondary concentration standard, which is omitted for O_3, CO and SO_2 as their standard values exceed the upper bound of the axis scale.

The analysis of monthly concentration data from 2020 to 2024 (Figure 5) revealed a characteristic U-shaped distribution, with the highest value occurring in January and December. This pattern reflects the combined influence of three key factors: winter heating demand, increased coal combustion emissions, and frequent temperature inversion events.

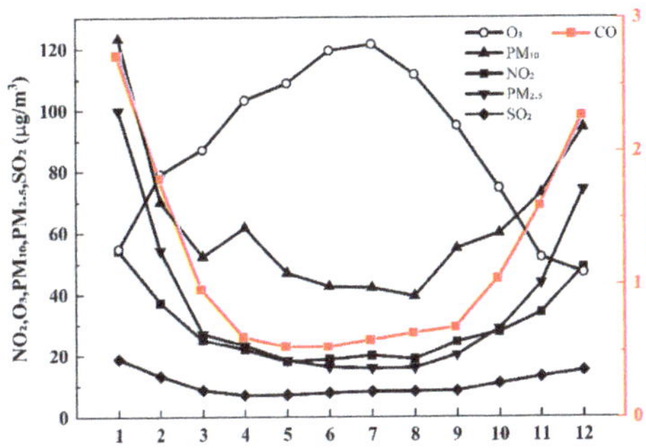

Figure 5. Monthly variations in six criteria pollutants in Yining City (2020–2024).

During the study period, $PM_{2.5}$ concentrations reached an annual maximum in winter (mean 76 $\mu g/m^3$) and gradually declined to their lowest levels in July–August (mean 16 $\mu g/m^3$). Precipitation and other meteorological factors were identified as the primary drivers of this seasonal variation. The data showed consistently high concentrations in December each year. PM_{10} exhibited similar seasonal trends, with two notable exceptions: an anomalous concentration increase in April and an accelerated rebound in September. These anomalies are likely associated with transient particulate pollution events resulting from a transported dust storm. The monthly distributions of SO_2, NO_2, and CO showed strong correlations with particulate matter trends. This correlation is attributable to increased emissions from coal-fired heating during winter months. In contrast, O_3 displayed a distinct inverted U-shaped pattern, peaking at 119–121 $\mu g/m^3$ in June–July. These summer maxima represent typical ozone formation conditions under elevated temperatures and strong solar radiation.

3.2. Effects of Meteorological Factors on Pollutants and Inter-Pollutant Correlations

The monthly variation patterns of pollutant concentrations in Yining City reveal that meteorological factors are a primary determinant of concentration fluctuations. Seasonal changes in meteorological conditions significantly influence air pollution levels, driving characteristic temporal patterns across different pollutants [27,28]. Key meteorological parameters, temperature and relative humidity mediate pollutant chemical transformation processes [29], while wind speed governs dispersion and transport dynamics [30]. Accordingly, this study performed correlation analyses between six pollutant categories and four meteorological variables (temperature, relative humidity, wind speed, and precipitation) utilizing 1461 valid samples. The results are summarized in Figure 6.

Regarding pollutant–meteorological relationships, CO concentrations exhibited significant negative correlations with temperature ($r = -0.74$), wind speed, and precipitation, but a strong positive correlation with relative humidity ($r = 0.61$). These patterns are likely attributable to the enhanced chemical conversion of CO at higher temperatures, resulting in lower concentrations.

NO_2 concentrations showed correlation patterns similar to CO, with a significant negative correlation with temperature ($r = -0.60$) and a distinct positive correlation with relative humidity ($r = 0.41$). These relationships suggest that both temperature and relative humidity substantially influence NO_2 concentrations, which consistently decrease with rising temperatures or declining humidity levels.

Similarly, SO_2 concentrations showed negative correlations with temperature, wind speed, and precipitation, but a positive correlation with relative humidity. Among the examined meteorological variables, temperature showed the strongest association with SO_2 concentrations ($r = -0.47$).

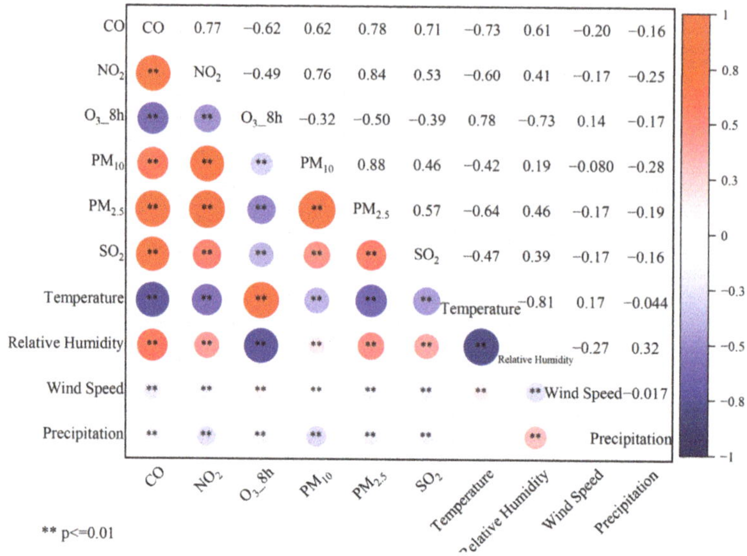

Figure 6. Pollutant–meteorology and pollutant–pollutant correlations.

Both $PM_{2.5}$ and PM_{10} displayed similar correlation patterns with meteorological factors, showing significant relationships with temperature and relative humidity. Specifically, PM_{10} concentrations were moderate negative correlations with temperature (r = −0.42) and precipitation (r = −0.28). $PM_{2.5}$ exhibited stronger associations, with a pronounced negative correlation with temperature (r = −0.64) and a positive correlation with relative humidity (r = 0.46). These findings indicate that elevated humidity promotes particulate matter accumulation through hygroscopic growth, whereby water vapor condensation increases particle size, while also enhancing secondary aerosol formation from gaseous precursors and reducing dry deposition efficiency. Collectively, these mechanisms contribute to higher ambient particulate matter concentrations. In contrast, precipitation facilitates the wet deposition of fine particles. Furthermore, summer meteorological conditions generally favor particulate dispersion, as higher temperatures intensify atmospheric turbulent mixing and enhancing dilution effects [31].

Unlike the other five pollutants, O_3 concentrations showed a strong positive correlation with temperature (r = 0.78) and a negative correlation with relative humidity (r = −0.73). This pattern primarily from temperature-enhanced photochemical activity that drives O_3 production The observed negative correlation with relative humidity stems primarily from its strong inverse relationship with temperature; consequently, high temperatures that favor O_3 formation coincide with attendant low relative humidity. Existing studies show that when relative humidity drops below 60%, the attenuation of solar radiation by water vapor is weakened, thereby accelerating photochemical reactions [32]. Furthermore, stagnant synoptic conditions, such as uniform-pressure fields and weak baric gradients, inhibit pollutant dispersion and thereby facilitate the accumulation of ozone precursors, notably NO_x and VOCs. For instance, in southern Xinjiang, meteorological systems like the Central Asian vortex or the periphery of a high-pressure system often induce surface wind speeds below 3 m/s, creating conditions conducive to localized ozone pollution [33].

Overall, low-temperature and high-humidity conditions favor elevated concentrations of CO, NO_2, PM_{10}, $PM_{2.5}$, and SO_2, aggravating surface-level pollutant accumulation—consistent with established research. During winter months, the low-temperature significantly weakens the atmospheric dispersion capacity, leading to frequent particulate pollution episodes [34]. Although wind speed and precipitation show limited direct influence on pollutant formation, pollution events occur more frequently under low-wind, low-precipitation conditions.

Regarding inter-pollutant relationships, this study systematically analyzed correlations among six criteria air pollutants using their daily monitoring data from 2020 to 2024. Following quality control screening, 1461 complete daily datasets meeting integrity criteria were retained for analysis. The inter-pollutant correlation results are presented in Figure 6. Statistical analysis confirmed significant correlations ($p < 0.01$) among Yining's primary air pollutants. With the exception of O_3, all pollutants exhibited positive inter-correlations. The particularly strong correlation between PM_{10} and $PM_{2.5}$ (r = 0.88) suggests shared pollution sources. The analysis revealed NO_2 as the predominant precursor influencing particulate concentrations (r = 0.84 with $PM_{2.5}$), followed by CO (r = 0.78) and SO_2 (r = 0.57). NO_2 reacts with water vapor in the atmosphere to form nitric acid (HNO_3), which subsequently reacts with ammonia (NH_3) to produce ammonium nitrate (NH_4NO_3) [35]. As a key $PM_{2.5}$ component, NH_4NO_3 formation is accelerated under high humidity conditions, leading to increased fine particulate concentrations. O_3 and CO demonstrate a strong inverse relationship ($r = -0.62$, $p < 0.01$), reflecting the consumption of CO as a precursor in photochemical reactions that generate ozone, thereby explaining their significant negative correlation.

3.3. Transport Trajectory Simulation Analysis of Typical Pollution Episodes

Wind field characteristics critically determine pollutant dispersion, thereby directly influencing transport efficiency and the atmospheric self-cleaning capacity [36]. The study first analyzed wind field characteristics using meteorological data, summarized in the wind rose diagrams (Figure 7a). Subsequent analysis explored the relationships between wind speed and the concentrations of particulate matter and ozone. In the corresponding visualization, Yining City is positioned at the pole, where azimuthal angles denote wind directions, radial distances represent wind speeds, and color gradients reflect pollutant concentration levels, enabling an integrated assessment of wind-driven dispersion effects. Figure 7b specifically depicts seasonal O_3 variations in relation to wind conditions.

Wind direction analysis reveals a consistent annual prevalence of northeast winds, which maintain stable directions with moderate speeds of 2–5 m/s. During spring, summer, and autumn, southwest winds occur at a relative frequency of 10–20%, often at higher speeds ($\geq$8 m/s) and frequently associated with short-duration, strong wind events, dust uplift, or localized meteorological disturbances. In winter, northeast winds prevail overwhelmingly, with a sharp decline in southwest occurrences and consistently lower wind speeds. This results in a clearly defined seasonal wind regime in Yining, characterized by persistent northeasterly flows punctuated by intermittent yet influential southwest winds.

The results of the relationship between O_3 concentrations and wind patterns reveals distinct seasonal dependencies. In spring, elevated O_3 levels are primarily associated with two wind regimes: (1) prevailing westerly winds at 3–4 m/s, and (2) northerly winds at 2–4 m/s. During summer, high O_3 concentrations show a strong correlation with northwesterly winds at 3–5 m/s. In autumn, O_3 peaks occur mainly under easterly to southeasterly winds at 2–3 m/s. Cross-seasonal data reveal that the highest O_3 concentrations consistently occur under low to moderate wind speeds (2–4 m/s) across all seasons. Notably, northwesterly winds contribute substantially to summer O_3 pollution, likely due to the regional transport zone and its precursors from upwind areas such as Huocheng County.

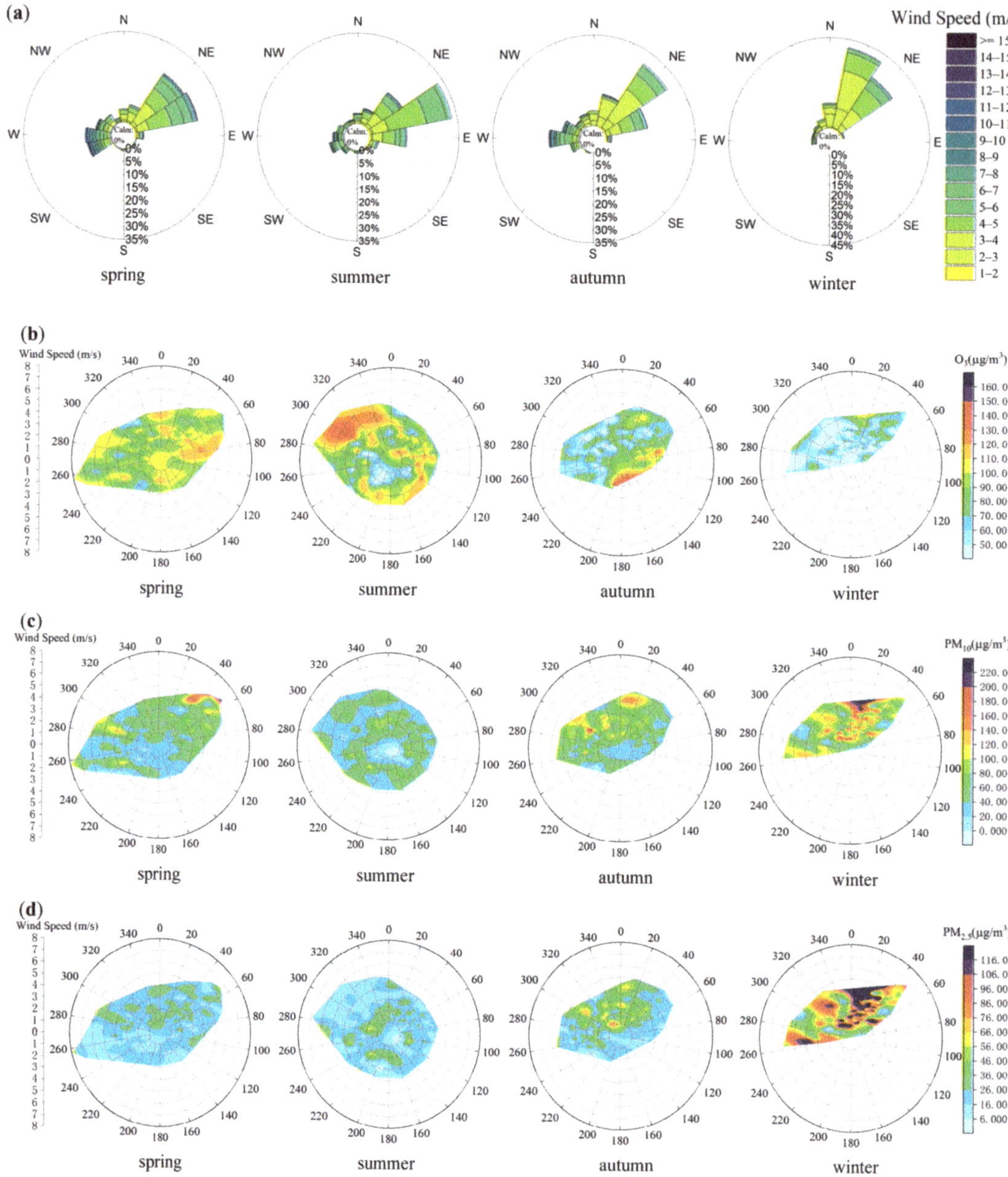

Figure 7. (**a**) Wind rose diagrams for Yining City by season; (**b**) Seasonal polar plots of wind fields versus O_3 concentrations in Yining City; (**c**) Seasonal polar plots of wind fields versus PM_{10} concentrations in Yining City; (**d**) Seasonal polar plots of wind fields versus $PM_{2.5}$ concentrations in Yining City.

Figure 7c,d illustrate the seasonal characteristics of PM_{10} and $PM_{2.5}$ concentrations in relation to the wind field parameters in Yining City. The analysis reveals distinct seasonal patterns in particulate matter concentrations. Spring peaks predominantly occur during northeast wind conditions at 3–5 m/s, while autumn shows particulate maxima under two different wind regimes: northwest winds at 2–4 m/s and stronger northeast winds at 4–5 m/s. During winter, the highest particulate concentrations consistently correlate with northerly winds at 3–4 m/s, demonstrating the clear seasonal dependence of air quality on local wind patterns. Overall analysis indicates that elevated particulate matter concentrations in Yining City predominantly occur under northeasterly wind conditions, suggesting that local emissions and regional transport from the northeast direction serve as the primary pollution sources. Notably, upwind areas such as Yining County contribute substantially to particulate pollution in the region.

Based on the identified pollution patterns, this study further investigated a severe $PM_{2.5}$ pollution episode in Yining City during the period 9–12 January 2020, the period with the highest recorded concentrations that month, using HYSPLIT backward trajectory analysis and WRF-model. To evaluate the reliability of the simulation results, WRF simulated meteorological data were compared with observational data from ground monitoring stations during the same period. The model performance was validated using the Mean Bias (MB) and Root Mean Square Error (RMSE) metrics (Table 4). The results demonstrate good agreement between simulated and observed values, indicating that the model effectively reproduced the variations in surface meteorological elements during the pollution episode.

Table 4. Evaluation of the WRF model performance for meteorological factors.

Meteorological Factor	Inspection Indicators	2020-01	2020-07
T2	MB	2.78 K	3.05 K
	RMSE	4.14 K	3.87 K
WS10	MB	3.59 m/s	2.30 m/s
	RMSE	4.24 m/s	3.20 m/s
RH2	MB	0.35%	−9.44%
	RMSE	13.18%	15.07%

Meteorological factors: T2-temperature at 2 m, WS10-wind speed at 10 m, RH2-relative humidity at 2 m.

To elucidate the potential source regions and transport pathways of a notable $PM_{2.5}$ pollution event (concentration increase from 164 $\mu g/m^3$ to 256 $\mu g/m^3$), a WRF-HYSPLIT analysis was performed. Figure 8 illustrates the backward trajectory of the polluted air mass, while Figure 9 shows the simulated meteorological conditions, including wind direction and speed, during the pollution episode.

Trajectory analysis reveals that on 9 January, the polluted air mass originating from the southeast significantly influenced on Yining City. This is corroborated by WRF simulation results, which show prevailing southeasterly winds on that day. From 10 to 12 January, air masses consistently originated from the southwest, deflection near 43.5°N latitude, and eventually approached Yining City from southeastern pathways. By integrating topographic data from Figure 2 with WRF-simulated wind fields in Figure 9, distinct orographic wind patterns around Yining City were identified during the study period. The analysis revealed that southwesterly winds prevailed south of the Halqitu–Narat mountain range, while southeasterly winds dominated the northern part of these topographic barriers. The trajectories of polluted air masses aligned closely with these dominant wind directions, indicating significant topographic influence on pollutant transport pathways during the pollution episode.

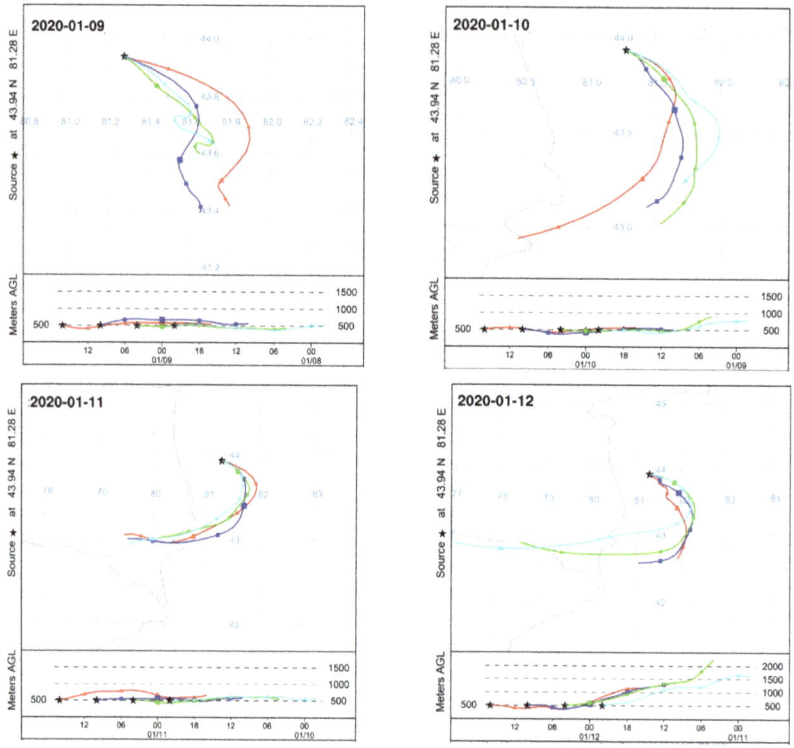

Figure 8. Twenty-four hour backward trajectories of polluted air masses during the period 9–12 January 2020. Colors correspond to different time points: red (16:00), dark blue (10:00), green (04:00), and light blue (22:00).

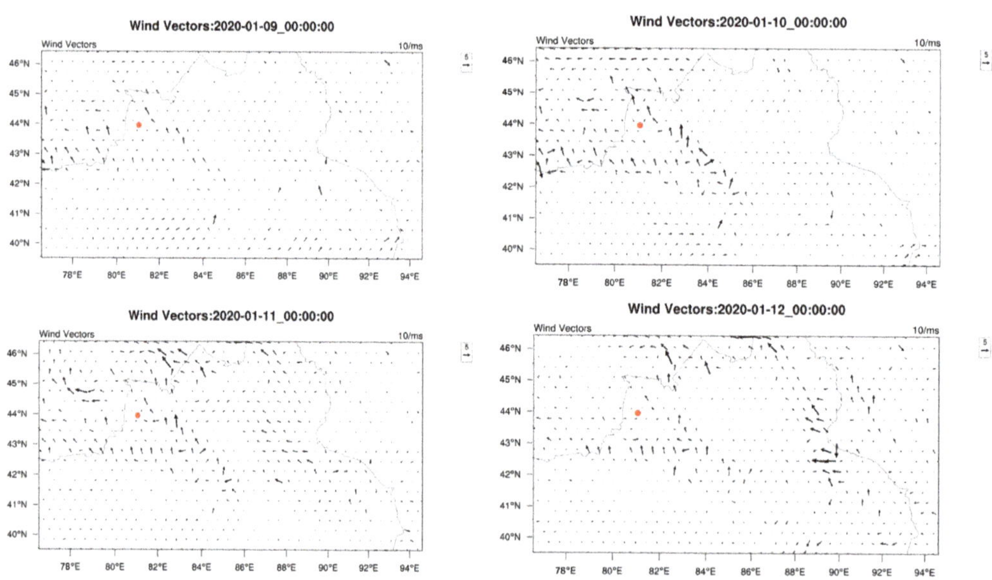

Figure 9. Simulated wind vectors during the period 9–12 January 2020. The red point represents Yining.

The results demonstrated that $PM_{2.5}$ concentrations in Yining City were significantly influenced by regional transport under specific meteorological conditions. When calm conditions prevailed in the southern mountains alongside southeasterly flow, enhanced regional transport led to noticeable increases in fine particulate matter levels. More notably, when southwesterly winds dominated the southern part of the mountain range, the transport mechanism incorporated pollutants not only from nearby counties (Yining, Chabuchar, and Gongliu) but also from transboundary sources including Kazakhstan. Under these conditions, particulate pollution intensified considerably, often reaching severe levels.

This study also examined a representative ozone pollution episode in Yining City during the period 11–15 July 2020 using numerical simulations. The analysis identified a characteristic variation pattern in O_3 concentrations, which rose from 116 $\mu g/m^3$ on 11 July to a peak of 150 $\mu g/m^3$ on 14 July before a subsequent decline. Figure 10 presents the backward trajectory analysis of the polluted air mass, and Figure 11 shows the corresponding WRF-simulated wind direction and speed during this episode.

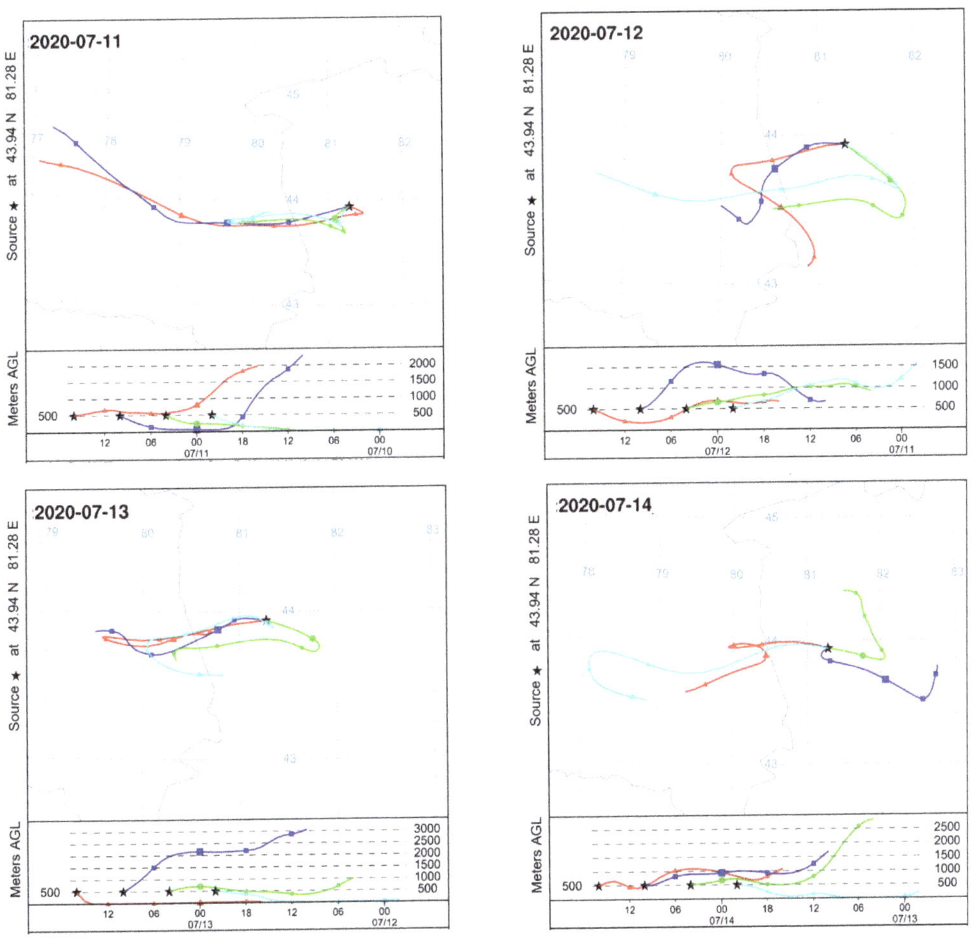

Figure 10. Twenty-four hour backward trajectories of polluted air masses during the period 11–14 July 2020. Colors correspond to different time points: red (16:00), dark blue (10:00), green (04:00), and light blue (22:00).

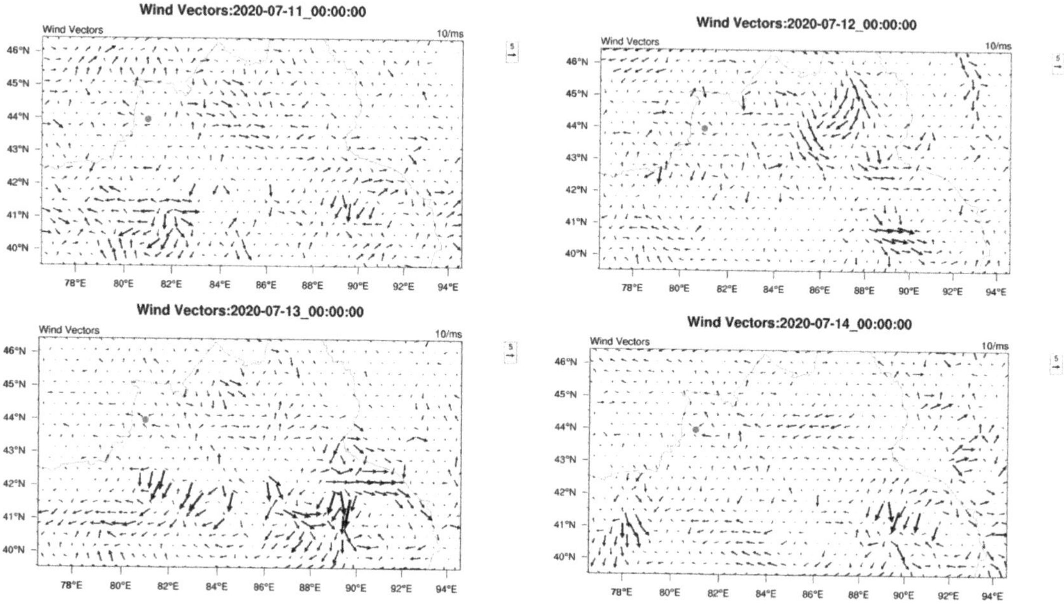

Figure 11. Simulated wind vectors in Yining (11–14 July 2020). The red point represents Yining.

The trajectory analysis revealed distinct transport patterns during the ozone pollution episode. On July 11th, all air masses originated from western directions; however, low wind speeds (<2 m/s) limited their influence on pollutant transport. During the period 12–13 July 2020, while the main trajectories came from the southwest, a secondary transport pathway from the southeast was identified, aligning with the simulated southeast wind patterns on those days. On the peak pollution day (14 July), air masses originated predominantly from the east, coinciding with intensified southeast winds (4–5 m/s) in the meteorological simulations, which enhanced regional transport. The analysis further indicated that air mass transport differentially influenced O_3 concentrations depending on prevailing wind directions. Southwesterly winds were associated with a moderate transport effect on ozone levels, whereas easterly wind conditions led to a more pronounced influence.

These characteristics suggest that ozone pollution in the study area results from the combined influences of emissions from southwestern and eastern adjacent regions, primarily Huocheng County, Yining County and Gongliu County, with additional contributions from transboundary air masses, including those originating in Kazakhstan.

4. Conclusions

This study presents a systematic analysis of atmospheric pollutant monitoring data from Yining City during the period 2020–2024, revealing distinct temporal variations in ambient air quality. By integrating meteorological monitoring data, the study identified the principal factors controlling air pollution in the region. Using coupled WRF and HYSPLIT modeling approach, the study established a pollution source-apportionment simulation system tailored for Yining and its surrounding areas. Numerical simulations were performed to characterize meteorological fields and pollutant transport pathways during representative monthly periods, yielding the following key findings:

1. Yining City showed marked improvement in overall air quality during the 2020–2024 period. The proportion of days meeting national air quality standards for all six criteria pollutants (SO_2, PM_{10}, $PM_{2.5}$, NO_2, O_3, and CO) increased from 82.51% in

2020 to 95.6% in 2024. Three-year average concentrations of SO_2, NO_2, O_3, and CO consistently met the Grade I standards of China's National Ambient Air Quality Standards (GB 3095-2012) [20], while PM_{10} and $PM_{2.5}$ met Grade II standards. Statistically significant reductions were observed for CO, PM_{10} and $PM_{2.5}$, indicating substantial air quality enhancement. Distinct temporal patterns emerged: NO_2 and SO_2 remained relatively stable, whereas O_3 increased significant in recent years (2022–2024), suggesting enhanced regional photochemical activity. Seasonal variations showed winter peaks for all pollutants except O_3, which peaked in summer. Monthly profiles revealed a U-shaped trend (January maximum) for SO_2, $PM_{2.5}$, NO_2, and CO; a W-shaped pattern for PM_{10}; and a distinct inverted U-shaped profile (July peak) for O_3;

2. Meteorological conditions critically influence air pollution in Yining. All six pollutants showed significant correlations with temperature and relative humidity. With the exception of O_3, pollutants mainly accumulated under cold, humid conditions, whereas O_3 formation was favored by warm, dry weather. Summer northwesterly winds (2–4 m/s) substantially enhance O_3 pollution, with Huocheng County identified as the primary upwind source region. During winter, northerly winds (3–4 m/s) significantly increased particulate matter levels, largely originating from Yining County;
3. Significant correlations were identified between gaseous precursors (SO_2, NO_2, CO) and secondary pollutants ($PM_{2.5}$, PM_{10}, O_3). $PM_{2.5}$ and PM_{10} showed a strong correlation coefficient ($r = 0.88$), indicating the same sources. Among the gaseous precursors, NO_2 exhibited the strongest positive correlations with particulate matter ($r = 0.84$ with $PM_{2.5}$; $r = 0.76$ with PM_{10}). In contrast, CO showed a strongest negative correlation with O_3 ($r = -0.62$), suggesting its consumption in photochemical processes. These results highlight NO_2 as the dominant precursor for particulate formation, while CO plays a key role in ozone depletion;
4. WRF and HYSPLIT modeling results indicated that regional transport significantly affects pollution in Yining. $PM_{2.5}$ pollution events were predominantly influenced by southwestern air masses from Chabuchar County, whereas O_3 episodes were mainly driven by eastern transport from Yining County and Gongliu County, reflecting distinct spatial patterns for different pollutants.

Author Contributions: Conceptualization, W.W. and X.L.; methodology, X.M.; software, X.L. and X.M.; validation, D.Q. and W.Z.; formal analysis, W.W., X.L., and S.W.; investigation, W.W. and X.L.; writing—original draft preparation, W.W. and X.L.; writing—review and editing, X.L., W.W. and X.M.;. All authors have read and agreed to the published version of the manuscript.

Funding: This work was supported by the Science and technology project of Ili Kazak autonomous Prefecture (YJC2024A02).

Institutional Review Board Statement: Not applicable.

Informed Consent Statement: Not applicable.

Data Availability Statement: The datasets used and/or analyzed during the current study are available from the corresponding author upon reasonable request.

Conflicts of Interest: The authors declare no conflicts of interest.

References

1. Kim, K.H.; Kabir, E.; Kabir, S. A review on the human health impact of airborne particulate matter. *Environ. Int.* **2015**, *74*, 136–143. [CrossRef]
2. Li, T.N.; Qiu, J.X.; Fang, C.S. A brief analysis of the hazards of ozone in the environment and relevant prevention and treatment. *World Environ.* **2020**, *5*, 16–18.

3. Li, H.S.; Hou, X.L.; Xue, J.; Guo, T.F.; Zou, T.S.; Zhang, H.F.; Guo, X.; Li, M.X.; Hao, J.M. Practices and Empirical Insights from the National Research Program for Key Issues in Air Pollution in Beijing-Tianjin-Hebei and Surrounding Areas. *Engineering* **2023**, *30*, 20–26. [CrossRef]
4. Yu, W.X.; Wang, Y.; Wang, H.L.; Zhu, S.Q.; Wang, P.; Zhang, H.L. Response of ozone to meteorology and atmospheric oxidation capacity in the Yangtze river Delta from 2017 to 2020. *Atmos. Environ.* **2024**, *332*, 120616. [CrossRef]
5. Mai, B.; Diao, Y.; Yang, H.; Deng, T.; Zou, Y.; Wang, Y.; Lan, W.; Liu, X.; Deng, X. Assessing atmospheric CO_2 concentrations and contributions from biogenic and anthropogenic sources in the Pearl River Delta region. *Urban Clim.* **2024**, *54*, 101–864. [CrossRef]
6. Han, X.; Liu, Y.Q.; Gao, H.; Ma, J.M.; Mao, X.X.; Wang, Y.T.; Ma, X.D. Forecasting PM2.5 induced male lung cancer morbidity in China using satellite retrieved PM2.5 and spatial analysis. *Sci. Total Environ.* **2017**, *607–608*, 1009–1017. [CrossRef]
7. Shrestha, A.; Luo, W. An assessment of groundwater contamination in Central Valley aquifer, California using geodetector method. *Ann. GIS* **2017**, *23*, 149–166. [CrossRef]
8. Min, Y.F.; Huang, W.L.; Ma, M.J.; Zhang, Y.N. Simulations in the Topography Effects of Tianshan Mountains on an Extreme Precipitation Event in the Ili River Valley, China. *Atmosphere* **2021**, *12*, 750. [CrossRef]
9. Zhou, C.H. Analyzing atomospheric environment pollution character and improving measure of Yining City. *Environ. Prot. Xinjiang* **2006**, *28*, 39–42.
10. Li, Y.; Song, X.M. The temporal and spatial distribution characteristics of heavy fog in Yining. *Desert Oasis Meteorol.* **2008**, *4*, 34–36.
11. Zhou, C.H.; Liu, S.H.; Jiapaer, Y.; Wang, S.Y.; Guo, X.Y.; Ding, F.; Luo, J. Analysis of Air Pollution Characteristics and Countermeasures in Yining City During the "13th Five-Year Plan" Period. *Arid Environ. Monit.* **2023**, *37*, 33–38.
12. Xie, Y.X.; Tang, X.; Guo, Y.H.; Lin, C.Y.; Wu, H.J.; Lu, M.M.; Wang, Z.F. Spatial and Temporal Distribution of Atmospheric Particulate Matter in Xinjiang. *Environ. Monit. China* **2019**, *35*, 26–36. [CrossRef]
13. Yining City People's Government. Yining City Ambient Air Quality Deadline Compliance Plan (2021–2025). Available online: https://www.yining.gov.cn/yining/c116600/202206/f0410c9a361d418aa7c0f8fe7e9314d7.shtml (accessed on 5 May 2025).
14. Zhang, S.X.; Zhang, Z.Z.; Li, Y.; Du, X.H.; Qu, L.L.; Tang, W.; Xu, J.; Meng, F. Formation processes and source contributions of ground–level ozone in urban and suburban Beijing using the WRF–CMAQ modelling system. *J. Environ. Sci.* **2023**, *127*, 753–766. [CrossRef] [PubMed]
15. Ge, S.; Wang, S.; Xu, Q.; Ho, T. CAMx simulations of the control of anthropogenic emissions on the reduction of ozone formation in Southeast Texas of USA. *Atmos. Pollut. Res.* **2021**, *12*, 101–114. [CrossRef]
16. Lang, J.L.; Liang, X.Y.; Li, S.Y.; Zhou, Y.; Chen, D.S.; Zhang, Y.Y.; Xu, L.T. Understanding the impact of vehicular emissions on air pollution from the perspective of regional transport: A case study of the Beijing-Tianjin-Hebei region in China. *Sci. Total Environ.* **2021**, *785*, 147–304. [CrossRef]
17. Wen, W.; Shen, L.Y.; Sheng, L.; Ma, X.; Wang, J.K.; Guan, C.G.; Deng, G.; Li, H.Q.; Zhou, B. Impact of meteorological uncertainties on PM2.5 forecast: An ensemble air quality forecast study during 2022 Beijing Winter Olympics. *Atmos. Environ.* **2025**, *344*, 121027. [CrossRef]
18. Yang, X.L.; Wang, Y.; Li, B.; Zhang, J.X.; Lai, X.L. Influence of Meteorological Elements on Pollutant Concentration in River Valley Terrain. *Res. Environ. Sci.* **2018**, *31*, 34–41. [CrossRef]
19. Zhang, L.L.; Sun, H.L.; Yang, Y.H.; Lu, B.B.; Liu, T.Y.; Lan, X.L.; Cao, L.J. Transport characteristics and pollution sources of PM2.5 in Yining City in winter. *J. Atmos. Environ. Opt.* **2022**, *17*, 294–303.
20. *GB 3095-2012*; Ambient Air Quality Standards. China Environmental Science Press: Beijing, China, 2012.
21. Stein, A.F.; Draxler, R.R.; Rolph, G.D.; Stunder, B.J.B.; Cohen, M.D.; Ngan, F. NOAA's HYSPLIT Atmospheric Transport and Dispersion Modeling System. *Bull. Am. Meteorol. Soc.* **2015**, *96*, 2059–2077. [CrossRef]
22. Zhou, Y.C.; Zhao, Y.Z.; Gao, X.; Lei, J.Q.; Dong, D.W. Dust weather in Tarim Basin derived from a dust event on April 18–21, 2023. *Atmos. Pollut. Res.* **2025**, *16*, 102605. [CrossRef]
23. Haseeb, M.; Tahir, Z.; Mahmood, S.A.; Batool, S.; Tariq, A.; Lu, L.; Soufan, W. Spatio-temporal assessment of aerosol and cloud properties using MODIS satellite data and a HYSPLIT model: Implications for climate and agricultural systems. *Atmos. Environ. X* **2024**, *21*, 100242. [CrossRef]
24. Wang, Y.; Stein, A.F.; Draxler, R.R.; de la Rosa, J.D.; Zhang, X. Global sand and dust storms in 2008: Observation and HYSPLIT model verification. *Atmos. Environ.* **2011**, *45*, 6368–6381. [CrossRef]
25. Wei, X.J.; Zhao, X.M.; Wang, Q.; Xiao, M.M. The Analysis of the Influence of Meteorological Factors on PM2.5 Concentration Based on Characteristic Index. *Environ. Monit. China* **2022**, *38*, 90–100. [CrossRef]
26. *HJ 633-2012*; Technical regulation on Ambient Air Quality Index (on Trial). Ministry of Environmental Protection of the People's Republic of China: Beijing, China, 2012.
27. Yu, P. The Memory of Air Pollutants in ChinaRegion. Master Thesis, Kunming University of Science and Technology, Kunming, China, 2023.
28. Yuval; Tritscher, T.; Raz, R.; Levi, Y.; Levy, I.; Broday, D.M. Emissions vs. turbulence and atmospheric stability: A study of their relative importance in determining air pollutant concentrations. *Sci. Total Environ.* **2020**, *733*, 139–300. [CrossRef]

29. Wang, H.; Li, J.H.; Peng, Y.; Zhang, M.; Che, H.Z.; Zhang, X.Y. The impacts of the meteorology features on PM2.5 levels during a severe haze episode in central-east China. *Atmos. Environ.* **2019**, *197*, 177–189. [CrossRef]
30. Yi, K.; Liu, J.F.; Wang, X.J.; Ma, J.M.; Hu, J.Y.; Wan, Y.; Xu, J.Y.; Yang, H.Z.; Liu, H.Z.; Xiang, S.L.; et al. A combined Arctic-tropical climate pattern controlling the inter-annual climate variability of wintertime PM2.5 over the North China Plain. *Environ. Pollut.* **2019**, *245*, 607–615. [CrossRef] [PubMed]
31. Chen, Y.L.; Wang, C.; Zhou, Y.Y.; Liu, S.Q.; Xiao, T.G.; Wen, X.H. Study on the Characteristics of Air Humidity and Its Effects on Air Quality in Chengdu. *Front. Environ. Prot.* **2020**, *10*, 569–576. [CrossRef]
32. Chen, C.; Zhang, M.X.; Liu, J.; Fang, J.L.; Guo, Y.F.; Li, T.T.; Shi, X.M. Acute impact of ambient fine particulate matter and ozone on daily outpatient visits and its seasonal differences in Beijing-Tianjin-Hebei and surrounding areas. *Acta Meteorol. Sin.* **2022**, *80*, 366–374. [CrossRef]
33. Zhang, J.L.; Li, R.Q.; Li, N.; Li, H.H.; Shi, J.J. Preliminary Analysis of the Water Vapor Characteristics of the "July 19" Heavy Rain in 2021 in the Tarim Basin, Xinjiang. *Chin. J. Atmos. Sci.* **2023**, *47*, 567–584.
34. Gao, S.N. Analysis of the Correlation Between Air Quality and Meteorological Elements in Xi'an City. *J. Xi'an Univ. (Nat. Sci. Ed.)* **2018**, *21*, 104–110.
35. Zuo, M.; Li, G.H.; Wu, J.R. Synergetic effects of NO_2 and SO_2 on air particulate matter pollution in the Guanzhong Basin (GZB), China. *J. Earth Environ.* **2022**, *13*, 571–587.
36. Cui, J.M.; Wang, T.J.; Gao, L.B.; Cao, Y.Q.; Wang, Q.G. Observational analysis on air pollution characteristics in Nanjing during winter of 2016. *J. Meteorol. Sci.* **2020**, *40*, 427–437.

MDPI AG
Grosspeteranlage 5
4052 Basel
Switzerland
Tel.: +41 61 683 77 34

Toxics Editorial Office
E-mail: toxics@mdpi.com
www.mdpi.com/journal/toxics

www.ingramcontent.com/pod-product-compliance
Lightning Source LLC
Chambersburg PA
CBHW061510210926
51850CB00008B/1870
* 9 7 8 3 7 2 5 8 7 7 1 3 3 *